I0130311

JOINT COMMITTEE ON ATOMIC ENERGY

SOVIET ATOMIC ESPIONAGE

Hold For Release
MONDAY A.M
April 9, 1951

Fredonia Books
Amsterdam, The Netherlands

Soviet Atomic Espionage

by
Joint Committee on Atomic Energy

ISBN 1-58963-134-X

Copyright © 2001 by Fredonia Books

Reprinted from the 1951 edition

Fredonia Books
Amsterdam, The Netherlands
http://www.fredoniabooks.com

All rights reserved, including the right to reproduce
this book, or portions thereof, in any form.

FOREWORD

The Joint Committee on Atomic Energy, from its inception almost 5 years ago, has continually given painstaking attention to the task of defending this Nation's atomic enterprise against Soviet agents. Since mid-1946, when the law creating the joint committee and the Atomic Energy Commission was enacted, American espionage defenses, so far as is known, have not been breached. The Federal Bureau of Investigation and other interested agencies have reported no successful act of atomic espionage committed against the United States from mid-1946 onward.

Before that time, however, Soviet agents did successfully penetrate the joint American-British-Canadian atomic projects. After mid-1946, moreover, further security breaches have occurred in the British program, through Dr. Klaus Fuchs and through the disappearance of Dr. Bruno Pontecorvo.

The individuals who had access to classified atomic information and who are definitely known to have conveyed such information to the Soviet Union while employed on the project are three in number. To list them in the estimated order of their importance, they are: (1) Dr. Klaus Fuchs, the German-born British scientist who worked both upon the key Oak Ridge, Tenn., process and upon weapons at the Los Alamos, N. Mex., laboratory during World War II; (2) Dr. Allan Nunn May, the British scientist, who was arrested and convicted in connection with the Canadian spy exposés of 1946; and (3) David Greenglass, an American citizen, who, as an Army technical sergeant, performed weapons work at Los Alamos during World War II.

A special situation is presented by Dr. Bruno Pontecorvo, the Italian-born British scientist, who worked at Canadian atomic energy centers and visited American laboratories during World War II. In October 1950 he fled eastward and disappeared behind the iron curtain. Pontecorvo may not have compromised information prior to his flight. But, for reasons developed later in this study, it must be taken for granted that every scrap of information known to Pontecorvo is today known to the Soviet Union. Thus the consequences are the same whether or not he did in fact serve as a Soviet agent while working in the west. Pontecorvo is therefore grouped with Fuchs, May, and Greenglass as a betrayer of atomic secrets. In terms of his contribution to the Soviet project, he is estimated to rank behind Fuchs but ahead of May and Greenglass.

Part I of this study seeks to develop factual, authentic information about the four known betrayers who themselves had access to atomic data. Also included in part I is material relating to Harry Gold, who served as an espionage courier and relayed information from both Fuchs and Greenglass to Soviet officials. Part I is entitled "Fuchs, Pontecorvo, May, Greenglass, and Gold."

Part II of the present report, in contrast to the proven cases which mainly comprise part I, deals only with allegations not tested in a court of law. These allegations all pertain to persons whose atomic spying or service in Russia's behalf are alleged to have occurred during World War II. Several of the individuals concerned are currently being prosecuted for perjury or contempt of Congress. The various allegations center particularly upon acts of espionage said to have been committed by Dr. Clarence Hiskey and Dr. Joseph Weinberg, two wartime members of the American atomic enterprise, and upon a Communist cell said to have been active in the wartime Berkeley Radiation Laboratory.

For the sake of clarity and completeness this report contains a part III relating to individuals who committed serious security breaches but who were not Soviet agents and whose actions did not assist the Russian atomic project. A leading example is the case of Dr. Sanford Simons. After working at Los Alamos during World War II he departed with a small sample of plutonium as a souvenir. Part III is entitled "Nonespionage Cases."

The purpose of the present study, prepared and compiled by the joint committee staff at the request of the chairman, is to gather in one place the salient facts about the various individuals within the categories above outlined. It is hoped that the study will assist committee members in assessing the atomic-espionage damage inflicted upon the United States.

All statements have been meticulously screened so as to include unclassified and publishable information only.

THE GEOGRAPHICAL FOCAL POINTS OF ESPIONAGE

Hanford, Washington, center of plutonium production. Both May and Pontecorvo knew critical Hanford secrets.

Los Alamos, New Mexico, atomic weapons laboratory. Fuchs and Greenglass betrayed the innermost Los Alamos secrets, 1944-1946.

Chicago "Met" Laboratory. May and Pontecorvo visited here, 1944-1945.

Chalk River, site of Canadian heavy water pile. May and Pontecorvo had detailed knowledge of this installation.

McGill University, Montreal. Here May and Pontecorvo performed atomic research, 1943-1945.

Columbia University research laboratory. Fuchs worked here upon U-235 productions, 1943-1944.

Oak Ridge, Tennessee, center of U-235 Production. Fuchs had complete access to research work upon the main production technique, 1943-1944.

ALASKA

CANADA

UNITED STATES

Birmingham University. Here, in 1942, Fuchs first became connected with atomic research.

The Cavendish Laboratory, Cambridge. Here May performed atomic research in 1942.

Helsinki, Finland. Pontecorvo and family were last seen here in October 1950.

Harwell, the major British atomic research laboratory. Here Fuchs worked from 1946 to 1949 and Pontecorvo from 1949 to 1950.

Fuchs was born in Germany and studied physics at Kiel University, fleeing to Britain in 1933.

Paris, France. Here Pontecorvo worked with Joliot-Curie on atomic problems from 1938 to 1941.

Pontecorvo was born in Italy, studied physics in Rome, and performed atomic research during the early 1930's.

FINLAND

UNITED KINGDOM

FRANCE   GERMANY

ITALY

THE GEOGRAPHICAL FOCAL POINTS
OF ESPIONAGE

The Library of Congress
Legislative Reference Service
Hubert L. Westrick, 3-21-50
No. 606

CONTENTS

PART I

FUCHS, PONTECORVO, MAY, GREENGLASS, AND GOLD

PART II

CHARGES NOT PROVEN IN A COURT OF LAW

PART III

NONESPIONAGE CASES

PART I

FUCHS, PONTECORVO, MAY, GREENGLASS, AND GOLD

1. A Synopsis of the Known Facts

(a) *Klaus Fuchs*

Dr. Fuchs was a member of the British atomic-energy mission which came to the United States during World War II, in accordance with the 1943 Quebec agreement between President Roosevelt, Prime Minister Churchill of the United Kingdom, and Prime Minister King of Canada. The British certified to the Manhattan Engineer District, which then had responsibility for American atomic development, that Fuchs could be trusted; and this certification was accepted without further inquiry.

During 1944 Fuchs lived in New York and worked at Columbia University, participating intimately in efforts to develop the gaseous diffusion U–235 separation process—a process now embodied in the K–25–27–29–31 plant complex at Oak Ridge, Tenn., and to be embodied in the major new plant complex under construction at Paducah, Ky. He was one of three Britishers who had complete access to all phases of the Columbia University work. It is little appreciated that Fuchs is not only the great betrayer of weapons data but also the great betrayer of the theory underlying the only Oak Ridge production method in use today.

In August 1944 Fuchs moved to Los Alamos with other members of the British mission and worked there until June 1946. He took part in the making of the earliest atomic bombs; he was privy to ideas and plans for improved atomic weapons, and he possessed insight into the thinking of the period as regards the hydrogen bomb. Fuchs returned to England in mid-1946 and soon became Chief of the Theoretical Physics Division at Harwell, Britain's principal atomic-research laboratory.

Some 3 years later American security authorities advised Britain of a lead developed in the United States, and this brought about Fuchs' arrest, conviction, and imprisonment in early 1950. He had shipped to Russia the most sensitive information, including extensive quantitative data in written form, regarding the Oak Ridge gaseous diffusion process, the weapons work at Los Alamos, British activities at Harwell, and other projects located in the United States, Canada, and the United Kingdom.

(b) *Bruno Pontecorvo*

The reactor of most advanced design and performance operating in North America is still the Canadian NRX heavy-water pile at Chalk River, Ontario, and Dr. Pontecorvo contributed to the nuclear aspects of this device. In addition, he had knowledge of nuclear problems

1

involved in building the plutonium-production piles at Hanford, Wash. Pontecorvo is considered by some of his scientific associates to be more outstanding as a physicist than Klaus Fuchs.

An Italian by birth, Pontecorvo studied at Rome and Paris and, when Nazi Germany invaded France in 1940, escaped through Spain to the United States. For a time he was employed by an American oil firm, and then, in early 1943, he joined the Anglo-Canadian atomic energy team at McGill University, Montreal. The next year he moved to Chalk River, Ontario, and—during the same year—engaged in classified discussions at the Chicago Metallurgical Laboratory. Later the main emphasis in Pontecorvo's efforts shifted from reactors to cosmic-ray research, although he continued to enjoy access to secret atomic energy data. Also his more recent studies included work upon tritium, a substance intimately related to the hydrogen bomb. Pontecorvo remained at Chalk River through 1948, the year in which he was naturalized as a British citizen. Thereupon he became a senior principal scientific officer at the Harwell Laboratory in England. Pontecorvo, his wife, and three children disappeared behind the Soviet iron curtain, fleeing via Sweden and Finland, during September 1950.

While Pontecorvo had no direct contact with weapons work, it is possible to speculate that he may have betrayed reactor data from 1943 onward—supplementing the bomb details and the U–235 processing information divulged by Fuchs and thereby furnishing Russia with a particularly well-rounded picture. In any event, as of September 1950, the Soviets acquired in Pontecorvo not only a human storehouse of knowledge about the Anglo-American-Canadian atomic projects but also a first-rate scientific brain.

(c) Allan Nunn May

Shortly before World War II ended, Dr. May, a natural-born British scientist, met a Russian military officer in Montreal, Canada, and gave him laboratory samples of U–235 and U–233. This dramatic betrayal attracted wide attention in the press when May was arrested in early 1946, but still more helpful to Russia, no doubt, was the information which he also compromised. He worked closely with the wartime Metallurgical Laboratory at Chicago, visiting there on three occasions; and as a result, he understood a number of the problems overcome in constructing the Hanford, Wash., plutonium piles. May, in addition, was familiar with the Canadian wartime atomic energy effort; and he knew miscellaneous facts about Oak Ridge, Tenn., and Los Alamos, N. Mex. He confessed to writing an over-all report on atomic energy as known to him and transmitting it to the Soviets.

A nuclear physicist of considerable ability, May joined the British atomic project during the spring of 1942. He worked first at the Cavendish Laboratory, Cambridge, England, and then, in January 1943, joined the Anglo-Canadian research team at McGill University, Montreal. In Canada he became a senior member of the nuclear physics division and remained until September 1945. Thereafter, until his arrest, May was a lecturer on physics at Kings College, London.

Igor Gouzenko, the Russian cipher clerk at the Soviet Embassy in Ottawa, Canada, helped bring about the arrest when he defected to the Canadian authorities in late 1945, taking with him papers which led to the exposure of widespread espionage activity. The wartime

Canadian spy ring is primarily associated, in many minds, with divulgement of atomic secrets; but Allan Nunn May is the only member of the ring who gave Russia information in this category. He was 34 years old when arrested and had become a Communist before the war. May is now serving a 10-year sentence in Wakefield Prison, Yorkshire, England.

(d) David Greenglass

Of the four betrayers who themselves had access to secret atomic information, only David Greenglass is American-born; only he is an American citizen; and he also stands alone as the only nonscientist in the group.

According to his own testimony in open court, Greenglass was assigned to the Los Alamos, N. Mex., weapons laboratory in the summer of 1944 and there worked as a machinist upon high-explosive lens molds, (a crucial nonnuclear phase of the atomic bomb). He gave Harry Gold (who was also a courier for Fuchs) sketches and detailed written descriptions of this work, as well as a list of scientific personnel associated with Los Alamos and a further list of Los Alamos individuals whom he thought might be willing to serve as Russian agents. In addition, Greenglass testified that he conveyed to Russia a diagram of the atomic bomb, along with a detailed explanation and related materials in writing.

Greenglass grew up on the lower East Side of New York City, attending the Brooklyn Polytechnic Institute and Pratt Institute. In 1938, at the age of 16, he joined the Young Communist League. At Los Alamos, as an Army technical sergeant, he eventually became foreman of a machine shop concerned with preparing weapon apparatus. Greenglass' testimony indicates that he was able to secure information far beyond the scope of his assigned duties.

He was arrested in June 1950 as the result of information supplied by Harry Gold and others, and he pleaded guilty to the charges against him. Although his espionage activities had terminated by 1946, when he left the Army, the case is significant because Greenglass was in a position to furnish Russia with mechanical details of bomb gadgetry and weaponeering that might have supplemented the data divulged by a theoretical physicist such as Klaus Fuchs.

(e) Harry Gold

Since the American population still contains several tens of thousands of Communists, according to official estimates, it is not surprising that Russia found means of conveying atomic secrets from Los Alamos, N. Mex., to her own consulate in New York City—once she had induced Fuchs and Greenglass, who worked behind the guarded Los Alamos gates, to provide these secrets in the first instance. The contact men, the couriers, the go-betweens, the spy recruiters, and Russian officials in New York—such individuals have played a lesser role than a Fuchs or a Greenglass, though an indispensable one from the Soviet viewpoint.

(1) Harry Gold met Fuchs in Boston, New York, Santa Fe, and elsewhere, and received both oral information and packets of papers for delivery to Russian officials. In addition, Gold served in the same capacity for Greenglass—and also for others who transmitted data having no relation to atomic energy. Arrested in May 1950, after

Fuchs gave FBI agents his description, Gold plead guilty and was sentenced to 30 years' imprisonment.

(2) Julius Rosenberg and his wife, Ethel, who were convicted on March 29, 1951, in New York City, of conspiring with Gold and others to procure national defense secrets for Russia. This couple engaged in widespread espionage activities, although none relate directly to atomic energy except as they focus through Greenglass, who is Ethel Rosenberg's brother. Greenglass testified that the Rosenbergs helped induce him to betray Los Alamos secrets.

(3) Ruth Greenglass, wife of David Greenglass, was named as a co-conspirator in the indictment against the Rosenbergs but has not herself been prosecuted. She testified to having helped her husband compromise Los Alamos information.

(4) Anatoli Yakovlev: This individual was formerly a Russian vice consul in New York. From 1944 to 1946 he supervised Harry Gold, receiving from the latter written and oral information furnished by Fuchs, Greenglass, and others.

(5) Semen M. Semenov: As a Russian official of the Amtorg Trading Corp., Soviet purchasing agency with American headquarters in New York City, Semenov directed Harry Gold's activities as an espionage courier from 1940 to 1944.

(6) Other individuals connected or allegedly connected with the same espionage network include Abraham Brothman and Miriam Moskowitz, convicted of persuading Harry Gold to give false testimony before a grand jury; Alfred Dean Slack, convicted of conveying to Gold secrets concerning a new type explosive; Morton Sobell, accused of betraying radar and electronics data; and William Perl, an expert on aerodynamics charged with perjury.

2. The Nature of the Information Betrayed

The extent of the espionage damage known to have been inflicted upon the atomic energy position of the United States is indisputably severe.

The Nation's three most important installations in this field are Hanford (plutonium), Oak Ridge (U–235), and Los Alamos (weapons). From Klaus Fuchs the Soviets learned critical information involved in two of these installations, Oak Ridge and Los Alamos; and David Greenglass supplied additional weapons data. Allan Nunn May and Bruno Pontecorvo were in a position to betray essential secrets about Hanford. The central atomic energy installation in Canada is Chalk River, and the main British research laboratory is Harwell. May and Pontecorvo each had extensive knowledge as to both. Moreover, all the four Soviet helpers—possibly excepting Greenglass—had opportunity to give Russia significant data concerning a number of added atomic energy facilities and processes in the United States, Britain, and Canada.

Whereas our own wartime Los Alamos laboratory grappled with a series of abstruse and exquisite weapons problems, finally arriving at practical solutions after great expenditures of effort, money, and technical talent, the fruits of this struggle were largely available to the Soviets at an early stage of their rival enterprise. Much the same may be said of the Oak Ridge, Hanford, and Chalk River projects. Whereas the wartime atomic partners, America, Britain, and Canada, overcame immense obstacles to construct reactors and to produce precious fissionable materials, a major share of their experience—thanks to the spies—was at hand for Russia to exploit without the independent exertion on her part otherwise necessary. Our own country, striking into the unknown, felt compelled to build three separate plants for U–235 production, each based upon a different process. One of these, the gaseous diffusion method, proved to be far superior and since the war has been used almost exclusively. It is the same method to which Klaus Fuchs had access during the wartime research and development phase. Here again the Soviets, from an early point in their effort, could avoid making many of the mistakes and following many of the costly false leads that inevitably attended the pioneering days of the American program. The same point may be made as regards the heavy water reactor at Chalk River, Canada. This is all apart from Fuchs' knowledge of American plans for postwar development both as to atomic weapons and as to the hydrogen bomb.

Thus the conclusion seems reasonable that the combined activities of Fuchs, Pontecorvo, Greenglass, and May have advanced the Soviet atomic energy program by 18 months as a minimum. In other words, if war should come, Russia's ability to mount an atomic offensive against the West will be greatly increased by reason of these four men. It is hardly an exaggeration to say that Fuchs alone has influenced the safety of more people and accomplished greater damage than any

5

other spy not only in the history of the United States but in the history of nations. This is not to imply that Russia could never have broken the American atomic monopoly through her own unaided labors. But if, for example, the United States had known early in World War II what Russia learned by the end of 1945 through espionage, it appears likely that our own project would today be at least 18 months ahead of its actual level of attainment.

The relative importance of the four betrayers is a matter of judgment and, in part, speculation. It seems crystal clear that Fuchs was the most damaging because no information surrounding the wartime Los Alamos weapons center and, likewise, no information involving what is now the sole Oak Ridge production process were withheld from him; and the evidence is plain that he effectively placed in Soviet hands the data at his command. Pontecorvo, May, and Greenglass all rank well below Fuchs in importance, and how they rank among one another is particularly speculative. It may be pointed out, however, that Pontecorvo is considered by some to be an even abler scientist than Fuchs and further that Pontecorvo successfully fled the West and is working behind the iron curtain today. Moreover, he has devoted himself to atomic energy problems since 1934, and the classified data known to him is fresh and recent. For these reasons Pontecorvo may be plausibly rated as the second deadliest betrayer, though considerably less deadly than Fuchs.

This rating of Pontecorvo is supported by the logic that the Soviets, in order to achieve the atomic-bomb stockpile now at their disposal, were required to solve two broad problems: (1) Quantity production of fissionable material, and (2) design and assembly of practical weapons. Fuchs contributed to the solution of both problems; but as to the first—production of fissionable materials—his own special knowledge focused particularly upon U–235 rather than plutonium. Therefore, unless Russia could locate an informant who would provide detailed information about nuclear reactors (the devices which produce plutonium), her range of understanding would contain a broad gap until such time as Soviet scientists filled it through independent effort. Pontecorvo may perhaps have served as an informant in this area, from 1943 onward. Certain it is that Russia today possesses nuclear reactors. But whether or not Pontecorvo in fact betrayed secrets before disappearing behind the iron curtain late last year, his recollection of those secrets is now available to Russia and his unusual scientific mind is also available for Soviet reactor development.

The question of whether Dr. Allan Nunn May or David Greenglass ranks as the next deadliest atomic spy depends partly upon an interpretation of May's own confession. On the one hand, he admitted giving Russia a written report on atomic energy as known to him; and, on the other hand, he claimed that this information was mostly of a kind published soon thereafter or about to be published. It so happens that the data to which May had access far exceeds the scope of any material in the public domain even today, 6 years later. Therefore, as to the one statement or the other in his confession, May lied. Since he was a fervent Communist, since Canada's Royal Commission considered the case to be serious in its consequences, and since May himself had so little repented at the time of his confession that he refused to identify a Soviet accomplice, it would be prudent to assume he told Russia everything or nearly everything of importance that he knew.

Such an assumption ties in with the fact that May stole specimens of U-235 and U-233, probably from our own Chicago laboratory, and handed them to his Soviet contact man—a deed suggesting extreme devotion to Russia's interest.

David Greenglass was a 23-year-old machinist, with no real scientific training, during the period of his main espionage activity at Los Alamos. The diagrams and written explanation of the Nagasaki-type atomic bomb that he gave to courier Harry Gold have a theatrical quality and, at first glance, may seem the most damaging single act committed by any of the main betrayers. But, on one occasion, Greenglass met with a Russian official in New York and was asked to supply a mathematical formula concerning high-explosive lenses used in the Nagasaki-type weapon. Not being a scientist, Greenglass lacked capacity to furnish this information—although, in all probability, Klaus Fuchs could have and did furnish it thereafter. By the same token, the bomb sketches and explanations that Greenglass—as a virtual layman—could prepare must have counted for little compared with the quantitative data and the authoritative scientific commentary upon atomic weapons that Fuchs transmitted.

When Harry Gold had once traveled West and received papers from both Fuchs and Greenglass on the one trip, his Russian superior ordered him to repeat the journey for purposes of again contacting Fuchs. Gold suggested that he also contact Greenglass, as before, and the Russian superior rejected this suggestion as not being worth while—an episode implying that after the Soviets had gained experience in what Greenglass was capable of telling them, they lost some of their interest. A like inference may be drawn from the proposals made to Greenglass, after the war, that he become a science student and thereby better equip himself to serve as a Soviet agent.

Greenglass' value to Russia in generally corroborating Fuchs' statements and perhaps in supplying miscellaneous information which Fuchs omitted is not to be discounted. It is even possible moreover, that Greenglass—in the narrow but important field of his own work upon high explosive lens molds—was able to convey practical data and know-how beyond Fuchs' understanding. Yet, everything considered, Greenglass appears to have been the least effective of the four spies, ranking behind Allan Nunn May in this regard. Had there been no Klaus Fuchs, Greenglass would take on far greater importance. Needless to say, this evaluation does not detract one iota from the horror of this man's crimes nor lessen his legal and moral guilt.

In writing of the 1945 Potsdam conference, former Secretary of State James F. Byrnes describes how Stalin reacted when President Truman told him about the atomic bomb, as follows:

At the close of the meeting of the Big Three on the afternoon of July 24 [1945], the President walked around the large circular table to talk to Stalin. After a brief conversation the President rejoined me and we rode back to the "Little White House" together. He said he had told Stalin that, after long experimentation, we had developed a new bomb far more destructive than any other known bomb, and that we planned to use it very soon unless Japan surrendered. Stalin's only reply was to say that he was glad to hear of the bomb and he hoped we would use it. I was surprised at Stalin's lack of interest. I concluded that he had not grasped the importance of the discovery. I thought that the following day he would ask for more information about it. He did not. Later I concluded that, because the Russians kept secret their developments in military weapons, they thought it improper to ask us about ours. (Reprinted by permission from Speaking Frankly by James F. Byrnes, 1947.)

3. Comment on Security Defenses, Espionage Techniques, and Motives

(a) Security defenses

The British Government was aware, in 1941, that Gestapo reports had named Klaus Fuchs as a German Communist, but had no way of checking the accuracy of this charge. Fuchs was sympathetic to communism from early college days in Germany, and several of his relatives were either members of the party or fellow travelers. According to Fuchs' confession, he associated mainly with Marxians for 6 years in England before the war. The pro-Soviet pressures operating upon Bruno Pontecorvo are evidenced by the fact that at least one brother and one sister have been lifelong Communists, and a cousin is today a Communist member of the Italian Parliament. Pontecorvo worked for 3 years in Paris under Frederic Joliot-Curie, himself a notorious Communist long surrounded by many Communist associates. When Pontecorvo visited the Chicago "Met Lab" in 1944, he had not yet acquired British citizenship. Allan Nunn May had been a devoted, though secret, prewar Communist in England and was a member of the Canadian Association of Scientific Workers, an organization infiltrated by Communists and having a Communist as its president. David Greenglass joined the Young Communist League in New York at the age of 16, and associated extensively with Communists until the time he entered the Army and was assigned to Los Alamos.

Why, then, were these four men—now known to be betrayers—granted access to atomic energy information?

The answer centers partly upon the fact that the FBI had no responsibility for security investigations during the wartime period. Not until 1947, when the present law controlling atomic energy was enacted, did the FBI become responsible for investigating project personnel. Previously the Army Manhattan Engineer District had full charge of atomic development and handled security matters exclusively through its own officials.

Another obvious reason why the four spies gained access lay in the fact that Russia was an ally of the United States during World War II. Although the Manhattan District attempted to exclude Soviet agents from the ranks of its employees, a great part of the effort was also devoted to excluding agents of Germany, Italy, and Japan.

In addition, the Quebec Agreement of 1943 stipulated generally that the United States, Britain, and Canada would collaborate as partners in the field of atomic energy. Methods of security clearance were omitted from the agreement, being worked out by the operating agencies concerned. The Manhattan District, having received written assurances from Britain that Fuchs, Pontecorvo, and May were loyal and trustworthy, did not make inquiries—because the United Kingdom is a sovereign nation—regarding the evidence and the security procedure that supported such assurances. Fuchs and Pontecorvo, moreover, could not be comprehensively investigated because each had passed most of his life in an enemy country—Germany and Italy, respectively. Pontecorvo spent 1941 and 1942 in the United States and may have revealed his true status through Communist activity during these years, but the FBI was excluded from responsibility for checking upon such possible activity.

Today it is often asserted that, during the war, atomic energy was perfectly concealed from the American people but that it was well-known to Joseph Stalin. This characterization is unfair in the sense that atomic energy was not only hidden from the American people but, in large degree, from the Axis Powers as well. The necessity of attempting to keep Germany and Japan totally in the dark meant that security efforts had to be so used as to shroud all information and to keep each project employee, however little he knew, from alerting the Axis. The mere fact that an atomic project existed was secret. As a result, security efforts were widely dispersed and could not be adequately concentrated upon screening the small numbers of people who would gain extensive knowledge and who could most assist Russia if so inclined.

Not to be overlooked, either, were the extreme conditions of urgency which underlay the wartime project. Tens of thousands of persons had to be employed in short order, and the job could not wait until exhaustive security procedures had been fully carried out. Our own qualified scientists are almost unanimous in believing, too, that only the participation of their British and Canadian colleagues made possible the achievement of the atomic bomb within the time available and that, without such participation, the success of the project would have been materially delayed. The sheer stresses and strains and urgencies of a war situation involving a three-nation atomic partnership probably constitute the greatest single factor accounting for the security lapses that gave entrance to Fuchs, Pontecorvo, May, and Greenglass.

(b) Espionage techniques

On the order of 200,000 people took part in the wartime American-British-Canadian atomic efforts, but only a few hundred—a fraction of 1 percent—were in a position to give Russia sweeping data of vital significance. This was due partly to a security compartmentation system whereby the individual received only such information as he needed to do his job. A large share of the two-hundred-thousand-odd wartime project members were not aware that their work related to atomic bombs. Moreover, the average employee could not have substantially helped Russia, no matter how disloyal to the United States he may have been, simply because his brain lacked the technical education to understand and convey the main secrets at stake. Soviet espionage, to be successful on a major scale, required three conditions: (1) A trained scientist or specialist, (2) having critical access to information about the American-British-Canadian project, and (3) willing to sacrifice his own country in behalf of Russia.

These conditions were all met in the cases of Fuchs, Pontecorvo, May, and Greenglass. Each man was among the few hundred—the fraction of 1 percent—possessing the mental equipment and holding the key positions which alone made possible, if combined with disloyalty, the betraying of essential secrets.

In at least two of these cases, it appears to be no coincidence that the very men now proven to have been spies were also among the few who, by reason of their training and duties, could best serve Russia as spies. Fuchs, as soon as he discovered the nature of the wartime work assigned him, personally took the initiative in contacting a Soviet agent and arranging to funnel information toward Russia. But Allan Nunn May was sought out by Russian military officers operating from the Soviet Embassy at Ottawa, Canada. They knew that May was both an ardent Communist and a key atomic scientist before establishing liaison with him. Greenglass worked upon a phase of the atomic bomb—the actual gadget itself—without realizing what it was until his wife told him. She had been informed by American operatives under the control of the Russian New York consulate as part of a successful scheme to recruit her husband into the espionage network. These operatives themselves chose Greenglass for their purpose, seeing that he had past Communist connections. The circumstances of Pontecorvo's flight to Russia from England via Sweden and Finland strongly indicate that the trip was prearranged, perhaps at the instigation of Soviet agents. In other words, the Russian intelligence system, so far as known, seems to have slighted the ordinary atomic employee who possessed little information and to have settled upon relatively few insiders, some of whom could and did deliver data of priceless value.

The fact that three of the four spies are physicists does not indicate that men of science, as a group, are more vulnerable to communism than other groups. It suggests, instead, that physicists alone fully understood many of the secrets which counted most and that physicists therefore attracted the concentrated attention of Soviet intelligence directors.

Some moral, however, may be drawn from the fact that both Fuchs and Pontecorvo had established reputations for their scrupulous obedience of formal security regulations, for the care which they seemed to exercise in guarding their language when noncleared persons were present, and in similar matters. Likewise, each man avoided conversations about politics, although not to the point of being conspicuous. Neither is known to have received any Communist literature such as the Daily Worker. Greenglass testified to the precautions which all Soviet agents took in shunning any appearance of disloyalty. Personal friends and associates of the three scientist spies were shocked and incredulous when the true facts became known. It may also be worth noting that the drastic espionage damage was not effected by men who broke into atomic energy centers through force and violence. The history of the espionage cases again demonstrates the oft-repeated maxim that personnel security is primary and that other forms of security such as physical barriers and documents control, however important, are secondary.

It is an interesting sidelight that May and Pontecorvo knew one another at Montreal, Canada, and Fuchs and Pontecorvo were acquaintances at Harwell, England. However, in all probability, none of these three men had knowledge that the others were also spies. The evidence shows that Russia maintained severe compartmentation within its espionage apparatus, at least so far as major sources of information were concerned.

Fuchs, May, Greenglass, and—presumably—Pontecorvo all received money from the Soviet Union. But the sums involved were relatively small and did not constitute an important inducement to the four spies (except that in Greenglass' case money may have been a distinct factor). It was the standard Russian technique to force token cash amounts upon espionage agents, for purposes of further corrupting them, helping to assure that they could be blackmailed in the future, if necessary, and signifying their complete subservience to the Soviet Union. Fuchs repeatedly accepted small bills "for expenses" and at one point received $400; Allan Nunn May was made to take $500; and Greenglass willingly took a total of $850—$150 on the first occasion he transmitted secret information. Shortly before his arrest last year, Greenglass was given $5,000 along with detailed instructions to flee the United States and go to Russia via Mexico.

(c) Motives

As the Canadian Royal Commission commented in 1946, "Membership in Communist organizations or a sympathy toward Communist ideologies was the primary force" which caused the spies to disregard their solemn oaths, to violate the trust reposed in them, and to undertake espionage for Soviet Russia. Each of the atomic agents, before World War II began, before he gained access to secret information, and as a young man, had been conditioned to what the royal commission calls "an atmosphere and an ethic of conspiracy." As a secret participant in Communist activities, he had been gradually brought to a state of mind where the dictates of honor, duty, loyalty, and integrity could be overcome. There are also traces of evidence, especially in May's case, that the spy helped justify his crimes to himself by reasoning thus: Russia is an ally and therefore deserves all possible help; the divulgement of information to the Soviets is in the tradition of free interchange of ideas among scientists; atomic data might promote industrial progress in a comparatively backward nation like the Soviet Union; and so forth. Greenglass may possibly have been influenced by similar thoughts. In other words, the spies— during their formative years—had been pulled into a Communist apparatus which systematically destroyed their sense of moral values and substituted the facile capacity for rationalization found in the code of totalitarian dictatorship.

Just as adherents of such a code sometimes tend to be arrogant and humorless, so there may also have been a powerful element of ego gratification in the actions of the atomic spies. Allan Nunn May has been described as "a bald, mousy little man" and Klaus Fuchs as "slight, shy, retiring, and very studious." Both were bachelors with few friends and scant interests outside science and communism. Such individuals can perhaps be imagined as relishing the secret knowledge that they, despite the seemingly prosaic pattern of their lives, were actually trafficking in information which affected the destiny of nations. David Greenglass, for his part, smiled while testifying about the atomic bomb before a crowded courtroom in New York, and he seemed to enjoy discussing his own insight into weapons data which others present regarded as mysterious and lying at the heart of American security. Bruno Pontecorvo, on the other hand, is recalled as a likeable extrovert, fond of telling jokes and socially inclined.

Dr. James Conant, president of Harvard University, has furnished another clue (although in a different context) that may help explain the conduct of the spies. In his book, Science and Common Sense, he writes:

Would it be too much to say that in the natural sciences today the given sociological environment has made it very easy for even an emotionally unstable person to be exact and impartial in his laboratory? The traditions he inherits, his instruments, the high degree of specialization, the crowd of witnesses that surrounds him, so to speak (if he publishes his results)—these all exert pressures that make impartiality on matters of his science almost automatic. Let him deviate from the rigorous role of impartial experimenter or observer at his peril; he knows all too well what a fool So-and-So made of himself by blindly sticking to a set of observations or a theory now clearly recognized as in error. But once he leaves the laboratory behind him he can indulge his fancy all he pleases and perhaps with all the less restraint because he is now free from the imposed discipline of his calling. One would not be surprised, therefore, if, as regards matters beyond their professional competence, laboratory workers were a little less impartial and self-restrained than other men, though my own observations lead me to conclude that as human beings scientific investigators are statistically distributed over the whole spectrum of human folly and wisdom much as other men. (Reprinted by permission of the Yale University Press, New Haven, Conn.)

The two most important betrayers, Fuchs and Pontecorvo, labored under especially powerful emotional pressure to use naive and irrational standards when thinking about politics; for both had personally suffered under fascism and both were refugees from that form of dictatorship. It is also noteworthy that both were born and reared on the Continent of Europe, the one in Germany and the other in Italy. All the four spies are alike in that their educational backgrounds reflect an unusual lack of contact with the liberal arts disciplines.

There may possibly be still another partial explanation for the warped mentalities of the spies; namely, an almost diseased yearning to remold the world after the image of their own work in physical science. The study of subatomic particles, the behavior of neutrons, the nature of fission—an attack upon these problems, however difficult, might seem orderly and rationally satisfying compared with the intangible complexities of moral and political issues. To an immature mind such as Fuchs' communism may have had special appeal because of a seeming resemblance between the regulated order it would impose upon society and regularities in his own laboratory research.

In any event, it is evident that a lack of moral standards, combined with an overweening and childlike arrogance—all induced by exposure to Communist recruiting techniques during early manhood—characterizes the atomic spy.

4. The Fuchs Case—Details

In early 1947, a disturbing headline on the front page of the New York Times foreshadowed the breaking of the Fuchs case. "Our Atom Secrets Tapped by Soviet, Baruch Believes," this headline began. Excerpts from the accompanying story, dated February 3, 1947, are as follows:

(By Anthony Leviero)

WASHINGTON, February 3—Bernard M. Baruch was reliably reported tonight to have told the Joint Congressional Committee on Atomic Energy that Russia evidently had tapped United States atomic-bomb secrets.

Clear indications of a penetration of some phases of the world's greatest secret, it was said, were unwittingly betrayed by Soviet delegates to the United Nations.

In endless discussions within the United Nations Atomic Energy Commission, in which the subject of international control was argued in the greatest detail, the Russians were said to have used phrases and asked questions that could have been based only on inside information, classified secret.

Mr. Baruch was chief of the United States delegation that implemented American control policy in the United Nations Commission. He testified on security in a closed session after the committee had heard him in an open hearing. The committee is considering the appointment of David E. Lilienthal and of his four associates on the United States Atomic Energy Commission.

Presumably the contact reaching beyond the security screen involved not only someone employed by Russia but others sympathetic with its aims * * *

Some Members of Congress linked the reported Soviet success in getting atomic secrets with the Russian spy system broken up in Canada last year. It was recalled that Allan Nunn May, British physicist and convicted spy, had obtained "general knowledge of the construction of the atomic bomb." * * *

Three years after this article appeared, Klaus Fuchs was arrested in England, and the reason for Mr. Baruch's fears became clear. Everyone who had concerned himself with the atomic affairs of the United States immediately experienced a deep and abiding sense of shock. Typical was the reaction of William L. Laurence, science writer for the New York Times. Recalling one of his own wartime visits to Los Alamos, N. Mex., he wrote:

And there in our midst stood Klaus Fuchs. There he was, this spy, standing right in the center of what we believed at the time to be the world's greatest secret. As he confessed 5 years later, he betrayed to the Soviet the most intimate details he learned—as a member of the innermost of inner circles—not only about the A-bomb, but about the H-bomb as well.

He was a trusted member of the theoretical division, the sanctum sanctorum of Los Alamos. This select group, behind doubly and triply locked doors, discussed in whispers their ideas about the "superdooper." His associates at Los Alamos today sadly admit that Fuchs made it possible for Russia to develop her A-bomb at least a year ahead of time. It is my conviction the information made it possible for the Russians to attain their goal at least 3 and possibly as much as 10 years earlier.

News of Fuchs' arrest disrupted a meeting of the American Physical Society underway in New York at the time, and several of the scientists present described the case as "an incredible shock." Dr. J. Robert Oppenheimer, wartime Director of the Los Alamos Laboratory, said, "We were not a very happy group of people." Later he indicated that Fuchs' espionage work might well have advanced the Russian project by as much as 1 year. Dr. Hans Bethe, a distinguished nuclear physicist and an expert on problems connected with the hydrogen bomb, headed the Theoretical Physics Division at wartime Los Alamos; and in that capacity, he was Fuchs' immediate superior. Excerpts from an interview with Dr. Bethe, in the February 5, 1950, Washington Star, read as follows:

Dr. Bethe said he was one of the few people Dr. Fuchs associated with at Los Alamos.

"We were very friendly together," he recalled, "but I didn't know anything about his real opinions.

"If he was a spy, he played his role absolutely perfectly."

The physicist recalled that when the Los Alamos scientists talked about the international exchange of scientific information, "Fuchs was not prominent in urging it, nor was he reluctant about it. He was like one of the rest of us. He didn't attract any attention."

Fuchs was one of the members of the atomic team of British scientists who worked with the Americans at Los Alamos. He worked under Dr. Bethe in the Theoretical Physics Division. "He made an extremely great contribution," said Dr. Bethe. "He was one of the most valuable men in my division. One of the best theoretical physicists we had." Dr. Bethe said Fuchs "knew everything we did." Fuchs did not work on the H-bomb which was then being considered, "but he could read any of the documents; he knew the•principles," said the American scientist.

"Everybody liked him," Dr. Bethe emphasied, "Everybody thought of him just as a quiet, industrious young man who would do everything he could to help our project."

"Fuchs was and is a bachelor. At Los Alamos he was neither aggressively social nor a hermit. He would visit at the homes of his fellow scientists on an average of twice a week."

The White House press secretary announced that President Truman had known about the Fuchs case before it reached the public but that he had not been informed at the time of deciding, a few days earlier, to order an intensified hydrogen program. According to another announcement, the case was discussed at a Cabinet meeting. The Joint Committee on Atomic Energy immediately held four executive sessions to canvass Fuchs' activities as then known and heard the testimony of the Atomic Energy Commissioners; Gen. Leslie R. Groves, commanding officer of the wartime Manhattan project; and J. Edgar Hoover, Director of the Federal Bureau of Investigation. During this period Senator McMahon, in addition to expressing his own concern about Fuchs, made public an unclassified paragraph taken from a top-secret letter written by Dr. Oppenheimer before the case broke. Herein the former Los Alamos Scientific Director states:

* * * it will be clear that members of the United Kingdom mission at Los Alamos played an integral part and a very important and responsible part in the actual wartime development of atomic weapons. It needs to be emphasized that even in those areas of laboratory work in which the mission played no direct part it, nevertheless, had complete access to all information and all reports.

On the day of Fuchs' arrest, February 3, 1950, General Groves had issued a similar statement:

Dr. Fuchs was at Los Alamos as a member of the British mission which worked at Los Alamos during the development of the atomic bomb. He was not the head of the mission nor was he one of the top members, but he was in the next rank. He had an important and responsible position.

In this position, he was necessarily afforded access to a great deal of information, both as to our development and as to future possibilities and proposed developments. Like all other members of the British mission, his responsibility, discretion, and loyalty were vouched for by his Government.

General Groves, like Dr. Oppenheimer, has estimated that Fuchs alone may have set ahead the Soviet project by 1 year.

The Atomic Energy Commission in its statement (also issued on February 3, 1950) mentions two facets of the case not emphasized elsewhere—Fuchs' work upon the Oak Ridge gaseous diffusion process during late 1943 and 1944 and his visit to the United States in 1947. This statement reads:

Dr. Karl Fuchs was a member of the British Atomic Energy Commission sent to this country during the war. He was engaged in project work from December 1943 until June 1946. He was first engaged in the early work done on the gaseous diffusion process in New York City.

In December 1944 he went to Los Alamos with the British scientific team assigned to weapons work and in this position had access to a wide area of the most vital weapons information.

Dr. Fuchs returned to England in June 1946, and has not been employed in the United States Atomic Energy project since that time.

In November 1947 he visited the United States as a member of the British delegation to participate in discussions on the use of the declassification guide, which had been developed during 1945 and 1946 and has been used by Canada and the United Kingdom since April 1946.

The declassification conference, announced November 14, 1947, was limited to fields of information shared by the three countries, as a result of their combined wartime efforts in the development of atomic energy. The conference was arranged to insure uniform application of declassification policy by the three governments.

In addition to the declassification conference in Washington, Dr. Fuchs was permitted to visit the Argonne National Laboratory at Chicago for discussions limited to nonsecret scientific material.

Tass, the official Soviet news agency, broadcast a dispatch denying that Fuchs was known to the Russian Government and declaring that the charges against Fuchs were a "rank invention."

Fuchs' confession.—When Fuchs confessed to espionage shortly after his arrest and before trial, a transcript was taken. He read the transcript when it had been typed, made corrections, and signed his name. Likewise, Dr. Michael Perrin, Deputy Controller, Department of Atomic Energy, Ministry of Supply, interviewed Fuchs regarding the nature of the technical information which he conveyed to Russia; and Fuchs' revelations on this score were then carefully organized and presented in written form. Copies of both documents—the confession and the interview—were made available by the British to the United States Government; and the complete texts were read, in executive session, to the Joint Committee on Atomic Energy.

During this committee meeting, it was suggested that each document be made public in toto, so that the people of the free world might know the full extent of the espionage which Soviet Russia had successfully accomplished through Fuchs. The suggestion has not yet been followed, however. Although the technical data about atomic energy set forth in the two documents assuredly reached Russia, there is no proof positive—so the argument runs—that every last shred of the information was effectively and accurately delivered or that all of it has been completely put to use by Soviet authorities. Thus, the argument continues, if the confession and interview were to be published, Russia might conceivably gain a few added details which her agents who contacted Fuchs had garbled, or which the Soviets had somehow neglected to exploit, or which Fuchs himself thought he had given but which actually he overlooked giving. Also involved is the possibility, however remote, that Fuchs might have deliberately "overconfessed," claiming to have betrayed secrets which actually he did not betray—all in hopes that such secrets would be published and thereby reach Russia via the daily newspapers. For these and related reasons, the confession and interview are still considered as secret. At some future time, the full contents will be declassified, along with additional papers concerning the case; and such writings will then at least rival the 1946 report of the Canadian Royal Commission as a documented study in Communist espionage. One excerpt from the Fuchs confession, however, is today in the public domain; for it was read to the British court which tried Fuchs on February 10, 1950. This excerpt is as follows:

After my release [from internment] I was asked to help Prof. R. E. Peierls in Birmingham on some war work. I accepted it without knowing at first what the work was but I doubt whether it would have made any difference to my subsequent action if I had.

When I learned about the purpose of the work I decided to inform Russia and I established contact through a member of the Communist Party. Since that time I have had continual contacts with persons completely unknown to me, except that they would give information to the Russians. At this time I had complete confidence in Russian policy and I had no hesitation in giving all the information I had.

I believed the western allies deliberately allowed Germany and Russia to fight each other to death. I tried to concentrate on giving information on the result of my own work. In the course of this work I began naturally to form bonds of personal friendship and I had to conceal from them my own thoughts.

I used my Marxian philosophy to conceal my thoughts in two separate compartments. One side was the man I wanted to be. I could be free and easy and happy with other people without fear of disclosing myself because I knew the other compartment would step in if I reached the danger point. It appeared to me at the time I had become a free man because I succeeded in the other compartment in establishing myself completely independent of the surrounding forces of society. Looking back on it now the best way is to call it a controled schizophrenia. In the postwar period I had doubts about Russian policy, but eventually I came to the point when I knew I disapproved of many actions of the Russians. I still believed Russia would build a new world and that I would take part in it. During this time I was not sure I could give all the information I had, however. It became more and more evident that the time when Russia would spread influence all over Europe was far away. I had to decide whether I could continue to hand over information without being sure I was doing right.

I decided I could not do so. I did not go to one rendezvous because I was ill at the time and I decided not to go to the following one.

Shortly afterward my father told me he might be going to the eastern zone of Germany. He disapproved of many things in Eastern Europe and had always done so, and he knew that when he went there he would stay there. I could not bring myself to stop my father from going there. However, it made me face at least some of the facts about myself. I suppose I did not have the courage to fight it out myself, and therefore took it out of my hands by informing the authorities that my father was going to the eastern zone.

A few months passed and I become more and more convinced that I had to leave Harwell. I was then confronted with the fact that there was evidence I had given away information in New York. I at first denied the allegations made against me. I decided I would have to leave Harwell, but it became clear that in leaving Harwell in these circumstances I would deal a great blow to Harwell and all the work I had loved and also leave suspicions against friends whom I had loved and people who thought I was their friend.

I had to realize that the controlled mechanism had warned me of danger to myself, I realized that the combination of the three ideas which made me what I was was wrong. I realized that every single one was wrong and that there were certain standards of moral behavior that are in you and which you cannot disregard. I find that I myself was made by circumstances. I know that I cannot go back on that and I know that all I can do now is to try and repair the damage I have done. The first thing is to make sure that Harwell will suffer as little as possible.

Before I joined the project most of the English people with whom I had made contact were left-wing and affected by a similar philosophy. Since coming to Harwell I have met English people of all kinds and I have come to see in many of them a deep-rooted firmness which enables them to live a decent life.

Family background.—Fuchs' 76-year-old father, Emil Fuchs, is today a professor of theology at Leipzig in the Soviet zone of Germany. He moved there from Frankfurt, in the American zone, 2 weeks before his son's arrest. Well known as a religious pacifist, he has written a number of books, including two postwar volumes entitled "Leonhard Raga, a Prophet of Our Time" and "Christianity and Socialism." The elder Fuchs visited the United States in 1949, making a lecture tour under Quaker auspices. Following the arrest, Professor Fuchs was contacted by telephone and expressed amazement at the charges against his son. The father is quoted as saying:

If he [the son] did it, it was not because of money but only because of his idealism and regard for communism.

Fuchs, senior, further said:

My son is like a child in everything outside his work—an absolutely childlike, good, helpful person, as all his friends can testify, an honest man through and through.

The elder Fuchs also predicted that east-west tension "may relax before long" and stated that, while he himself is without political affiliations, he has "respect for the Soviet experiment in building a new social system."

Rebecca West, a frequent British writer upon the subjects of espionage and treason, has this comment:

To understand Dr. Fuchs we must note that this is no case of the godless scientist cradled in materialism. Klaus Emil Fuchs came from a pious home. His father, Emil Fuchs, was a preacher well known in Germany since the beginning of the century; first as a Lutheran pastor and then as a Quaker. He was a true mystic, illumined by the love of God, and his courage in earthly affairs was superb. He was the first pastor to join the Social Democratic Party, and between the wars he was well known as a speaker for a group known as the Religious Socialist. He defied the Hohenzollern rule and defied Hitler. He was also a loving husband who made a delightful home for his sons and daughters. But in the opinion of some of those who liked him best he was not very intelligent, and his wirtings show that he was intensely egotistical and self-satisfied. His virtues are so great that it would be foolish to mention his failings, were it not that they have a bearing on his son's career.

Another writer, Kurt Singer, also observes that Fuchs' father "hated war * * *, standing in the shadow of Tolstoy and Gandhi." Mr. Singer adds:

His father's house had emphasized values of brotherhood, duty, internationalism, peace, religion, but there was little flexibility or humor in the teaching. Life was gray, grim, earnest, boring, and there was no time for carefree joy or laughter.

Klaus Fuchs was the youngest child of the family. His mother, who died in 1931, had relatives afflicted with insanity. One sister, Elizabeth aged 40, today lives in Cambridge, Mass., with her American husband. A brother, George Karl, aged 41, lives in Davos, Switzerland. Another sister, an artist, committed suicide before the war, jumping under a subway train in Berlin after an unsuccessful attempt to flee Germany and reach Czechoslovakia. All members of the family were persecuted by the Nazis, and the father spent 9 months in a German concentration camp.

Fuchs' early years.—Klaus Fuchs was born in 1911. Mr. Kurt Singer describes Fuchs' youth as follows:

The boy Klaus in the provincial town of Russelsheim, near Frankfurt, was strictly forbidden to join the cheering of the soldiers off for the front. Little Klaus began life as the outsider, the observer, the nay-sayer. He had no close boyhood friends and except for his three elder brothers and sisters he lived in virtual isolation, shielded from the contagion of hysterical patriotism and living in an aseptic world of his father's making. * * *

The first war was followed by the annihilating inflation, and the roots of nazism flourished in the economic swamp which Germany had become. Even in the primary school, politics was an urgent reality, and the pacifist's son Klaus was the butt of soldiers' sons, who made fun of the timid, studious boy. The troubles of Germany turned his father inward to reflection and religious experience. He became a Quaker in 1925. Klaus found no comfort there. Instead. it was clear to him that the boys who fought back and did not fear the violent little nationalists were the Reds, the Communists.

Later, at Kiel University, when the Nazis were already a major political force, Klaus joined the Young Communist League. Against his father's quakerism he embraced the doctrine of the class struggle. But he was never a great reader of Marxist literature. His field was science and, like so many brilliant mathematicians and physicists, the experimental and analytical techniques he used so scrupulously in the lecture halls and laboratories, he abandoned completely when confronted with political argument. He accepted all the worn clichés of Communist propaganda. Russia was the worker's fatherland; all weapons were permissible in the class struggle; the Communists were fighting for a classless society; there was no such thing as absolute truth or objective science: art and science were class weapons: the artist and scientist who believed in communism were in uniform and must take part in the world struggle. * * *

His father urged Klaus to escape from Germany so that he might continue his studies abroad, but Klaus remained working in the Communist underground movement. It was not activity that appealed to him. The disorganized life, being hunted from pillar to post, the need to abandon organized studies, did not suit the young student-scientist.

After a short while he crossed the frontier into France and from there he came to Britain. He went to Bristol University, where he specialized in mathematics and physics, and was awarded a doctorate in philosophy. His lodgings in Hampton Roads, Redland, Bristol, were the typical student's retreat, untidy, strewn with papers and books. It was a simple life and a happy one, on the whole. Too happy, perhaps, for Klaus to justify his conscience, for his father was in Germany where he had chosen to stay, although American Quakers had offered him a chance to get out. Emil Fuchs had replied to them that his place was in Germany in the fight against Hitler. Where, then, was the place of his son, Klaus?

Somehow the student had to justify to his father that his departure from Germany was not a flight from fear, but a tactical withdrawal to a place from which he could renew his role in the struggle. For the first time he was living in conditions of freedom and reasonable stability. Politics in Britain did not have the violence or the upsets that he had known in Germany. His fellow students were not consumed by bitterness nor deeply involved in doctrinal debate. Klaus, quiet and sensitive, emotional to an etxent which his poker-faced appearance belied, was attractive to certain types of girls.

Lonely and abstracted, he aroused the maternal impulse, and during his years in Britain he was never without female friends who admired and fussed about him. At the same time his studious, ingrown personality did not make him an exciting friend; his conversation did not often go beyond scientific small talk and university gossip. To his friends, Klaus was frankly a bore, but a nice bore.

Then his field of research widened. In 1938 he went to Edinburgh University, where he took his degree as doctor of science. His original researches in atomic and nuclear physics were placing him in the forefront of the younger scientists, and he published papers in the Proceedings of the Royal Society. The refugee-immigrant was making a name for himself in scientific circles.

Rebecca West depicts Fuchs' early years in this fashion:

He studied at Leipzig University, and then, when his father was made professor of religious science at a teachers' training college in Kiel, moved to the university there. At both places he was deeply involved in the useless and silly and violent political activities by which German undergraduates did so much to destroy the coherence of their own country and the peace of the world. College is a grand place for political discussions and a terrible place for political action. When college students go in for deeds, not words, cold-blooded adults get hold of them and without mercy use them as cat's-paws.

In Germany at that time the Communists were indulging in a campaign against the Social Democratic Party, although they should for obvious reasons have joined with the Social Democrats and the various schools of liberals in an unbroken popular front against the Nazis. Their secret reason for this was a tragic and ridiculous miscalculation; they wanted Hitler to come to power, in the mistaken belief that the Nazi regime would collapse immediately and leave Germany ripe for capture by communism. But they put up a noisy and hypocritical pretense that they were attacking the Social Democratic Party not because it was doing too much against the Nazis to suit them but too little.

Klaus Emil was completely taken in by this fraud, and very active under its influence. The political follies committed in the dying Weimar Republic are as unpalatable as yesterday's melted ice cream, but Klaus Emil's career must be followed because it led him and us to our present situation. He ran about with the high-speed inconsistency characteristic of German political life. He joined the students' section of the Social Democratic Party, but left it because the party supported a policy of naval rearmament, and he had been brought up to be a pacifist. But very soon afterward he joined a society with a mixed membership of Social Democrats and liberals, which was in fact a semimilitary organization with a taste for street fighting.

Then he moved to Kiel and went back to the Social Democratic Party, but presently left it again and offered himself as a speaker to the Communist Party without joining the party, and at the same time became a member of an organization, much frowned upon by the Social Democrats, in which rebel members of their party joined with Communists in the dangerous game of fraternizing with those students belonging to the Nazi Party whom they thought "sincere" and possible converts.

This was a nasty organization in which everybody was trying to double-cross everybody else. Then, when the Communists had so greatly weakened the Social Democratic Party that it could do nothing to fight the Nazis, Klaus Emil left it in disgust at the impotence and joined the Communist Party. The record reads like a receipe for mincemeat, but produced nothing wholesome.

When Hitler came into power in 1933 Klaus Emil was engaged in a complicated and futile campus intrigue, in which he showed a great deal of courage, particularly considering that he was of feeble physique, but little sense and even less fastidiousness. When he was in the train on his way to Berlin to attend a secret conference of anti-Nazi students he read of the burning of the Reichstag, saw that the hunt of the Communists had begun, took the hammer and sickle badge out of his coat, and went into hiding.

He was presently drawn into the operations of a mechanism which was one of the most brilliant achievements the Communist Party has to its name. After they had helped the Nazis get into power, they worked to get control of the organizations set up to care for the refugees from Nazi tyranny in all the countries to which they fled. They then saw to it that the Communist refugees received preferential treatments, that the non-Communist refugees were exposed to Communist propaganda and learned to look on Communists as their benefactors, and that the Communists and non-Communists refugees alike served the ends of the Communist Party.

All this they did with a pickpocket ingenuity, covering up their activities from the observation of the non-Communist members of these organizations, who were merely furnishing the bulk of the money and the personal service. Klaus Emil was told by his party that he must go abroad and finish his studies, because when the Nazis had been thrown out the Communists would need members with high technical qualifications to build up Soviet Germany; and he was first sent to France and then to England, where he was befriended by the Society for the Protection of Science and Learning, a body consisting almost entirely of non-Communists.

Here, it may also be recalled, Fuchs stated in his confession that, before he joined the atomic project, "most of the English people with whom I had made contact were left-wing and affected by a similar philosophy." This remarks appears to mean that many or most of Fuchs' prewar friends in England were, like himself, Communists.

Fuchs internment.—In 1940, after World War II had commenced, Fuchs was interned as an enemy alien of German nationality, along with hundreds of others. He was first sent to the Isle of Man in the Irish Sea and later to a detention camp in Canada. The reason for interning nearly all enemy aliens at this time was the crisis produced by the German invasion of the Low Countries and France. The British Government evidently felt it had best detain the aliens to be on the safe side and then release the trustworthy ones as soon as screening could be accomplished. The British Enemy Aliens Tribunal examined Fuchs in 1941 and released him from detention, although the British Government knew that Gestapo reports, which could not be verified, had named Fuchs as a member of the Communist Party. Sir Hartley Shawcross, Attorney General of Great Britain and Fuchs' prosecutor, made this statement at the trial:

> The investigations which were undertaken at that time [1940–41] had not shown that he [Fuchs] had any association whatever with British members of the Communist Party, and the investigations all tended to show that he was quite immersed in his academic studies and his work as a research worker, and was taking no active role in politics.

Mr. Kurt Singer gives the following version of the 1940–41 period:

> On September 1, 1939, Hitler invaded Poland and the war was on. Klaus Fuchs suddenly found that he was, finally, regarded as a German enemy alien. A few months later, despite his feelings against Hitler, his antipathy to nationalism, his years in British universities, he was told to pack a bag and get ready for internment. To him, British tolerance was a sham, as his Communist friends had told him. In the show-down, the British ruling classes were ruthless, heartless, barefisted—Fascist. The effect of internment on Klaus, the trip as an internee in the North Atlantic through waters infested by submarines, was to revive the Communist allegiance which had become quiescent. It also added the excitement of martyrdom to his essentially adolescent nature.
>
> In his Nissen hut, in the Canadian camp, it was not difficult for Klaus to imagine that "fate" had pointed out to him the error of his backsliding ways. It is certain that he emerged from internment with his Communist faith renewed. Separated from his friends in Britain, surrounded by many of his countrymen who were grieved that though anti-Nazi they were treated as enemies. Klaus looked again toward the distant, greener fields of the Soviet paradise.
>
> When he was able to resume his work, his old convictions were firmly fixed. In 1941, he was released from internment to continue his work, research which was to help in the development of the atom bomb. Although it was known that he was communistically inclined, so high was his qualification that he was allowed into the most secret consultations. Security officers, after careful screening, had reported that there was no danger he would become a foreign agent.
>
> Meanwhile, British Military Intelligence was receiving reports of an extensive German plan to build a new weapon, an atom bomb, which would be decisive in the war.
>
> In occupied Norway, secret underground agents reported the construction of strange, heavy-water plants, where hundreds of German scientists had been put on special duty. British-Norwegian commando teams went into action to cause as much physical destruction as possible for the new German production centers.
>
> At the same time, a meeting was called in London to lay the plans for an answer to German atomic research. To this, the deepest secret of the war, Klaus Fuchs, the Communist, was given access. The Communist, now ready to conduct espionage for Russia, was given material to work with. At his trial, Mr. Curtis-Bennett, his defense attorney, said: "Anybody who had read anything about Marxist theory must know that a man who is a Communist, whether in Germany or Timbuctoo, will react in exactly the same way. When he gets information, he will automatically and unhappily put his allegiance to the Communist idea first."
>
> Amazing enough, although it was on record at the Home Office that he was a member of the German Communist Party, a year after being released from interment, Klaus Fuchs was naturalized as a Briton. The superb resistance of the Red Army to the German invaders, the atmosphere of allied amity, all made it easy for Fuchs to submerge the vestiges which remained of his British "conversion" while accepting its citizenship.

His work with Prof. Rudolph Peierls, one of the outstanding atomic research scientists, during 1941, showed that he was clearly a genius in his field, "more a candidate for a Nobel peace prize or membership of the Royal Society," as Mr. Curtis-Bennett said at his trial, than a likely traitor. Fuchs lived happily with Professor Peierls and his family in a large, detached house in Birmingham. The young scientists was a favorite with the children.

Dr. Klaus Fuchs was now close to the pinnacle of the atomic pyramid: the abstruse and most vital theoretical side of the bomb. The information which he acquired in this position, as well as his own brilliant discoveries, meant years of toil to a nation still young in atomic problems like Russia. * * *

Rebecca West adds this:

It must be emphasized that at no time did Klaus Emil have grounds for complaint against Great Britain. He never found it niggardly, or on the side of reaction. He was sent to Bristol University, where he got his doctorate of philosophy in mathematics and physics, and then to Edinburgh University, where he got his doctorate of science, and was given a Carnegie research fellowship. When war broke out between Great Britain and Germany the aliens tribunal, before which he appeared to show cause why he should not be interned, accepted his membership in the Communist Party as proof that he was anti-Nazi.

It is true that in 1940, when the Germans invaded the Low Countries and France, he was interned and taken to Canada. But this was the treatment which was applied both to refugees who were thought especially suspicious and those who were thought specially meritorious, and in his case it was certainly a proof that the authorities believed him worthy of being saved from a possible German invasion of Britain.

In 1942 he was allowed to return to Great Britain, where a position was waiting for him at Glasgow University. Soon afterward he was asked by Professor Peierls, a very eminent German-born refugee physicist, to come to Birmingham University to help him in some war work. This proved to be atomic research. In June of that year he signed the usual security undertaking, and applied for naturalization as a British subject a month later, taking the oath of allegiance to the king in due course, while at the same time he made arrangements to hand over all particulars of the research to couriers who he knew would deliver them to the Soviet authorities.

Fuchs' work on the gaseous diffusion process.—When Fuchs first undertook atomic research in 1942 at Birmingham University, England, he found himself working upon the gaseous diffusion process which later evolved into the great Oak Ridge, Tenn., production plants. He immediately contacted a Communist acquaintance, who put him in touch with Soviet intelligence. The acquaintance, according to Sir Hartley Shawcross, Fuchs' prosecutor, was a foreign Communist "not recognized by the British authorities as a person who would be a Communist." In Britain, from 1942 to December 1943, Fuchs met with Russian agents at intervals of 2 or 3 months and transmitted documents, simultaneously receiving instructions for future meetings. At first he merely conveyed papers prepared by himself, but soon began to divulge all the information in his possession. The same contact procedure was followed in the United States from December 1943 to August 1944 while Fuchs lived in New York City and worked at Columbia University, although the meetings became somewhat less frequent.

The famous Smyth report (Atomic Energy for Military Purposes, by Henry D. Smyth) makes known that the scientists working upon the gaseous diffusion process at Columbia University, during the time Fuchs participated there, were divided into two groups. One operated directly under the Manhattan District and, supervised by Nobel prize-winner Harold Urey, was known as the SAM unit. The other group was connected with Kellex Corp., which had contracted to undertake studies in the gaseous diffusion field. Fuchs was intimately connected with both groups. Of the two letters reprinted below, one was written by a scientist who played a prominent and distinguished role in the SAM group and the other by a scientist who played a like role in the Kellex group. The letters follow:

THE H. K. FERGUSON Co., INC.,
New York 6, N. Y., March 19, 1951.

Hon. BRIEN MCMAHON,
   Chairman, Joint Committee on Atomic Energy, Congress of the United States,
   Washington, D. C.

DEAR SIR: Mr. William L. Borden, executive director of the staff of the Joint Congressional Committee on Atomic Energy has requested that I write a letter to you giving an estimate of the knowledge of the gaseous diffusion project acquired by Klaus Fuchs, with particular emphasis on the period 1943–44 when he was a frequent official visitor to the SAM Laboratories at Columbia University. Recognizing the importance of this infomation to the joint committee, I am pleased to accede to this request. My recollection of this period is naturally somewhat imperfect, but I believe I can reconstitute the situation with a fair degree of accuracy.

I should like to emphasize that although Fuchs became cognizant of a great many details of the gaseous diffusion project as developed in this country by the SAM Columbia group, 'the Kellex Corp., and many others, there is absolutely no question of impropriety in the disclosure of information by the many Americans involved. Fuchs was a fully accredited member of a British mission, whose assignment was to assist the American scientists and engineers working on the various phases of the atom bomb project. The mission was part of the wartime British-American cooperation on the atom bomb, the history of which is doubtless familiar to you. In every case discussions with the British mission were held within restricted areas, admission to which required the presentation of a visitor pass issued by the security division of the Manhattan District, which specified the people to be contacted, the topics of conversation, and limited the visit to specific dates.

Prior to Fuchs' arrival in this country, toward the end of 1943, he was chief theoretical aide to Professor Peierls, of Birmingham, who was one of the leading spirits in the entire British Tuballoy project and also an ardent advocate of the gaseous diffusion process. Fuchs' name appeared on theoretical papers on the gaseous diffusion process to my certain knowledge in 1942, and I believe as early as 1941. Because of visits to this country of Peierls and others in early 1942, when the relative merits of the Birmingham and Columbia versions of the diffusion process were discussed at length, and the established (though inefficient) channels of Anglo-American interchange of technical information, it is clear that before Fuchs' arrival he had good knowledge of the American plans for the gaseous diffusion plant. It is important to bear in mind that because of Fuchs' grasp of the theoretical principles involved, which interrelate the process variables so that the choice of a few determines the remainder within narrow limits, he would be able to reconstitute our whole program from only scattered pieces of information. Thus even before his arrival in New York when he obtained full and detailed information, he could have transmitted a very good outline of the American gaseous diffusion project.

Another point, which may be of some interest, is that compartmentation of information, which in this country isolated workers in the individual projects (plutonium project, calutron project, gaseous diffusion project, atomic weapons project, etc., etc.) from knowledge of other projects, was not followed in Britain. The British theoretical physicists in particular worked on all phases of the project. By the end of 1943, Fuchs undoubtedly had an over-all picture of the Manhattan District operations comparable to that of the most responsible members of the American project.

During 1943 Fuchs (in England) worked on the theory of the control of the gaseous diffusion plant. He was principally interested in two questions. The first was how the flow of gaseous uranium hexafluoride in the plant could be made to conform on the average to process specifications; the second was the effect on the plant production of fluctuations in the flow about the average flow rate. He made substantial contributions on both problems. Toward the end of 1943, the question of the proper method of control of the gaseous diffusion plant was being actively discussed. Work on the control problem was being carried out by the SAM Theoretical Division (under my direction) and by a section of the Kellex Corp. (under Manson Benedict). The future operating company, Carbide & Carbon Chemicals Corp. was also interested. Fuchs arrived in New York—I believe it was on December 7—and a series of meetings were set up, to be held alternatively at the Kellex Corp. offices in the Woolworth Building and at the SAM Laboratories in the Nash Building, wherein the two American groups and the British (Peierls and Fuchs) would compare results.

After several meetings in December a division of work was adopted, Fuchs' part of which was to calculate numerically for the plant being actually designed, the effects of fluctuations on production rate. All phases of the control problem depend on the intimate details of plant construction, and in the course of his assigned task Fuchs obtained from the Kellex Corp. complete knowledge of the process design of the K-25 plant.

It does not necessarily follow from his assignment that he would be familiar also with the mechanical design of K-25 or with the method of fabrication of the diffusion barrier. However, he was present at meetings where both these items and, in addition, schedules for barrier production, equipment delivery, and plant construction were discussed. For example, on December 22 he was present at a meeting at Kellex presided over by General Groves, attended also by Conant, Tolman, Colonel (now General) Nichols, a Columbia contingent headed by Urey, a Kellex contingent headed by Keith, Felbeck (Carbon & Carbide); and a British contingent consisting (besides Fuchs) of Akers, Simon, Curtin, Peierls, and Pfeils. The entire question of the feasibility of the K-25 program was reviewed at this meeting—going down the line of pumps, valves, control, barriers, and so forth. At another meeting on December 28, Peierls and Fuchs conferréd with SAM representatives on diffuser designs. On December 29 he attended a meeting at Kellex on the mechanical design of diffusers. On January 5, 1944, another meeting was held at Kellex largely on barriers, again with Groves as chairman, Conant, representatives from Columbia and Kellex, and a British contingent. I believe Fuchs was present: The minutes of the meeting will confirm this. At this meeting the status of barrier materials was reviewed with special reference to meeting plant schedules.

Fuchs continued active work on the control of the K-25 plant through January, February, and March of 1944. For example, on March 31, 1944, he reported at SAM on the results of his calculations. During this period much discussion of barriers took place, both theoretical and practical, and Fuchs must have known of the major policy decisions made at that time, particularly as the physical properties of the barrier entered into his calculations.

On May 1, 1944, I left SAM to accept another position and consequently have no knowledge of Fuchs' activities on the gaseous diffusion project after that date. However, between December 1943, when he arrived and at least until April 1944, he was a frequent visitor at both Kellex and SAM Laboratories, and participated in discussions covering a wide range of practical and theoretical problems. My knowledge and conviction is that during this period he had intimate and detailed knowledge of all phases of the design of the K-25 plant, including methods of fabricating the barrier, the assembly of the diffuser, and the planned production rate.

Looking back on this period in the light of recent disclosures of Fuchs' activities as a Russian agent, he possessed to an unusual degree the attributes of an efficient spy: Knowledge of the general scope of the work, access to detailed information, and an appreciation of its significance. In my opinion his contribution to Soviet progress in the production of atomic weapons, and the corresponding damage to the security of the United States and to world peace, was enoromus; one would have to search diligently in the records of treason to find a case with more far-reaching and permanent effects.

Compared to these consequences, Fuchs' betrayal of the personal integrity of scientists is of minor importance. Nevertheless, it is a blow which all scientists bitterly resent. For this, and many other reasons, distasteful as it may be to rake over this sordid affair, you may be assured of my cooperation in your investigation.

Very truly yours,

KARL COHEN,
Director, Atomic Energy Division.

HYDROCARBON RESEARCH, INC..
New York 6, N. Y., March 19, 1951.

Hon. BRIEN MCMAHON,
United States Senate, Washington, D. C.

DEAR SIR: I am happy to comply with the request of Mr. William L. Borden that I send you a brief, unclassified account of the extent of knowledge regarding the K-25 gaseous diffusion plant for concentrating U-235 obtained by Dr. Klaus

Fuchs during the period in which he assisted the Kellex Corp., the designers of this plant.

I was employed by the Kellex Corp. from 1943 to 1946 as head of the process development division. I was present at a number of the meetings and conferences authorized by the Manhattan District which were attended by Dr. Fuchs and other British scientists. My impression of the extent of Dr. Fuchs' knowledge regarding the diffusion plant is based on my recollection of the topics discussed at these meetings and of the contents of reports prepared by Dr. Fuchs.

Prior to his visit to this country in 1943, Dr. Fuchs was a member of the group of British scientists and engineers who had been designing a diffusion plant for construction in England. During this work in England Dr. Fuchs made important contributions to the theory of the diffusion process. He wrote a number of excellent reports which were of value to us at Kellex; one report on the theory of the control of a diffusion cascade was outstanding for its originality and usefulness.

In 1943 the British decided not to build a diffusion plant of their own. The leading engineers and scientists who had been working on the British plant, including Dr. Fuchs, were sent to the United States to be of assistance in the design of the American diffusion plant.

In December 1943, at the request of the Manhattan District, a series of meetings were held at the offices of the Kellex Corp. with the British group, including Dr. Fuchs, to acquaint them with the general features of the American diffusion plant and to obtain the advice, suggestions, and criticism of the British group. Another purpose of these meetings was to determine in what way the British could be of more specific assistance in the design of the plant.

It became apparent that the British workers could be of little help in the mechanical engineering of the plant, because the plans for the American plant were so different from those the British had intended to use. On the other hand, it was clear that the British workers could be of assistance in examining theoretical problems connected with the diffusion plant, because these were substantially the same in the British and American plants.

Consequently, a group of three British physicists, Dr. R. E. Peierls, Dr. Fuchs, and Mr. Skyrme, remained in New York to collaborate with the staff of the Kellex Corp. and the SAM laboratories in investigating theoretical design problems. For a short time this group was headed by Dr. Peierls, but when Dr. Peierls left New York, Dr. Fuchs took charge of the work. Dr. Fuchs remained during December 1943 and the early part of 1944.

The main problems investigated by the British group were—

(1) The control of the American diffusion plant.

(2) The efficiency of alternative arrangements of diffusion stages for separating uranium isotopes.

(3) The amount of separating efficiency lost when air leaked into the diffusion plant.

(4) The amount of separating efficiency lost when flow through the plant was disturbed at periodic intervals.

None of these investigations was vital to the success of the plant, but it was helpful to have this work done by the British group.

To do this work, the British group, and Dr. Fuchs, needed to have specific knowledge of the size and number of stages in the K–25 plant, the temperatures and pressures at which the stages operated, and the purity and amount of U–235 which the plant was being designed to produce. In addition, the general meetings Dr. Fuchs attended when he first came to the country undoubtedly acquainted him with the types of diffusion barrier which were being developed, and with the types of mechanical equipment, such as pumps, seals, valves and coolers, which were to be used in the plant.

Dr. Fuchs' contact with the diffusion process ended long before the plant went into operation in 1945, so that it is unlikely that he knew the U–235 production rate of the actual plant, and he may not have known which of the various types of barriers under development were actually used. Nevertheless, the comprehensive knowledge of the development status of the diffusion process early in 1944 which Dr. Fuchs acquired through these contacts with the Kellex Corp., all duly authorized by the Manhattan District, must have been of great value to him in his treasonable enterprise. Although Fuchs did not have detailed plans and specifications of the plant, he was in possession of information which, if transmitted to the Russians, would have saved them years of development effort.

Very truly yours,

MANSON BENEDICT.

Los Alamos

The extremely sensitive data to which Fuchs had access at Los Alamos, from August 1944 until June 1946, was highlighted in every statement made about him at the time of his arrest. To cite a further example, Senator McMahon said:

there can be no doubt * * * as to the general extent of Fuchs' information and knowledge both as to atomic weapons and so-called hydrogen bombs.

One wife of a Los Alamos scientist is quoted as recalling Fuchs thus: "A very quiet, rather sweet, reticent little guy." Another describes him as "a mild, unobtrusive, pleasant little man who never talked politics." Along with British friends he made one trip to Mexico. Harry Gold served as espionage courier for Fuchs during most of the Los Alamos period, contacting him two or three times at Santa Fe, N. M.; but during February 1945, Fuchs met once with a Russian official in Boston. Many of the atomic scientists were accompanied by security officers wherever they went, for purposes of assuring their safety. Fuchs did not happen to be among those so protected.

After returning to England in mid-1946, he became head of the Theoretical Physics Division at Harwell Laboratory and continued passing information to Soviet agents. Harry Gold has testified to the arrangements made for Fuchs' first meeting in Britain. Describing a conversation with Yakovlev, his Russian superior. in New York, Gold said:

The final item I reported to Yakovlev were the details of an arrangement which Fuchs and I had arrived at, which arrangement concerned the means by which someone would get in touch with Fuchs when he returned to England. The exact details were these: Beginning on the first Saturday of every month after it had been determined that Fuchs had returned to England, at a stop on the British subway, underground in London called Paddington Crescent, possibly Teddington Crescent, 8 p. m., Fuchs was to be on the street at the underground stop, the street level. He was to be carrying five books bound with strings and supported by two fingers of one hand; he was to be carrying two books in another hand. His contact, whoever that would be, was to be carrying a copy of a Bennett Cerf book, Stop Me If You Have Heard This.

Referring to the same conversation with Yakovlev, Harry Gold also testified as follows:

I reported to Yakovlev that Fuchs had told me about being present at the first atomic explosion at Alamogordo, N. Mex. Fuchs had said that the flash had been visible some 200 miles away. I told Yakovlev that Fuchs was very worried about one matter: This concerned the fact that the British had gotten to Kiel, Germany, ahead of the Nazis—ahead of the Russians, and Fuchs was very worried, very much concerned over whether the British Intelligence might not discover the Gestapo dossier upon him. I told Yakovlev that Fuchs had said that he, Fuchs that is, had been the leader of the student group, the German student group at the University of Kiel and had fought the Nazis, Nazi storm troopers in the streets of Kiel. Fuchs had said that there was a very complete dossier by the Gestapo upon him and he was greatly troubled by the fact that should the British Intelligence come upon it they would become aware of his very strong Communist background and ties.

Speaking of Fuchs' departure from the United States and his work at Harwell, Kurt Singer comments:

In 1946, Dr. Klaus Fuchs returned to Britain, carrying the prestige of his considerable achievement in the atomic project. He was given the high post of head of the theoretical physics division of the Atomic Energy Establishment at Harwell. He was a scientist's scientist, devoted to the welfare of his colleagues, a steady contributor to the Proceedings of the Physical Society and of the Royal Society. He apportioned jobs, passed on the qualifications of applicants, selected people for promotion. As chairman of the Staff Association Committee at Harwell, he presided over matters affecting personnel with a fine impartiality, liked by his employers and associates. * *. *

He participated in Harwell's social life, a little stuffily, unbending, awkwardly, but then genius has its mannerisms. It is nonsense to assume that his unmarked, repressed personality was a pose to assist his espionage. It was, however, a very useful weapon in the Soviet network. Fuchs, lonely, engrossed, inhibited, was actually alive only to a very small circle of intimates, who accepted the "flatness" as the hallmark of so many great scientists.

1947 visit

Fuchs visited America in 1947 to attend a declassification conference as between the United States, Britain, and Canada. This conference related to information held in common by the three countries as a result of their wartime collaboration. The question before the conference was what security classification should be assigned such commonly held information and whether or not any of it could properly be made public. Dr. Robert F. Bacher, a leading physicist at wartime Los Alamos, later one of the original Atomic Energy Commissioners, and now chairman of the physics department of the California Institute of Technology, wrote the following letter in regard to Fuchs and the 1947 declassification conference:

FEBRUARY 11, 1950.

Senator BRIEN MCMAHON,
   Chairman, Joint Committee on Atomic Energy,
        Senate Office Building, Washington, D. C.

DEAR SENATOR MCMAHON: As you have requested through Mr. Willam Borden, I am writing you to give you a brief summary of the joint declassification meetings which were held with representatives of the United Kingdom and Canada in Washington in November 1947 and to give you my recollection of the participation in those meetings of Dr. Klaus Fuchs who was one of the United Kingdom representatives.

Since I have no records available to me in Pasadena, I have refreshed my memory somewhat by talking on the telephone with representatives of the Atomic Energy Commission Secretariat who referred to minutes of these meetings. Due to the nature of the material, it was possible to refresh my memory only in part on the telephone. I found, however, that I seem to recall the meetings rather well and the following information is the best of my recollection of these meetings.

The classification conferences were held in Washington on Friday, November 14, 1947, and Saturday and Sunday, November 15 and 16. I attended these meetings as a guest in order that I might be able to assure the other members of the Commission directly that the terms of reference were carefully followed. I do not believe that I stayed through each of the meetings but according to the minutes and also according to my recollection I attended at least a part of each session.

The purpose of these declassification conferences was to discuss the interpretation of the declassification guide which all parties were trying to follow. It had become apparent that different interpretations of the guide might be made and it was essential to have a uniform interpretation if security of classified information was to be achieved. At the first meeting Dr. Warren Johnson, of the University of Chicago, acted as chairman and I gave a brief word of welcome and said what we hoped could be achieved along the lines of uniform classification policy as stated above.

On Friday, November 14, and Saturday, November 15, the discussions were mostly on general questions. On Sunday, November 16, meetings were held of two subcommittees. One of these subcommittees discussed questions of classification pertaining to weapons and associated work at Los Alamos. Since it was the policy in these discussions as stated in the terms of reference that "no classified information not already known to all parties concerned would be discussed," talks on weapons were attended only by representatives of the United Kingdom and the United States. To the best of my recollection, I attended this session in full. During the general sessions on November 14 and 15 as well as during the subcommittee sessions on November 16, I believe that the terms

To the best of my recollection, Dr. Fuchs was present at the subcommittee meeting at which topics associated with Los Alamos were discussed. I am quite sure, however, that no information was discussed which was not known to him already from his work at Los Alamos. In fact, the nature of these discussions was so general and so much concerned with the interpretation of words and phrases, that the informational content was negligibly small compared to that which any worker would have assimilated in a short period at Los Alamos.

I have refreshed my memory on the above subcommittee meeting by learning as well as I could over the telephone the specific topics which were discussed. The nature of these specific topics confirms my recollection and also the statements made above.

It is my opinion that the agreement achieved in the discussion of the declassification guide and its intepretation, added materially to the security of information which we were trying to keep classified. Furthermore, I believe that these objectives were achieved strictly within the bounds of the terms of reference of those talks, which called for no discussion of information which had previously not been available to the parties concerned.

I hope that this brief summary may be of use to you. If I may help you further in any way, I shall be very happy to do so.

Sincerely yours,

ROBERT F. BACHER.

While in the United States during 1947, Fuchs also paid a visit to the Argonne National Laboratory (successor to the Chicago "Met Lab"). The eminent scientific director of that laboratory, Dr. Walter H. Zinn, describes this visit in the following letter:

ARGONNE NATIONAL LABORATORY,
Chicago, Ill., February 6, 1950.

Hon. BRIEN MCMAHON,
Chairman, Joint Congressional Committee on Atomic Energy,
Senate Office Building, Washington, D. C.

DEAR SENATOR MCMAHON: In reply to the request made to me by Mr. Borden I am setting forth herein the details of Dr. K. Fuchs' visit to the Argonne National Laboratory in 1947. The visitor's pass at the laboratory shows that this gentleman arrived at the pass desk at 2:50 p. m., November 28, 1947, and departed from the same point at 4 p. m. The total duration of his visit therefore was 1 hour and 10 minutes.

In accordance with established procedure, access to the laboratory was not given to this visitor without previous permission and instruction by the Security Department of the Chicago office of the AEC. This was done by the transmittal to the director of the laboratory of three teletypes from AEC Washington to AEC Chicago concerning this visit. Two excerpts, as follows, from these teletypes give their essential content:

"Skinner and Fuchs are cleared to discuss unclassified and declassified aspects of neutron spectroscopy. No access to restricted data is to be afforded."

"Drs. H. W. B. Skinner and Fuchs, British scientists, who will visit your area are cleared to inspect the crystal spectrometer and mechanical velocity selector."

Again, in accordance with established procedure, I took the necessary steps to insure that the visit of Dr. Fuchs was concerned only with unclassified or declassified matters. This meant that he was conducted by a member of the security guard force from the pass gate to the director's office and that his pass was countersigned by the director or other officer of the laboratory who would have knowledge of the basis on which the visit was arranged. For this particular visit in November 1947 I cannot remember whether or not I personally conducted the visitor to the instruments which he had been cleared to see or whether or not this was delegated to a member of the scientific staff. I countersigned his pass. The presumption is fairly strong that I personally conducted him since these instruments had been constructed and used by me and the group working with me.

The crystal spectrometer referred to was completely described in the Physical Review, volume 71, page 752, June 1, 1947, in an article by W. H. Zinn. The mechanical velocity selector was described completely in an article in the Physical Review, volume 72, page 585, October 1, 1947, by T. Brill and H. V. Lichtenberger.

It is quite likely that upon arriving at the place where these instruments were in use, the scientists working on them participated in the conversation concerning these devices. Inquiry so far has not revealed who these scientists may have been, but it is also quite possible that this would be a difficult matter to determine since these instruments have been used from time to time by a considerable number of persons, some of whom are no longer at the laboratory. In any case, I am positive that precautions were taken to guarantee that the visit concerned only the unclassified matters for which clearance had been granted from Washington.

Yours truly,

W. H. ZINN, *Director.*

About the time of Fuchs' 1947 visit to the United States, he began to entertain doubts about the wisdom of Russia's policy. Shortly before his arrest he used sickness as an excuse to miss one scheduled appointment with a Soviet contact man, and he may also have held back some information which, except for these doubts, he would have transmitted. Kurt Singer comments:

At the peak of his career, Dr. Fuchs examined his course and decided that there was the possibility of a doubt creeping into his faith in communism: the sin of pride before the party in the Communist book of rules and regulations.

As a pledge of his subservience, Fuchs accepted a few hundred pounds payment from the Soviet agent. There had never been a road back for Fuchs; this was his way of demonstrating that he did not want one.

The fact is, however, that at this same time the first real doubts were creeping into his mind. He confessed later: "In the postwar period I began to have doubts about the Russian policy. During this time I was not sure I could go on giving information I had."

The arrest

Prime Minister Attlee has summarized the circumstances under which Fuchs came to be arrested. On March 6, 1950, he advised the British House of Commons that in the fall of 1949 a "tip" was received from the United States authorities and that an investigation disclosed Fuchs' true role as a spy. The Prime Minister's entire statement on this occasion is as follows:

I want to say one word about a matter which has caused a good deal of writing in the press, and that is the Fuchs case. It is a most deplorable and unfortunate incident. Here we had a refugee from Nazi tyranny, hospitably entertained, who was secretly working against the safety of this country. I say "secretly" because there is a great deal of loose talk in the press suggesting inefficiency on the part of the security services. I entirely deny that. Not long after this man came into this country—that was in 1933—it was said that he was a Communist. The source of that information was the Gestapo. At that time the Gestapo accused everybody of being a Communist. When the matter was looked into there was no support for it whatever. And from that time onward there was no support. A proper watch was kept at intervals. He was a brilliant scientist. He was taken on in 1941 for special work by the ministry of aircraft production. He was transferred to the department of scientific and industrial research. He went to America. He came back to Harwell. On all those occasions all the proper inquiries were made and there was nothing to be brought against him. His intimate friends never had any suspicion. The universities for which he worked had the highest opinion of his work and of his character.

In the autumn of last year information came from the United States suggesting there had been some leakage while the British mission, of which Fuchs was a member, was in the United States. This information did not point to any individual. The security services got to work with great energy and were, as the House knows, successful. I take full responsibility for the efficiency of the security services and I am satisfied that, unless we had here the kind of secret police they have in totalitarian countries, and employed their methods, which are reprobated rightly by everyone in this country, there was no means by which we could have found out about this man.

I do not think there is anything that can cast the slightest slur on the security services; indeed, I think they acted promptly and effectively as soon as there was any line which they could follow. I say that because it is very easy when a thing like this occurs—it was an appalling thing to have happened—to make assertions. I do not think that any blame for what occurred attaches either to the Government of the right hon. Gentleman opposite or to this Government or to any of the officials. I think we had here quite an extraordinary and exceptional case. I mention that because of the attacks that have been made.

Three months before the arrest Fuchs was promoted to a higher position at Harwell Laboratory in order to prevent him from suspecting that he might be under investigation. He was named one of the 25 senior scientific officers, whereas previously he had been a deputy officer. His promotion involved a salary increase from $4,480 to $5,040 per year. During this period he lived on the Harwell site itself in a small prefabricated house.

Both shortly before and after the arrest British security officials conferred with Fuchs. At first admitting nothing, he suddenly decided to tell everything. Rebecca West has discussed the statement he prepared, as follows:

Some measure of his oddity is given by the opening of the statement he made to the security officers on his detention. He began by giving them the date of his birth and assuring them that he had had "a very happy childhood." Now, British policemen seem much milder than American policemen and are certainly more stolid. But it is unlikely that they looked at Dr. Fuchs in a manner suggesting that it would take a weight off their mind if they could learn that he had not been unhappy when he was a small boy. It is unlikely, too, that most people, charged with a crime involving long-standing and heartless fraud and certain to cause hideous consequences, would fail to recognize that society might have other anxieties which it would like to settle first. This is a strange bird * * *.

Here was one of the most gifted scientists of our time, with power to be part creator of lethal weapons transcending all the previous malice of mankind, and to be as dangerous in his work as a single-handed traitor, because of his rare and exalted gifts. And his statement reads like the ramblings of an exceptionally silly boy of 16.

He was 38 years old. He was suspected of an appalling crime. He began by assuring the special branch officers of the happiness of his childhood, and went on to relate how brave he had been when he was a boy. It appeared that there was once a celebration at his school on the anniversary of the foundation of the Weimar republic; and as a protest many of the pupils arrived wearing the imperial badge, so he had put on the republican badge, and the other children had torn it off.

He recalled that; and he recalled, in the minutest detail, all his foolish and futile political activities at his universities. And in the course of this merciless recapitulation, which must have made the security officers groan aloud, he betrayed an unusual degree of political ignorance.

Every student of contemporary history knows that Communist strategy in Germany during the early thirties aimed at splitting the Popular Front and letting Hitler in so that he could be got out again by a revolution which the Communists would turn to their profit. Indeed, it is so well known that it would be virtually impossible for a non-Communist to write of those times without taking it as established historical fact, or for a Communist to write of them without attempting to disprove that assumption. But it is plain that Klaus Emil had never even heard of this interpretation of the events in which he took part. He wrote of them as naively as if he were still 20 and they had never been discussed.

Some of these tedious fatuities of his youth he recounted to the security officers for the sake of their moral, rather than their political, implications: and that, too was a curious self-betrayal. Throughout the statement Klaus Emil expressed himself with extreme egotism and vanity. Even if we take into account the strong strain of self-satisfaction running through his father's writings, and remember also that he had spent all of his childhood in minor industrial towns where his father was the unchallenged intellectual and moral leader, his sense of being an elect being must be pronounced extraordinary, particularly in a man of 38.

But it worried him, when what he had been doing was brought out into the open and he had to discuss it, that such a perfect character as his own should have been capable of practicing the continued deception, which, as he admits with an air of being fair-minded, had been a part of his treachery. He explained to the security officers at enormous length that this was all due to a mildly dirty trick he had played on some Nazi students during his campus intrigue in 1933. He had, not given them fair warning that he was going to publish an attack on them for a course of action which, had they received such a warning, they might have abandoned. He had omitted to resolve this point in his mind, he said, and so he had set up a mental process which he described as "controlled schizophrenia." It was, in fact, plain lying and cheating, but these were too realistic terms to be used in the "Cloud-Cuckoo-Land" where he had made his home.

There was no limit to his sense that power should be his. At one point in his statement he rebuked the British authorities for not letting the internees in the Canadian camp read newspapers. He ignored the practical reason for this, which was the difficulty of keeping discipline and protecting the non-Nazi internees from the Nazi internees, had the news continued to be bad over any length of time. Gravely he complained that it had prevented him from learning the truth about the real character of the British; and it is implied that had he known more about them he might have spared them, might not have aided their enemies to drop A-bombs on them. Not for a moment did it cross his mind that perhaps it was not for him to smite them or to spare them.

The trial

A hearing was held on February 10, 1950, at Bow Street Court, London, and the following is a summary that appeared in next day's New York Times:

[New York Times, Saturday, February 11, 1950]

TESTIMONY AT FUCHS' HEARING IN LONDON

LONDON, February 10.—Following is a detailed record of the testimony of three witnesses—Wing Commander Henry Arnold, retired security officer at the Ministry of Supply's atomic plant at Harwell; William J. Scardon, Harwell security officer; and Michael Perrin, atomic scientist—at today's hearing in the case against Dr. Klaus Fuchs:

Commander Arnold, the first witness against Dr. Fuchs, said he had "impressed security regulations on Dr. Fuchs" and that he thought at one time that Dr. Fuchs was "an exceptionally security-conscious person."

T. Christmas Humphreys, the prosecutor, opened his interrogation of Commander Arnold by asking him point-blank:

"Did you ask Dr. Fuchs if he had disposed of information to Russia?"

"Yes," Commander Arnold replied.

"Did he tell you the technical information he disclosed?"

"In a broad sense," Commander Arnold replied.

"Was it technical information of the greatest value to Russia?

"Definitely." * * *

Then Magistrate Sir Laurence Dunne broke in to say that this information was so secret that "it will not be taken up at this point." He added that the director of the Harwell atomic-research laboratory would be called to give testimony "on that point."

GUIDED BY CONSCIENCE

During examination of Mr. Scardon, who said he had had a great many interviews with Dr. Fuchs, the following exchanges occurred with the prosecutor:

"Did the conversation touch upon his oath of allegiance?"

"Yes."

"What did he say about it?"

"He said he regarded his oath of allegiance, taken upon naturalization in 1942, as a serious matter, but he claimed the freedom to act in accordance with his conscience should circumstances arise in this country comparable to those which existed in Germany in 1932.

"He said he would feel free to act on the loyalty which he owed to humanity generally."

"At a later stage did you make it clear that you suspected him of passing information to the Soviet authorities?"

"Yes."

"What was his first reaction?"

"He seemed surprised and said, 'I don't think so.' "

"Did you make it clear to him that you were in possession of precise information on this matter?"

"Yes."

"What did he say?"

"He again replied, 'I do not think so.' I told him that that was an ambiguous reply and he said, 'I do not understand. Perhaps you will tell me what the evidence is. I have not done any such thing.'"

"Did that remain his attitude for some time?"

"Yes."

"On the grounds of his father's presence in Leipzig?"

"That is so."

"On January 24 did you see Dr. Fuchs at his own request at his private address in Harwell?"

"Yes."

"What did you say to him?"

"I said, 'you asked to see me and here I am.' He replied, 'Yes, it is rather up to me now.'"

"Did he once again tell you the story of his life but with no admission of these offenses?"

"Yes."

"What seemed to be his mental condition?"

"He was under considerable mental stress."

"What did you say to him?"

"I suggested that he should unburden his mind and clear his conscience by telling me the full story. It seemed to me that whereas he had told a long story providing a motive for his acts he had told me nothing about the acts themselves."

"What did he say to that?"

DECIDES TO SPEAK OUT

"He said, 'I will never be persuaded by you to talk.' There was then an interval for lunch and after lunch Dr. Fuchs said to me suddenly and voluntarily that he had decided it would be in the best interests to answer questions. He added that he had a clear conscience at present, but was very worried about the effect of his behavior upon the friendships which he had contracted at Harwell."

"Will you summarize what he said in answer to your questions?"

"He said he was engaged in espionage from the middle of 1942 until about a year ago. There was a continuous passing of information relating to atomic energy at irregular but frequent meetings. This illegal association commenced on his own initiative, and no approach had been made to him. He himself spoke to an intermediary who arranged the first rendezvous.

"Did he himself say something about the continuation of his work at Harwell or his possible resignation?"

"Yes; he said that since he was under suspicion he might, upon reflection, think it impossible to continue to work at Harwell and that if he came to that conclusion he would offer his resignation. He thought it would be perfectly simple for him to obtain a university post. He also foresaw that there would be no particular disadvantage in his doing so. It seemed to me to be quite clear that his great interest was in the work upon which he was then engaged."

"Was there a further interview on December 30 when you told Dr. Fuchs that the Ministry would undoubtedly decide to dispense with his services?"

"Yes."

"Thereafter future interviews were arranged at the current meeting when an alternative arrangement was made to meet every eventuality. For a long time the defendant confined his information to the product of his own brain. But as time went on this developed into something more. He said the talks were sometimes certainly with Russians, but others were with persons of unknown nationality. He had realized that he was carrying his life in his hands, but he had done this from the time of his underground days in Germany.

"He said there was a prearranged rendezvous, and recognition signals were exchanged. The association continued through 1944 in New York, for a period at Los Alamos and in London again on his return to England."

DOCUMENTARY INFORMATION

"Generally, the meetings were of short duration and consisted of his passing documentary information and with the other party arranging the next rendezvous. At times he was questioned, but the defendant thought it to have been inspired from some other quarter than his contact.

"For the last 2 years of his association with the Russians there was a gradual reduction in the flow of information which he imparted since he was beginning to have doubts as to the propriety of his actions.

"He said he still believed in communism, but not as practiced in Russia today. In this form he thinks it is something to fight against. He said he had never been a member of the British Communist Party. He said that he had decided fairly recently that he could only settle in England and that he had been terribly worried about the impact of his behavior upon his friendship with various people and in particular with Wing Commander Arnold at Harwell."

"What did he say about expenses or reward?"

"He said that in the early days of the relationship he had accepted expenses and admitted taking the sum of £100 shortly after his return to England in 1946 from his contact.

"He explained that he had discussed the acceptance by Dr. Allan Nunn May [British scientist who was sentenced to 10 years] of money from the Russians with a friend who knew him who said that he thought Dr. May had taken this money merely as a token payment.

"The defendant after thinking it over accepted the sum of £100, regarding this as a symbolic payment signifying his subservience to the cause.

"On January 26 of this year I saw him again at his own request. He was anxious that his position should be resolved as quickly as possible. He wondered whether the authorities would clearly understand his position and I asked him whether he would like to make a written statement, incorporating any details which he thought ought to be borne in mind. I suggested three possibilities: that he should write out a memorandum himself; that he should dictate a statement to a secretary; or that I should write down a statement at his dictation.

MEETINGS WITH AGENTS

He said he would like to avail himself of my services and we made arrangements to meet in London on January 27 for this purpose.

"I then asked him about meetings with agents, and he said the first was a private house in London where he had met a man whom he believed to be a Russian. That was early in 1942.

"He said that after that first meeting there were meetings at intervals of 2 or 3 months for about 6 months before he went to New York in December 1943.

"In New York three or four meetings took place following the first which had been arranged before he left England. There were further meetings between the time when he went to Los Alamos in August 1944, and his return to England in 1946. He said there was only one person at each contact.

"He thought his first meeting after his return to England was at the beginning of 1947 and thereafter meetings took place at roughly two monthly intervals, always with the same man in London.

"On January 27 this year I took the defendant to the War Office, having met him at Paddington station. I cautioned him and said 'I ought to tell you that you are not obliged to make a statement, and you must not be induced to do so by any promise or threat which has been held out to you.' He said 'I understand. Carry on' I then wrote down a statement at Dr. Fuchs' dictation.

"Dr. Fuchs read it over, corrected it in several places in his own handwriting and himself wrote the last line which read, 'I have read this statement and, to the best of my knowledge, it is true.' "

Prosecutor Humphreys passed the statement to Mr. Scardon for identification and then asked the court to treat it as a "secret document."

Mr. Scardon continued testifying:

"After making the statement Dr. Fuchs said he was most anxious to discover what his future was to be He said he did not want to waste any time in getting the matter cleared up.

"Dr. Fuchs offered to give all technical information to a technical expert and on January 30, after meeting Dr. Fuchs at Paddington station, I took him to the War Office where he met Mr. Perrin."

There was only one question by Dr. Fuchs' own lawyer when Mr. Scardon had ended his testimony for the prosecution:

"Would it be fair to say that since lunch time on January 24 he has helped you and been completely cooperative in every way?"

"Yes, sir," Mr. Scardon replied to the attorney, Thompson Halsall.

VALUABLE TO AN ENEMY

Mr. Perrin was then called. An atomic scientist whom the security officers had used to question Dr. Fuchs, Mr. Perrin said Dr. Fuchs had given him what the accused said were full details.

Mr. Perrin was asked when Dr. Fuchs said he had had his first contact with a Russian about work at Los Alamos.

"In February 1945, at Boston, Mass.," he replied.

"Did he tell you whether he had passed any information?" Mr. Perrin was asked.

"Yes, he did."

"Was the information of value to a potential enemy?"

"It was."

"During the course of his discussions with you did he deal with the year 1947?"

"Yes."

"Did he describe the information and the purport of it?"

"Yes."

"Was that technical information about atomic research, and of the greatest possible value to a potential enemy?"

"Yes"

"In general was all the technical information thus passed valuable to a potential enemy?"

"Yes, it was."

When Mr. Perrin finished, Mr. Humphreys asked the court for trial at Old Bailey's at the next session.

Dr. Fuchs stood in the dock while the magistrate read the charges against him. He then asked Dr. Fuchs if he wished to make any statement. Dr. Fuchs said he did not.

Mr. Halsall said on Dr. Fuchs' behalf:

"He has nothing to say at this stage, and will call no evidence in this court."

Dr. Fuchs was then committed for trial, and the hearing ended.

The Fuchs' trial was held on March 1, 1950, with Lord Chief Justice Rayner Goddard presiding. Attorney General Sir Hartley Shawcross was the prosecutor and Derek Curtis-Bennett, chief defense counsel. Ultimately, there were four counts in the indictment, charging that Fuchs communicated to unknown persons atomic research information calculated to be useful to an enemy. Two counts were added to the original charge at the time of Fuchs' arrest. The four offenses were said to have been committed in Birmingham, England, in 1943; in New York City between December 31, 1943, and August 1, 1944; in Boston, Mass., in February 1945; and in Berkshire, England, in 1947. There was only one witness, Mr. William J. Skardon, who had also testified earlier and who was called by the defense to state that when Fuchs made his confession of guilt he was not under arrest but was a freeman. The trial lasted 1 hour and 27 minutes; there was no jury and no evidence beyond that offered at the prior Bow Street proceedings.

Lord Goddard told the prisoner:

You have betrayed the hospitality and protection given to you with the grossest treachery.

Dare we now give shelter to political refugees who may be followers of this pernicious creed, who may well disguise themselves and bite the hand that feeds them?

You might have imperiled the friendship between this country and the great American Republic with whom His Majesty (the King) is allied.

You have done irreparable harm both to this land and to the United States of America, and you did it as your statement shows, clearly for the purpose of furthering your political creed.

Sir Hartley Shawcross, after telling the court that the information betrayed was "likely to be of utmost value to an enemy," made the following statement:

As to the value of the information, perhaps, it is not in the public interest to say more than this. There were, of course, many fields of atomic research and of the general experimental and developmental work in regard to atomic energy which were being carried on and which were unknown to him, and those fields were consequently protected.

On the other hand he was a scientist of the highest standing in his own particular field, and although, according to his statement, he did not disclose the whole of his knowledge as to that field, information he had admittedly disclosed would undoubtedly have been of the greatest assistance as to that particular field.

One must therefore, regard the disclosures as a very grave matter indeed. That gravity he cannot now, even if he would, mitigate, and the bitterness of his position must be made the more acute by his own belated realization that the cause to which he gave such unswerving devotion was itself a false cause.

Mr. Curtis-Bennett, the defense counsel, said of Fuchs, in part:

He came to Britain for the purpose of conducting his scientific investigation and study, and he said quite frankly, to fit himself out as a scientist in order to help the rebuilding of Communist Germany. He did not come to Britain to build atom bombs.

He pursued his peaceful studies, and if the war had not come he might have been more a candidate for a Nobel peace prize or a membership of the Royal Society.

Within the Joint Committee on Atomic Energy it had been suggested that the possibility of seeking to extradite Fuchs from Britain for trial in the United States should be explored. Under British law the maximum possible sentence which could be imposed was 14 years, whereas United States law would have allowed the death penalty. However, the applicable treaty between the United States and Great Britain did not permit extraditing Fuchs unless its provisions were to be construed with great liberality; and in this branch of international law a strict construction has been the rule. Also, where an individual has violated the laws of two countries, he is normally tried and punished in the country where apprehended before the issue of extradition is considered. Therefore Fuchs was sentenced to 14 years by the British court and is now serving this term at Brixton Prison, near London. He is permitted to devote some time, as a prisoner, to scientific studies.

A month after the trial it developed that Fuchs' name had appeared in Halperin's notebook, an exhibit in the Canadian spy trials of 1946. Israel Halperin, a member of the wartime espionage ring operating in and around Ottawa, Canada, had compiled many hundreds of names and addresses of people who, for one reason or another, he desired to remember. The fact that Fuchs had been so listed in Halperin's notebook was announced by the Lord Chancelor, Viscount Jowitt, before the British House of Lords on April 5, 1950. He said:

My Lords, I wish to correct a point of fact arising out of a speech which I made last week on the motion of the noble Lord, Lord Vansittart. I then said that there was no truth in certain statements which had been made in the press about the Fuchs case. The fact is that in a notebook belonging to a man who was one of those examined by the Canadian Royal Commission there did appear, amongst a long list of other names, the name of Klaus Fuchs. This notebook, together with all other relevant material, was promptly made available to us by the Canadian authorities. Subsequent events have, of course, attached a significance to that name which it did not then bear. As I was, when I made my speech, imperfectly informed on this particular, I thought I owed it to your Lordships and to the press to make this correction.

Another post-trial development had to do with the question of whether or not American FBI agents would be permitted to interview Fuchs in prison and seek to elicit from him information which might help bring about the arrest of United States confederates. The Department of State requested the United Kingdom to grant permission for such an interview, and the chairman of the Joint Committee on Atomic Energy subsequently made the strongest possible representations in favor of this step. On May 11, 1950, Mr. James Chuter Ede, in behalf of the Labor Government, made this statement to the British Parliament:

It is provided by the Prison Rules, 1949, that an officer of police may visit any prisoner who is willing to see him on production of an order issued by or on behalf of the appropriate chief officer of police, such visit being additional to the prisoner's normal entitlement of visits. This provision is intended to relate to visits by officers of British police forces and I am not aware of any precedent for such a visit by police of other countries. The Government of the United States has, however, recently made a request that a representative of the Federal Bureau of Investigation should be allowed to visit the prisoner Klaus Fuchs, and in the exceptional circumstances of this case the request has been granted. In accordance with the usual practice the visit will take place in the presence of a prison officer and will be subject to the usual conditions governing the interviewing of prisoners.

As a result FBI agents, Hugh Clegg and Robert Lamphere, visited at length with Fuchs in his prison cell in order to fill out the complete details of his activities.

The FBI eventually pieced together enough information to bring about the arrest of courier Harry Gold. Fuchs was able neither to give the FBI Gold's name nor to furnish an accurate description. But the FBI managed to single out Gold after sifting approximately 1,500 possible suspects. When photographs of Gold were located and shown to Fuchs, he could not recall having seen the individual pictured. But when the FBI produced motion pictures of Gold, the prisoner was then able to identify Gold with positive assurance. This occurred after Gold had been arrested and had confessed.

Fuchs is today 40 years of age, 5 feet 8 inches tall, broad-shouldered, sallow, dark-haired, and balding. He wears glasses and speaks English with a strong German accent. His last public statement at the trial is as follows:

There are also some other crimes which I have committed, other than the ones with which I am charged. When I asked my counsel to put certain facts before you I did so in order to atone for these crimes.

They are not crimes in the eyes of the law.

I have had a fair trial and I wish to thank you, my Lord, my counsel, and the governor and staff of Brixton Prison for their considerate treatment.

The following excerpt from an interview with Gordon Dean, Chairman of the United States Atomic Energy Commission, constitutes a kind of postmortem on the case—insofar as such a case can ever be considered closed:

[Reprinted by permission of the United States News Publishing Corp.]

*   *   *   *   *   *   *

Q. How long does it take to clear an individual?

A. The average today is 53 days.

Q. And also the money that it costs to do the clearing?

A. The cost is between $100 and $200 per person, and if you clear thousands of people for projects, that's a lot of money—and time.

We are taking, I might add, a very hard look now. As a matter of fact, we've just had a long session on what we might recommend to change the act, with the objective both of keeping secret those things which we should keep secret and at the same time getting on with the job. I think that probably by the time Congress comes back we will have some constructive suggestions to make as to how it can be done.

Q. What percentage of the people you examine fail to get clearance?

A. Very small. My guess is that it is less than half of 1 percent.

Q. Would you say that the arrangement or the setup you have with other governments for obtaining information from us is now satisfactory?

A. Do you mean, can we trust the certifications of other governments?

Q. Well, that is another way of asking it. The Fuchs case arose out of the fact that we didn't have it. Is the present arrangement satisfactory?

A. I think the present arrangement is generally satisfactory. We did have conferences, you know, with the British and the Canadian security officers immediately after this thing. They came over here and we had a 3-day session, largely to determine the comparability of our security standards, and I think it is reasonably safe.

Q. You don't feel so apprehensive of losing out in that direction?

A. No, although in the most perfect system there may be someone who will slip through.

Q. We now clear foreigners, don't we? We didn't clear Fuchs ourselves—we depended upon the British?

A. You still can't have the FBI running investigations through all foreign countries, making their own investigations. What we have to do is to delegate it to a competent security group, comparable to our own, to make sure that the investigations cover the same type of points we make here in the States. Of course, the FBI, in turn, does the same thing for other governments. When someone is over here that the foreign government wants a check on, the FBI will make the check for them.

Q. Can you evaluate the damage that was done to our country by Fuchs and his associates in terms of the Russian progress?

A. It is hard to do, but I don't think you would be taking too extreme a position if you said he had advanced them between a year and 2 years.

Q. To what extent did the British have access to our atomic information? I believe we were supposed to be partners with them in the original development of atomic energy?

A. During the war, it was a complete partnership. The British decided to give up trying on the gaseous-diffusion work and they came over to this country, and we had a complete partnership. As a matter of fact, there were about 30, I believe, in the military mission from Britain who went to Los Alamos. They knew everything. They helped us very much in the development of the weapon. Since the war, we have operated under an understanding with the British and the Canadians in several areas which are not weapons areas. We have exchanged some visits within those areas, but that is the extent of it.

Q. In weapons there is now no real exchange?

A. No.

Q. Has the Fuchs episode had any effect on those scientists who were inclined for a long time to pooh-pooh the need for security—American scientists who were a little bit annoyed and irritated by our desire to have security because they thought it was inconceivable that Russia could do what she has done?

A. I wouldn't limit it to scientists. I would say that the Fuchs episode has had a sobering effect upon everybody connected with the program.

Q. In that way it was a blessing in disguise?

A. I think so. Some good came from it. It certainly doesn't equal the bad, but some good did come from it.

Q. Have you any idea what is wrong with human beings or with our system in these democracies of ours that these people will do the things that Fuchs did? Does the scientist have less regard for loyalty to his country than other people? Is he a world citizen who wants to give everything away? What is the reason that Fuchs got into this thing?

A. I don't think that you can say that scientists are an entirely different breed in that respect. In Fuchs' defense, let me say that we have had some of them who were not scientists. Fuchs is the type of man who, while he might have been caught had there been a real security check on him, might never be caught by any kind of investigation, because apparently he owes his allegiance to nothing that ordinary humans owe theirs to. He is going to make his own decisions regardless of any rules he purports to operate under.

What do you do with a man like that? Usually he is a very intelligent man. He is an independent man. He is an idealist of some kind. He might be a Communist-idealist, but he is a man of ideals of some kind. You don't usually spot this type in a check.

Q. Was he an inventive type, or was he just a theoretician?

A. I would say the inventive type—he was a very bright man.

* * * * * * *

5. The Pontecorvo Case—Details

Sixteen years ago there appeared in the April 1935 Proceedings of the Royal Society of London what is now a famous scientific article. It is entitled, "Artificial Radioactivity Produced by Neutron Bombardment—II," and the six authors are E. Amaldi, O. D'Agostino, E. Fermi, B. Pontecorvo, F. Rasetti, and E. Segre. Several of these same authors, although not including B. Pontecorvo, had published an earlier article along similar lines in 1934. Both articles stemmed from research performed at Rome, Italy, and one of the authors, E. Fermi, is today known throughout the world as a Nobel price winner in physics and a foremost contributor to American atomic energy development both during and after the war.

In 1935 the six scientists filed a patent claim with the United States and other countries concerning the results of their work. United States Patent No. 2,206,634 was actually awarded in 1940 and is labeled, "Process for the Production of Radioactive Substances." This patent is today the subject of litigation, with the holders claiming $10,000,000 in damages for wartime and postwar governmental use of their processes. Bruno Pontecorvo would have stood to gain from the litigation, if it results in a damage payment and if he had not disappeared behind the iron curtain.

Dr. Fermi has written the following letter in regard to Pontecorvo:

THE UNIVERSITY OF CHICAGO,
INSTITUTE FOR NUCLEAR STUDIES,
Chicago 37, Ill.. March 13, 1951.

Senator BRIEN McMAHON,
United States Senate, Washington, D. C.

DEAR SENATOR McMAHON: This is the answer to the telephone call that I received this morning from William Borden, asking me to supply you with a statement about B uno Pontecorvo.

I knew Pontecorvo very well when he was my student and collaborator in Rome for 2 or 3 years, beginning about 1935. After that I have seen him much less frequently, perhaps on the average about once a year.

Personally Pontecorvo is a very attractive person, who makes friends easily and appears to be quite e troverted. Scientifically he is one of the brightest men with whom I have come in contact in my scientific career. He is responsible for a number of important contributions to physics. Since his student days he has always been working in the field of nuclear physics except for a short interlude when he was employed by an oil-prospecting firm in Oklahoma.

His family owned a small manufacturing concern in Pisa, Italy. He has a large number of brothers and sisters. One of his brothers is a noted biologist, who is now living in England I believe that one of his brothers and a brother-in-law are, or have been. high in the council of the Communist Party in Italy.

After his student days in Rome, probably about 1936, Pontecorvo worked in France in the Curie Laboratory. He came to this country at the time of the fall of France and found first employment in an oil concern in Oklahoma ; later on he joined the Canadian atomic project and worked there until about 2 or 3 years ago, when he moved to the Harwell Laboratory in England. I believe that he is a British citizen.

I do not know, of course, what are the reasons that prompted his alleged escape to Russia. My personal impression of his research activities has been that he did not have much interest in the atomic developments except as a tool for

38

scientific research. In particular I do not remember any instances in which he took up with me any subject connected with atomic technology and he did not seem to have any special interest in atomic weapons.

For these reasons my impression is that if he went to Russia he may not be able to contribute to their work by the things that he has learned during his connection with the Canadian and the English projects but rather through his general scientific competence. This naturally is only a surmise.

I do not remember that Pontecorvo seemed very much interested in politics and I do not remember ever to have had political discussions with him.

Sincerely yours,

ENRICO FERMI.

Another of the authors of the 1935 scientific article and one of the claimants in the patent suit is Dr. Franco Rasetti. He kindly consented to be interviewed about Pontecorvo, and this interview resulted in the following joint committee staff paper:

Memorandum for the file.
From: Harold Bergman, Deputy Director, JCAE.
Subject: Interview with Dr. Rasetti, May 14, 1951.

Following advance telephone arrangements, I this afternoon interviewed Dr. Franco Rasetti in his office at Johns Hopkins University, Baltimore, Md., on the subject of Bruno Pontecorvo. Dr. Rasetti has been a professor of physics at the Johns Hopkins University since October 1947. Previously, he was associated with the physics department of Laval University, Quebec City, Quebec, from 1939 to 1947; the University of Rome, Italy, from 1927 to 1939; and the University of Florence, Italy, from 1923 to 1927. His chief field is the physics of geology and, in 1936, Prentice-Hall published his Elements of Nuclear Physics.

According to Dr. Rasetti, the city of Pisa, Italy, was—before the war—a small town in which "everybody knew each other." Consequently, he knew the Pontecorvo family quite well, although his closest contact was with the oldest son, Guido, now about 46. Dr. Rasetti lists the Pontecorvo family as follows:

Massimo, father	Guiliana	Anna
Maria, mother	Bruno	Giovanni
Guido	Gilberto	
Paul	Laura	

The children were born approximately 2 to 3 years apart, and probably in the order above-named. Guido is now professor of genetics at the University of Edinburg, Scotland. Two of the other children are probably now in France and England.

Dr. Rasetti commented that the Pontecorvos were a respected and fairly wealthy Italian-Jewish family. Massimo, the father, and several brothers owned textile mills in and around Pisa until the depression of the 1920's, when they lost much of their money. The parents are still living in Milan, Italy

While Dr. Rasetti was working at the University in Rome, he received a letter from the oldest brother, Guido, in either 1931 or 1932. The letter indicated that Guido's younger brother, Bruno, was interested in the field of physics and desired to study at the University of Rome. Dr. Rasetti was asked to do what he could to help.

Bruno Pontecorvo, according to Dr. Rasetti's recollection, attended the University of Rome as an undergraduate for 2 years and continued there for perhaps three additional years as a graduate student. Dr. Enrico Fermi was then a professor of physics at the University of Rome, and it was there, during 1934, that he, Pontecorvo, and others, performed the work which lead to the publication of the famous scientific papers appearing in the Proceedings of the Royal Society of London.

Dr. Rasetti commented further that during his association with Bruno Pontecorvo, during 1934 and 1935, the latter was quickly recognized by his seniors as a "gifted youth." Furthermore "everybody liked him." Dr. Rasetti stated that this popularity has been true of Pontecorvo through the years. Recently Dr. Rasetti met some of Pontecorvo's Chalk River associates and, according to Dr. Rasetti, they commented on Pontecorvo's flight to Russia thus: "We just can't believe it." "He appeared to us in 1935 as a normal young man, only more gifted," Dr. Rasetti said.

He added that, during the Rome years, Pontecorvo showed no interest in politics and managed, "as most of us tried," to stay out of the Fascist organizations—with the possible exception of normal membership in "required" organizations.

In 1935, Mussolini—under pressure from Hitler in Germany—began emphasizing anti-Semitism. About this time, according to Dr. Rasetti, Bruno Pontecorvo felt that he could not progress in his scientific work if he remained in Italy. Therefore, Dr. Rasetti continued, Pontecorvo applied for and received a Government fellowship for study in the field of physics with Prof. Fredric Joliot-Curie in Paris. Dr. Rasetti observes that he himself may have signed a letter to help procure the fellowship, but he does not recall for sure. The duration of the fellowship was 1 year, but Pontecorvo secured its renewal for an additional year. Thereafter, he may possibly have received fellowship assistance from a French source. Dr. Rasetti stated that he doubts Pontecorvo had much actual contact with Joliot-Curie, since the latter was a recognized scientist and the other a student. However, Dr. Rasetti has heard that during Pontecorvo's stay in Paris, he did associate with "Leftist" elements, probably younger individuals than Joliot-Curie.

In Paris, according to Dr. Rasetti, Pontecorvo met and married a Swedish girl, named Mariana. They had a boy in Paris. Mariana is known to Dr. Rasetti, who later met her in Montreal. He states that she is "no intellectual"; showed no interest in politics or science, and appeared to be interested only in her husband and children. When the Nazis took Paris in June 1940, Pontecorvo, his wife, and child left Paris hurriedly by bicycle for unoccupied France and remained in Bordeaux until late 1940 or early 1941, when they came to the United States.

In the United States Pontecorvo worked during 1941 and 1942 for the Wells Survey, Inc., Tulsa, Okla., doing radiographic oil well logging. He did not invent the method, but he improved it. In the fall of 1942 Pontecorvo was offered a position with the National Research Council of Canada in Montreal. Rasetti was then teaching at Quebec City and saw Pontecorvo again for the first time since 1935. They subsequently met at scientific meetings perhaps a half dozen times between 1942 and 1949 in Canada and the United States. Rasetti advised that although he realized that Pontecorvo, while in Montreal, was working on "something classified" for the Government, he had no idea what it was, as "he never told me and I never asked." Two children were born to the Pontecorvos in Montreal.

Dr. Rasetti states Pontecorvo was transferred from Montreal to Chalk River with the rest of his group in either 1945 of 1946, and "after he went to Chalk River, I saw him perhaps three or four times at Physical Society meetings in the United States. The last time was in Washington, D. C., in April 1949. We talked primarily about work in cosmic rays." Dr. Rasetti further stated, "It was my understanding he was working at Chalk River in the unclassified field of cosmic-ray study." Queried as to the source of this understanding, Rasetti said he believes it derives from his conversations with Pontecorvo and his associates and from the fact that Pontecorvo later published papers on cosmic rays. Shortly after this Dr. Rasetti heard from a third party that Pontecorvo had gone to Harwell, the British nuclear research center.

Dr. Rasetti further stated he never heard Pontecorvo discuss politics or express any interest in political questions. He does not recall ever discussing the problem of citizenship with him. He does not believe they have exchanged more than a half dozen letters—the last, about 5 years ago—always on scientific subjects or relating to scientific meetings. He stated he knows definitely that Pontecorvo's parents and the two oldest brothers have shown no political interests. He thinks that if Bruno Pontecorvo has gone to Russia, it is "because he has sincere faith in their system" and wants to live with "congenial spirits" and "not for the express purpose of giving them atomic secrets." In fact, Dr. Rasetti doubts that Pontecorvo could have had access to "very valuable secrets" since 1945, when he left Montreal, and Dr. Rasetti also observed that 1945 secrets would be "no secrets" now.

He commented that he classes Pontecorvo as "a good nuclear physicist but not a first-rate one." By "first rate," he said he means men like Enrico Fermi, Edward Teller, George Gamow, Hans Bethe, etc. He stated "I can better express it by saying that in this country there are maybe 100 nuclear physicists in Pontecorvo's class."

He stated that while he is perhaps the best authority in the United States on Bruno Pontecorvo's "early period"—others know more about his later life.

Upon reaching the United States in 1940, Pontecorvo secured private employment with the Wells Survey Co., of Tulsa, Okla., in connection with work on the location of oil deposits through neutron sources. He took out first papers for American citizenship and, in order to keep his application valid, continued to pay American income taxes after he later moved to Canada.

At the beginning of 1943 Pontecorvo joined the British Government's atomic effort, briefly in New York City and then as a member of the Anglo-Canadian team in Montreal. For the following 2 years he was engaged in classified experiments on the development of the Canadian heavy-water piles, moving to Chalk River for that purpose. At the beginning of 1946 he took a position with the British Ministry of Supply Atomic Energy Organization, but remained at Chalk River, Canada, until January 1949. During 1948 he was granted British citizenship in absentia.

Pontecorvo made three visits to American atomic installations. In January 1944 he was one of a group of British scientists who participated in classified discussions at the metallurgical laboratory of the University of Chicago. Two years later he appeared at the General Electric installation in Schenectady and unsuccessfully sought employment. In November 1948 he paid an unclassified visit to the University of California at Berkeley.

In early 1949 Pontecorvo transferred to the Harwell Laboratory in England, becoming a senior principal scientific officer in the Nuclear Physics Division. After some months there he accepted appointment to a professorship at Liverpool University. Before he and his family moved to Liverpool they took a vacation on the continent of Europe, never returning.

On October 20, 1950, all except the Communist newspapers in Rome, Italy, banner-headlined stories to the effect that British officials were seeking the whereabouts of Dr. Bruno Pontecorvo, who had disappeared. Two days later the New York Times carried the following dispatch:

[New York Times, October 22, 1950]

STOCKHOLM, SWEDEN, October 21.—(Reuters).—Bruno Pontecorvo, naturalized British scientist from Harwell, Britain's biggest atom research center, has vanished after flying to Finland. He is believed to be in Moscow.

Scandinavian Airlines System officials said today that the Italian-born professor had checked out of Stockholm September 2 for Helsinki, en route to the Soviet capital.

Helsinki Airport officials said no one named Pontecorvo had flown back to Stockholm. The only other routes out of Finland by boat, train, or plane all lead to Russia.

The Finnish Ministry of the Interior announced tonight it had been unable to find any record that Mr. Pontecorvo or his family had left Finland by air, sea, or land. The name did not appear on any passenger list. Nor are they listed as foreigners staying in the country, the ministry added.

Fihnish officials did not discount the possibility the family might have left the country under an assumed name.

VISITED ITALY

Mr. Pontecorvo is known to have gone to Italy on vacation after resigning from Harwell in July. His Swedish wife and three sons—Gil, 12; Tito Nils, 6, and Antonio, 5—went with him. He was scheduled to become professor of experimental physics at Liverpool University in January.

From Rome his movements were traced today to Helsinki, where the Finnish Airlines said he and his family were checked off as they left the plane from Stockholm.

They had reached Stockholm by the night express from Copenhagen. A Stockholm Airport attendant said they caught the Helsinki plane after taking a taxi from the train.

On the flight from Rome to Copenhagen the 40-year-old scientist and his family had 110 pounds of registered luggage, not counting 20 pounds of hand luggage that Mr. Pontecorvo insisted on taking with him to his seat on the plane. He also was reported carrying a bulging brief case.

The family reached Copenhagen the evening of September 1 and went straight from the airport to the railway station to reserve sleepers to the Swedish capital.

VISITED UNITED STATES, CANADA

On arrival at Stockholm the morning of September 2 Mr. Pontecorvo went straight to the airline ticket office in the city center and bought tickets for Helsinki.

The home of Mr. and Mrs. Hans Nordblom, Mrs. Pontecorvo's parents, is a 15-minute streetcar r.de from the airlines office, but neither the professor nor his wife visited or telephoned them.

The flurry over the Pontecorvo family's travels began in Rome yesterday when Italian newspapers said he had "disappeared." They said British and Italian intelligence were searching for him.

Fellow scientists say they believe Mr. Pontecorvo never worked directly on atomic bomb projects. Dr. C. J. Mackenzie, head of Canada's National Research Council, said he did much independent research on cosmic rays.

He visited the United States with a team of British atomic scientists in 1943 or 1944, Dr. Mackenzie added, and later worked in Montreal and at the Chalk River project, Ontario, though he was never officially employed.

The Hartford Courant of October 23, 1950, carried this item:

[Hartford Courant, October 23, 1950]

ROME, Monday, October 23 (AP).—The Italian atomic scientist, Dr. Bruno Pontecorvo, now a British citizen, sailed on a Russian ship from Finland for Leningrad on September 2, the Stockholm correspondent of Il Tempo reported today.

The correspondent, Henrico Altavilla, said he flew from Stockholm to Helsinki to investigate the international mystery of the whereabouts of Pontecorvo, described by European scientists as one of the foremost researchers in tritium, basic element of the hydrogen bomb.

Since July 25 the scientist has been on a vacation from England where he worked for 2 years at the British atomic research plant at Harwell, Italian newspapers reported Saturday that he had skipped to Russia.

Il Tempo's correspondent said Pontecorvo, his Swedish-born wife and three children flew from Italy to Copenhagen where they left their Swedish plane "for fear of being followed" and took an express for Stockholm.

At Stockholm, he reported, the family spent the night of September 1 at a house occupied by the Soviet Embassy.

An attendant at Stockholm's Broome Airport recalled that they arrived at the airport about 11 a. m. on September 2 in a Soviet Embassy car with 12 suit-cases, the correspondent said.

Finland's Interior Minister Johannes Viroelinen flew to Helsinki on the same plane.

Altavilla said the scientist and his family refused to take the airline bus at the Helsinki airport and waited until an automobile from the Soviet Legation arrived and took them to the harbor.

The correspondent said the Russian ship *Bellostor*, scheduled to sail at 10:40 a. m., waited until 5 p. m. when the party arrived. He said the ship weighed anchor as soon as the scientist and his family were aboard. The ship was due in Leningrad September 5, he added.

Neither the British nor Italian legations here have heard from the scientist, who for 2 years was one of the top men at Britain's atomic research plant at Harwell along with Dr. Klaus Fuchs, now serving 14 years on conviction of having given atom secrets to Russia.

Meanwhile, British officials, both in the Ministry of Supply, which controls atomic research, and at Scotland Yard, said there had been no developments and 'no progress whatsoever" in the search for the scientist.

There was a report that luggage had been found at Pontecorvo's Harwell residence bearing the fresh label of a hotel in Spain. The original reports from Rome suggested that the scientist had flown direct from Italy to Czechoslovakia. Still another report states that Pontecorvo had talked with an unidentified Czech and an uniden-

tified Italian near Lako Como in Italy during August 1950, where-
upon he said to his wife, "I dare not go back. I should be sent to
prison if I did." The Hartford Courant, on October 25, printed this
further story:

[Hartford Courant, October 25, 1950]

PONTECORVO'S SON'S REMARK MAY BE CLUE

BOY WAS OVERHEARD ON PHONE SAYING "WE ARE GOING TO RUSSIA"

HELSINKI, FINLAND, October 24 (UP).—A passenger on the plane which
brought Prof. Bruno Pontecorvo, the missing British atom scientist, to Finland
said today the professor's 5-year-old son told him, "We are going to Russia."

Pontecorvo, his wife and three children have not been seen since September
2 when they landed at Malmi Airport outside Helsinki after a flight from
Sweden. The Italian-born physicist is employed at Britain's top secret Harwell
research station and fear has been expressed in the House of Commons that he
may have gone to Russia with priceless atomic bomb secrets.

Police reported today they had uncovered no leads on how the family left
Finland, if they did leave.

The plane passenger, who asked that his name be withheld, said Pontecorvo's
son, Antonio, was seated in front of him on the plane and that the boy said,
"We are going to Russia."

At another point the boy was said to have looked out of the plane window
as it flew over land and asked: "Is that Russia?" Later, on the bus which
brought the passengers from the airport to Helsinki, Antonio reportedly asked if
the houses beside the road were Russian houses.

Lord Lucas of Chilworth, parliamentary secretary, British Ministry
of Transport, gave an official account of the episode on November 7,
1950. Speaking before the House of Lords, he said:

My Lords, I must apologize for the length of this reply. Dr. Pontecorvo, a
senior principal scientific officer in the nuclear physics division at Harwell, and
a naturalized British subject, was granted leave of absence on July 25 last and
left this country with his family on the same day for a holiday in France and
Italy. He took his car with him. Before proceeding further, I must explain
that, under the British Nationality and Status of Aliens Act, 1914, a naturalized
British subject has to all intents and purposes the status of a natural-born
British subject. Even if there had been any reason at the time to suppose that
Dr. Pontecorvo might not return to this country—and there was no such indica-
tion whatever—there would have been no legal means of preventing his departure.

Dr. Pontecorvo's leave expired on August 31. On this date he had written a
note to Harwell, which was received on September 4, explaining that he had
had trouble with his car—the fact that there had been a car accident has been
confirmed through other channels—but hoped to be back in time for a conference
to be held between September 7 and 13. A message had also been sent to him
from Harwell asking him to visit and advise a team of scientists employed in
Switzerland on cosmic ray work, on which he had specialized. It was his failure
to pay this visit, or to communicate with Harwell after September 4, which
gave cause for concern, and security inquiries as to his whereabouts were started
on September 21. These were naturally conducted with considerable discretion, because there was little point in creating undue alarm about an absence
which might have been capable of a perfectly innocent explanation, and because
all the evidence at Harwell and elsewhere suggested strongly that it was Dr.
Pontecorvo's intention to return to this country. He had received in June last
an offer of appointment to a professorship at Liverpool University, had written to
the vice chancellor accepting the offer, after discussing it with his director, had
arranged to take a university flat and had booked a return passage for his car.

Dr. Pontecorvo and his family flew to Stockholm from Rome on August 31
and on September 2 they traveled to Helsinki. Since then, there has been no
definite information about Dr. Pontecorvo's movements, but His Majesty's Government have no doubt that he is in Russia. Dr. Pontecorvo, although employed in
the nuclear physics division at Harwell, was not engaged on secret work. For
several years past his contacts with such work had been very limited, and
he had no direct contact with work on atomic weapons.

Although this statement makes clear that, under British law, the authorities possessed no power to prevent Pontecorvo's trip to Europe, it nevertheless leaves much of the mystery unresolved; and that mystery remains largely intact today.   However, two discussions in the House of Commons tend to clear up several points in the case. The first, held October 23, 1950, is as follows:

SCIENTIFIC OFFICER, HARWELL (ABSENCE FROM DUTY)

Mr. ERROLL (by private notice) asked the Minister of Supply if he has any statement to make on the disappearance of the Harwell atomic scientist, Professor Bruno Pontecorvo.

Mr. G. R. STRAUSS. Dr. Pontecorvo is a senior principal scientific officer at Harwell. He was granted leave of absence on 25th July last and was due to return to duty on 31st August. He had accepted an appointment at Liverpool University and was shortly about to take up this position.

Dr. Pontecorvo was born in Italy. He left that country for France in 1936 and went from France to the U. S. A. in 1940. In 1943 he became a member of the Joint Anglo-Canadian atomic energy team at Montreal and was transferred to the Ministry of Supply atomic energy organization in January 1946. He remained in Canada as a member of that organisation until January 1949, when he was posted to Harwell. Dr. Pontecorvo became a naturalised British subject in March 1948. For several years past Dr. Pontecorvo's contacts with secret work have been very limited. I have no information about Dr. Pontecorvo's present whereabouts beyond what has appeared in the Press.

Mr. ERROLL. May I ask two questions? First, can the Minister state that the Professor has never had the opportunity of acquiring knowledge of atom bomb manufacture likely to be of value to a foreign power, and second, can he explain how reliable the British screening of this person was in view of the fact that, according to the "Daily Herald" his sister is the wife of a Communist official in Italy?

Mr. STRAUSS. Although Dr. Pontecorvo has not had direct access, except in a very limited way, to secret subjects for some time, it would be quite impossible to say that he has not been able to gather information while he was resident in Harwell or in Canada which might be of value to an enemy. On the second point I can only say that this individual has been screened several times during the last few years by our security officers.

Mr. C. S. TAYLOR. When a man like this is known to have relations who are Communists—[HON. MEMBERS: "Oh."]—when it is also known that he was a bosom friend of Dr. Fuchs, why is it he is allowed to continue in such a responsible position?

Mr. STRAUSS. I do not agree with the hon. Member in his last allegation. I do not think it is true. As I say, this man has been screened several times, and according to the security officers the screenings were particularly satisfactory.

Mr. R. A. BUTLER. Would the right hon. Gentleman tell us, in view of the profound disquiet created by this news, whether, since the Fuchs episode, the whole of this screening business has been tightened up, and whether there has been any investigation since this case into all the officers concerned?

Mr. STRAUSS. Yes, Sir, since the episode the whole matter has been looked at very carefully and there has been a certain tightening up of the system.

Mr. BUTLER. Does that apply to this case in particular?

Mr. STRAUSS. Yes, Sir, but this man was leaving Harwell anyhow and going to Liverpool.

Sir P. MACDONALD. Is not it a fact that he was leaving the country? Why was he allowed to take his family and car to Italy when it was well known that he was visiting his sister, who is the wife of a prominent Communist in Italy?

Mr. STRAUSS. As he was the holder of a British passport there was no means of retaining him in this country.

Mr. BLACKBURN. While entirely admitting the great difficulties of screening refugee scientists and welcoming the policy of the Government in accepting refugees, may I beg the Minister to bear in mind that this very distinguished scientist, as the Minister admitted, was only making £1,100 a year and that the salaries of scientists ought to be reconsidered?

Mr. LOW. In replying to my right hon. Friend the Member for Saffron Walden (Mr. R. A. Butler) the right hon. Gentleman seemed to indicate that, because this man was shortly leaving Harwell, the new, strict precautions for screening were not taken in this case. [HON. MEMBERS: "No."] I may have misunderstood the right hon. Gentleman, but is it not important that this screening should have been applied because of the opportunities the man had to get to know the most detailed secrets?

Mr. STRAUSS. I said that more rigid screening was applied after the Fuchs case. I continued that just at that time this man took up an appointment at the Liverpool University and was not going to be at Harwell more than a few months longer at the outside.

Mr. SUTCLIFFE. Has the right hon. Gentleman given any indication of the reason for his leaving Harwell to take up another appointment?

Mr. STRAUSS. This post was offered to him; it was suggested he would be doing more useful and more remunerative work at Liverpool than at Harwell.

Mr. BOYD-CARPENTER. In view of the fact that the right hon. Gentleman said this gentleman was due back from leave on 31st August, can he say when inquiries as to his whereabouts were set on foot?

Mr. STRAUSS. No, not at the moment. I am not sure when the inquiries started. I imagine quite recently, when he was overdue from his leave, but I am not certain about the exact date.

Mr. PICKTHORN. When the right hon. Gentleman says that it was suggested that this scientist would be more useful at Liverpool than elsewhere, can he tell us by whom it was suggested, and to whom?

Mr. STRAUSS. The deputy director at Harwell was taking up an important post in Liverpool and he wanted assistants. It was suggested by the people at Harwell that this man might well go with Dr. Skinner, the deputy director, and would be useful to him at Liverpool, and he agreed to take the post.

Mr. H. STRAUSS. Was any investigation made of what documents the scientist took with him, since such an investigation would have been within the law?

Mr. STRAUSS. No, sir, so far as I am aware no such investigation took place.

Air Commodore HARVEY. Does not all this suggest that the present method of screening is completely ineffective? Will the right hon. Gentleman go into it with his colleagues and overhaul the methods of screening?

Mr. STRAUSS. I do not agree that it is ineffective, but we are always looking for ways to improve the system.

Wing Commander HULBERT. Can the right hon. Gentleman say if, in view of the recent incidents, he is satisfied with the present security and screening arrangements at Harwell, and how many pseudo-Communists are still there?

Mr. STRAUSS. I am satisfied that the screening arrangements are very good. It is never possible to be absolutely certain that anybody who may have had any connection, either himself or through his friends or his relatives, with any Communist or Fascist organization, is working in a research establishment. We cannot be absolutely certain about that, but we believe that the screening arrangements are as good as they can possibly be devised, unless we go to limits which this House would not tolerate.

Mr. THURTLE. Can my right hon. Friend say whether the security officers have still complete faith in the loyalty of this gentleman to the British nation?

Mr. STRAUSS. I would like to wait a few days longer to see what happens.

Later—

Lord JOHN HOPE. On a point of order. I beg to ask leave to move the Adjournment of the House for the purpose of discussing a definite matter of urgent public importance, namely, the disappearance of Professor Pontecorvo and the failure of the Government to take adequate precautions to prevent it.

Mr. SPEAKER. The noble Lord has asked leave to move the Adjournment of the House for the purpose of discussing a definite matter of urgent public importance, namely the disappearance of Professor Pontecorvo and the failure of the Government to take adequate precautions to prevent it.

The noble Lord's Motion fails on the ground of urgency in regard to this particular case. Professor Pontecorvo is not in this country, there was no particular reason for stopping him when he left the country, and the Government have no power now to get him back or deal with him.

If the remedy sought by the hon. Gentleman is to impose some kind of exit permit on any persons employed on atomic research who wish to leave this country, then the matter fails to qualify as definite. I cannot therefore allow the noble Lord's Motion.

Lord JOHN HOPE. Further to that——

Mr. SPEAKER. The matter cannot be argued. I have given a Ruling and that must stand.

A later discussion in the House of Commons, which took place on November 6, 1950, is as follows:

DR. PONTECORVO

4. Mr. BOYD-CARPENTER asked the Minister of Supply when Professor Bruno Pontecorvo entered the employ of his Department; what was the nature of his work; and if he has any further information as to where this gentleman now is.

6. Mr. BAKER WHITE asked the Minister of Supply how many days elapsed between the date on which Professor Pontecorvo was due back from leave and the date on which his Department notified the security departments, on a high level, of his disappearance.

Mr. G. R. STRAUSS. Dr. Pontecorvo entered Government service in January 1943, and joined my Department in January 1946. He took up duty at Harwell in January 1949, where he was employed in the Nuclear Physics Division. Dr. Pontecorvo's leave expired on 31st August. On this date he had written a note to Harwell, received on 4th September, saying that he had trouble with his car but hoped to be back in time for a conference to be held between 7th and 13th September. A message was sent to him from Harwell asking him to visit and advise a team of scientists employed in Switzerland on cosmic ray work on which he had specialized. It was his failure to pay this visit or to communicate further with Harwell after the date of the conference which first gave cause for concern. Inquiries as to his whereabouts were started on 21st September. I have no conclusive evidence of his present whereabouts, but I have no doubt that he is in Russia.

Mr. BOYD-CARPENTER. Can the right hon. Gentleman say whether, at the time this gentleman's employment at Harwell began, it was known to his Department that he was closely connected with a well-known Italian Communist; and can he also say if he has any information as to whether, on his recent trip abroad, Professor Pontecorvo took with him any documents of a secret nature?

Mr. STRAUSS. No, Sir, it was not known at the time Professor Pontecorvo was first employed at Harwell that he had a relative abroad who was connected with the Communist Party. As to the second part of the hon. Gentleman's question, so far as is known he took no documents abroad with him.

Mr. BAKER WHITE. May I ask the Minister, although I see the difficulty in this case, whether it would be possible to extend to Harwell the system used in the Services, namely, that if anybody is absent and over-stays his leave machinery is set in motion immediately to find out where he is?

Mr. STRAUSS. Yes, Sir, but the reasons for this man overstaying his leave seemed quite normal. He had a motor car breakdown, and was asked to visit some people in Switzerland, and it was, naturally, only about a week afterwards that those at Harwell became worried about him.

Captain WATERHOUSE. In answer to a previous Question I understood the Minister to say that foreigners were carefully screened before they got these jobs. Does the right hon. Gentleman agree that we ought to have found out that Professor Pontecorvo had a Communist connection in Italy?

Mr. STRAUSS. One finds out a great deal by screening, but it so happens that the fact that he had a Communist relative abroad was not in the possession of the security officers.

Mr. NABARRO. Can the Minister say whether the three defections which have taken place in the last four years—Dr. Nunn May in 1946, Professor Fuchs in 1949 and now Professor Pontecorvo in 1950—arise from the fact that these three gentlemen were working in conjunction with one another in the Harwell atomic establishment?

Mr. STRAUSS. It is true that Dr. Fuchs and Professor Pontecorvo were working in the same major establishment, the one research establishment in this country. But they were not particular friends, and I do not think that the fact that they were working in a centre where research takes place is really a significant fact.

Mr. WALTER FLETCHER. Would not about the first series of questions asked in screening be about Communist connections, either relations or friends, of the person who is being screened?

Mr. STRAUSS. I do not think so, particularly in this case. The man had not been abroad for a long time. He had left Europe in 1940. The fact that he had relatives abroad who were members of the Communist Party was not suggested to the security officers, and they were not aware of that fact.

Captain CROOKSHANK. If these people are not asked about their friends and relations, what are they asked about?

Mr. STRAUSS. It must be remembered that the major screening of this man took place in 1943, when he joined the atomic energy organisation in Canada and the security services made the necessary inquiries. Similar inquiries were made at a later date and, as I have informed the House, this fact was not known when the man joined the Harwell organisation at the beginning of 1949. It was known later, but by that time he had made arrangements to leave Harwell anyhow, and go to Liverpool.

Mrs. MIDDLETON. Is my right hon. Friend aware that a "star turn" at the recent Conservative Party Conference was a former Communist? What screening would be appropriate in that case?

Several Hon. Members *rose*——

Mr. SPEAKER. We cannot debate this matter further. We have had at least five minutes on it already.

SECURITY MEASURES, HARWELL

9. Squadron Leader BURDEN asked the Minister of Supply if he is satisfied with present security measures at Harwell.

Mr. G. R. STRAUSS. Yes, Sir.

Squadron Leader BURDEN. Is not it a fact that a list in a telephone directory, giving the names and places of work of the people working at Harwell, is available to anyone? Would not the Minister agree that the first step to breaking down security is to know the names of those people and, in particular, to discover whether they have relatives living behind the iron curtain upon whom pressure can be brought to bear?

Mr. STRAUSS. I do not think there is any danger in someone discovering the names of people working at Harwell by looking at a list in one of the telephone booths at Harwell.

Sir WALDRON SMITHERS. Would the right hon. Gentleman consult with other Ministers to see if they are satisfied with the security measures at Broadcasting House?

Mr. GODFREY NICHOLSON. May I ask the Minister whether any steps are taken to keep a check on people abroad with whom these people may correspond?

Mr. STRAUSS. Certain checks are kept. Reasonable checks are kept on the people working at Harwell and other secret research stations. But it would be quite impossible, in view of the thousands of people who work there and elsewhere, to keep a check on people in other countries with whom they may communicate.

It is possible to hope that some innocent explanation accounts for Pontecorvo's disappearance. Conceivably he was kidnaped by the Soviets. Conceivably he became insane and is hiding in western Europe. If an innocent explanation should develop subsequent to the appearance of this report, apologies are herewith rendered in advance. At the same time, the facts as reported to the British House of Lords and House of Commons suggest that the possibility of an innocent explanation is extremely remote, and prudence dictates the assumption that Pontecorvo indeed went to Russia at his own volition.

By way of pure speculation it may be theorized that Pontecorvo—since he was about to lose all contact with classified information at the time of his European trip—had been ordered by Soviet superiors in an espionage apparatus to leave England and go to Russia. In the same vein it may be speculated that, just as a travel itinerary from the United States to Mexico to Czechoslovakia to Russia seems to be a standard escape route for American agents of the Soviet Union, so Pontecorvo's path from Britain to Italy to Sweden to Finland to Russia constitutes a standard escape route for English agents of the Soviet Union. It may also be speculated that Pontecorvo's flight is somehow related to the Fuchs case. Conceivably Pontecorvo feared that, since he was acquainted with Fuchs, he too would fall under suspicion. Still other speculations are to the effect that Pontecorvo was somehow blackmailed into fleeing or that he was lured by a Soviet offer of some description or even that he thought he might serve as an international liaison man between scientists of the East and West, looking toward a peaceful solution of differences. The psychology which apparently motivated Fuchs implies that the most far-fetched speculation cannot be ruled out as to Pontecorvo, but the mystery remains.

Mr. Gordon Dean, Chairman of the United States Atomic Energy Commission, indicated his attitude toward Pontecorvo during an October 24, 1950, press conference. The relevant portion of the transcript reads thus:

Question. Mr. Dean, is the Atomic Energy Commission taking interest in the disappearance of that British atomic scientist?

Chairman DEAN. Yes.

Question. Do you know if he has any relations in this country? It is reported he has a brother working on a project.

Chairman DEAN. As yet I don't know, personally.

Question. Mr. Chairman, when he came here with a British team, wouldn't he have had some clearance then from us?

Chairman DEAN. I believe the process then was for the British to certify that he was a reliable man, or words to that effect, and could be used in the program.

Question. Can you tell us what kind of work he did, then, sir.

Chairman DEAN. In Canada?

Question. No; while he was here.

Chairman DEAN. I am a little bit uncertain as to that period before he went to Canada so I wouldn't like to say. I just don't know at this point. We could probably find out for you.

Question. That was prior to 1946.

Chairman DEAN. It was prior to 1946; 1943 is when he went to Canada.

Question. Do you think this incident offers any serious period of threat?

Chairman DEAN. That depends entirely on the extent to which he had access to vital information, and that to me, would mean largely weapons information. I think the British will have to make the appraisal of the extent of that access, not ourselves.

Question. Will this have any effect on our relationship with the British Atomic Commission?

Chairman DEAN. Of course, it is always difficult when somebody goes sour. What the effect is, it is very hard, without being a good prophet, to forecast.

Of course, we are interested in the disappearance of anybody who has had access to restricted data or been in any way connected with the program. However, he is not our baby.

Question. There have been a good many reports that he was quite an expert. Is that true? If it is true that he has gone someplace, is it true that he is the man that could take away with him any important information?

Chairman DEAN. It is pretty hard for us to answer on this side of the water. He has been a senior scientist, I believe, at Harwell. Some sort of statement, which I don't have before me, was made by the British, indicating that his access was not particularly broad. I have forgotten the exact language of that. But I think they are in a better position to say what his standing was in Harwell than we are.

Question. Had he any kind of American access?

Chairman DEAN. I believe the record will show that he came over here with a British team during the war, and went shortly thereafter into the Canadian program, where he would have access to one phase, but a very limited phase. I think he has returned to this country on two occasions. In neither case did he come on a classified visit or have access to any classified material.

Question. So his connection, if any, with the American program would have been collateral through the Canadian work.

Chairman DEAN. Yes; except for a very brief period before he went to Canada at which time he was stationed in the British Mission probably in New York.

Mr. SALISBURY. I don't know precisely.

Question. Was he given clearance by the British or was he cleared here, sir?

Chairman DEAN. He was never cleared here, so if there was any kind of a certification of that group, it was from the British.

Question. The same way as Fuchs had the clearance. Have you asked for information from Britain, sir?

Chairman DEAN. I don't believe that we, as the Atomic Energy Commission, have made such a request. I imagine it has been made through other departments.

6. *Allan Nunn May—Details*

Dr. May was the first of the major atomic energy spies to be apprehended. He is today serving a 10-year prison sentence for espionage at Wakefield, Yorkshire, England. The London Times of March 20, 1946, describes an early phase of his trial as follows:

ATOMIC SECRETS CHARGE—SCIENTIST SENT FOR TRIAL—BAIL REFUSED

Dr. Allan Nunn May, 34, university reader, of Stafford Terrace, Kensington, was at Bow Street yesterday committed for trial at the Central Criminal Court on a charge of having, for a purpose prejudicial to the safety and interests of the State, communicated to some person unknown certain information calculated to be directly or indirectly useful to an enemy, contrary to the Official Secrets Act, 1911.

Mr. Anthony Hawke and Mr. H. A. K. Morgan conducted the case on behalf of the Director of Public Prosecutions; Mr. Gerald Gardiner, instructed by Mr. Harold Kenwright, defended. Dr. May pleaded "Not Guilty," and reserved his defence.

Mr. HAWKE, opening the case, said that Dr. May was a reader in physics at the London University, but from May 1942 until September 1945, he was a member of an organization, set up by the Government in November 1941, to investigate problems of atomic energy. The organization was part of the Department of Scientific and Industrial Research, and for security reasons was known at the time as Tube Alloys Research. It was under the direction of Sir Wallace Akers, and consisted of various teams of experts working in different university and industrial laboratories. Dr. May worked first at the Cavendish laboratory, Cambridge and in January 1943 went to Canada. While in Canada he was a senior member of the nuclear physics division of the organization. Not only did he have knowledge of all the physics work in connexion with this research, but also, owing to his ability—and there was no question that it was considerable—he was consulted on problems in connexion with physics relating to the researches in progress.

CONDITIONS OF SECRECY

He was on two committees which gave him access to secret reports and to the latest developments in connection with uranium. This indicated the vital importance and secrecy of his post. On his own admission Dr. May communicated, in Canada last year, to a person whose identity he refused to divulge, a written report on atomic research, as it was then known to him, and he also gave to the same person specimens of certain types of newly discovered material regarded as of peculiar importance and secrecy. Before accepting the post he was required to sign a document that he would not give information except to persons named. Conditions of secrecy and confidence were imposed upon him.

At the end of last year or early this year certain information reached the Intelligence Corps in this country. On February 15 last Deputy Commander Burt, head of the Special Branch at Scotland Yard, saw Dr. May at Shell-Mex House, where he had resumed his work. Dr. May was asked if he was aware that there had been a leakage of information in Canada relating to atomic energy. He replied that it was the first he had heard of it. He denied that any approach had been made to him while he was in Canada, and added that he was not prepared to answer any question if it related to counterespionage. Between February 15 and February 20 further information came to the notice of the authorities, and Deputy Commander Burt again saw Dr. May. He told him it was understood that shortly after his return from Canada he had an appointment to meet someone in the vicinity of the British Museum, and it was known that he did not keep that appointment.

49

Mr. Hawke added that the usefulness of the information disclosed by Dr. May could be measured in this way—that it might enable persons in possession of it to save a substantial period of time in arriving at conclusions connected with this research.

ALLEGED APPOINTMENT

Deputy Commander Burt gave evidence that when he spoke to Dr. May about not keeping an appointment near the British Museum his reply was: "No; I did not keep that appointment as when I returned I decided to wash my hands of the whole business." He was asked who, in Canada, had made an appointment for him to meet someone in London.

In cross-examination, Mr. Burt said he did not exercise any pressure on Dr. May, either by way of inducement, threat, or promise of favour. He did not tell him the authorities had a lot more information about him and that he was implicated. At the interview on February 20 he was not in a position to accuse Dr. May. Dr. May had been followed from February 15 until February 20.

Mr. GARDINER. Did you tell him you did not suggest he had done this for gain?—No. That was contrary to my instructions, which were that the question of gain had entered into it.

A statement alleged to have been made by Dr. May at this interview was handed to the Magistrate, and it was agreed that the contents should not be divulged at this stage.

Mr. Hawke said there was one passage in the statement which it was desirable should not be read at any time.

Sir Wallace Akers, on the board of Imperial Chemical Industries, Limited, and director responsible for scientific research by that concern, said that on December 1, 1941, he was appointed director to supervise a special organization set up to handle the problem of atomic energy. While in Canada Dr. May, as senior member of the Nuclear Physics Division, would be consulted on problems connected with physics. He was also a member of two important committees, which would give him access to secret reports and thus keep him informed of the latest developments in the methods of use and production of uranium. Everything in connection with that matter discovered by Canadian or British sources since the war had been kept secret.

Referring to the statement alleged to have been made by Dr. May to Mr. Burt, Sir Wallace Akers said that materials mentioned in it would enable scientists to determine important nuclear physical data earlier than if they had had first to prepare the material.

Cross-examined, he agreed that Dr. May was engaged on research on nuclear atomic energy, and not on atomic bombs.

Mr. GARDINER. Most of what was known about atomic energy some little time ago has been published?—Qualitatively, but not quantitatively.

Is there a strong feeling among scientists, rightly or wrongly, that contributions to knowledge made by them with respect to the benefits of atomic research ought not to be the secrets of any one country?—Yes.

Mr. Gardiner asked if Russia was a gallant ally in February last year.

Mr. HAWKE. I should like to know why Russia has been introduced. I made no reference to Russia or America.

The Magistrate, Mr. McKenna, refused an application for bail.

On May 2, 1946, the London Times carried this additional story:

ATOMIC SECRETS

SCIENTIST SENTENCED TO 10 YEARS

Dr. Allan Nunn May, 34, atom scientist and lecturer in physics at King's College, London, pleaded "Guilty" at the Central Criminal Court yesterday to communicating information contrary to the Official Secrets Act, and he was sentenced to 10 years' penal servitude. He admitted giving to an unknown person on a day between January 1 and September 30, 1945, for a purpose prejudicial to the safety and interest of the state, information about atomic research calculated to be useful to an enemy.

The Attorney General, Sir Hartley Shawcross, K. C., who prosecuted with Mr. Anthony Hawke, said that if certain information had got into the hands of scientists of other countries it would have shortened their researches by a considerable period.

Mr. Gerald Gardiner, defending, remarked that Dr. May told him that the person to whom he gave the information was Russian.

The Attorney General replied: "There is no kind of suggestion that the Russians are enemies or potential enemies. The court has already decided that this offense consists in communicating information to unauthorized persons."

Sir Hartley said that early in May 1942, Dr. May was invited to become a member of one of the staffs of scientists set up by the Government for research into atomic energy. In January 1943 he went to Canada with other scientists to continue their researches. He occupied a position there of considerable responsibility as senior in the Nuclear Physics Division. He remained in Canada until September 1945 and no suspicion had arisen. In February last he was seen by Lieutenant Colonel Burt, head of the Special Branch of the CID of the military intelligence authorities, who told him it was known he had an appointment after his return from Canada to meet somebody in the neighborhood of the British Museum, and that he did not keep the appointment.

Dr. MAY replied: "No, I did not keep the appointment. When I got back to this country I decided to wash my hands of the whole business." He made a statement in which he said: "About a year ago, while in Canada, I was contacted by an individual whose identity I decline to divulge. He called on me at my private apartment in Montreal and apparently knew that I was employed by the Montreal laboratory, and sought information from me concerning atomic research. * * * After this preliminary meeting I met the individual on several subsequent occasions while in Canada. * * * I gave the man a routine report on atomic research as known to me. This information was mostly of a character which has since been published or is about to be published."

The ATTORNEY GENERAL remarked that all of this information had by no means been made public.

The statement went on to say: "Before I left Canada it was arranged that on my return to London I was to keep an appointment with someone I didn't know. I did not keep the appointment because I had decided that this clandestine procedure was no longer necessary in view of official release of information.

"The whole affair was extremely painful to me. I only embarked on it because I thought this was a contribution I could make for the safety of mankind. I certainly did not do it for gain."

Mr. GARDINER, addressing the Judge, said: "Doctors take the view that if they discover something of benefit to mankind, they are under obligation to see it is used for mankind and not kept for any country or people. There are scientists who take substantially the same view, and Dr. May held that view strongly."

Mr. JUSTICE OLIVER, passing sentence of 10 years' penal servitude, said to May: "I cannot understand how any man in your position could have the crass conceit to arrogate to himself to do what you did, knowing it was one of the country's most precious secrets. I find you have acted not as an honourable man but as a dishonourable man. It is a very bad case indeed."

Additional documentation of the case is to be found in a letter sent by Gen. Leslie R. Groves to Senator Hickenlooper on March 12, 1946. The Senator read the letter to the Senate on March 19, 1946:

[From the March 19, 1946, Congressional Record]

WAR DEPARTMENT,
Washington, D. C., March 12, 1946.

Hon. B. B. HICKENLOOPER,
United States Senate, Washington, D. C.

DEAR SENATOR HICKENLOOPER: In reply to your letter of March 7, 1946, in which you requested certain information particularly as to any connection which Allan Nunn May, the British physicist whose arrest was recently announced in London, had with scientific research and development of atomic energy and the atomic bomb, and the seriousness of any possible disclosures by him, I submit the following information which I believe to be correct.

Dr. Allan Nunn May is a native-born English physicist of about 40 years of age. He holds a doctor of philosophy degree from Trinity College, Cambridge. He came to Canada some time in 1943 and was employed in the British group in the Montreal Laboratory of the National Research Council of Canada which was then embarking on research in the atomic energy field. He had been investigated, for security purposes, by the British Intelligence. That organization cleared him for access to any atomic energy work. It was not practicable nor was it our custom to look behind the approval of the British organization as to the trust-worthiness of any individual whom they had investigated. I am sure that they found no indication that he was not completely loyal and of unquestioned integrity.

On January 8, 1944, in the company of 12 other scientists from the National Research Council, Dr. May first visited the metallurgical laboratory of the University of Chicago, which was, as you know, engaged in work for the Manhattan District exclusively. May's visit had been approved by me in accordance with our previously established and approved arrangements for the interchange of information. My records indicate that I met May at that January 8 meeting, which was the first meeting between representatives of the metallurgical laboratory and of the National Research Council since I had been placed in charge of the project. At that meeting some phases of the work then in progress at the Chicago Laboratory were discussed, particularly with reference to phases of importance to the Montreal work.

On April 13, 1944, May returned to the Chicago Laboratory and stayed until April 27. He worked on a minor experiment at the Argonne Laboratory, where the original graphite pile was, and is, located, and where a small-scale heavy water pile had also been constructed. He came to Chicago again on August 28 and stayed through September 1, conferring with officials of the Chicago Laboratory on the construction and operation of the Argonne pile and the proposed Montreal pile.

His third and last visit occurred between September 25 and October 30, 1944. At that time he carried on extensive work in collaboration with our scientists in a highly secret and important new field. His work resulted in a research report in which he collaborated with an American scientist. May necessarily must have become familiar with the work then going on in the Argonne Laboratory He also, at this time, probably acquired knowledge of some technical problems which we encountered in the operation of the first Hanford pile.

During his first two visits Dr. May stayed at a Chicago hotel. On his last visit he stayed at an Argonne dormitory, except for week ends which he spent, with an American physicist, in the Chicago apartment of another American physicist who was temporarily out of town. He had few social contacts with the other scientists although he was generally well liked by them. They have described him as charming, shy, little man with a dry sense of humor. The American scientists with whom he was in most intimate contact are in my opinion men of unquestioned loyalty and integrity. The revelation of his activities came as a complete shock to them.

By this time (October 1944) May had spent more time and acquired more knowledge at the Argonne than any other British physicist. Although I had absolutely no reason to suspect him, I did not like to have him acquire such a wide knowledge of later developments. It is for that reason that in the spring of 1945 I declined to approve a proposed fourth visit of 1 month's duration. May never returned to the Chicago Laboratory and never visited any other Manhattan District installation.

It is very doubtful if May has anything but a general knowledge of the construction of the atomic bomb. He would not have been able to secure any such knowledge through legitimate channels. It would have required a breaking down of the compartmentalization rules in each instance where he secured such knowledge.

Dr. May has a rather wide knowledge of the Canadian effort. He understands the principles of design and construction of piles. He knows some important facts about the design, construction, and operation of the Hanford pile.

We do not know whether May was in the pay of any outside agency at the time of his visits to the United States. Summarizing, he could have furnished to an unauthorized person small samples of plutonium and U–235 of unknown purity and degree of enrichment since it would be a virtual impossibility to trace the theft of such a small amount; and he could have given information of varying degrees of accuracy with respect to—

(a) The bulk of the research and development carried on at the Argonne Laboratory of the Metallurgical Laboratory at the University of Chicago during 1944 and early 1945.

(b) Some of the technical problems involved in the design, construction, and operation of the Hanford Engineer Works.

(c) A very limited amount of information with respect to the materials used in the actual bombs.

(d) A very limited amount of information on which to base guesses as to our

I do not include in the above summary any knowledge secured from his work under British auspices.

Although one of the other individuals whom the Canadians have arrested has been announced as an employee of the National Research Council, I understand that he was not connected with the atomic energy project there and he did not visit the United States Manhattan District installations.

I do not know what other persons, Canadian or otherwise, may hereafter be publicly announced by Canada, the United States, or Great Britain as having been implicated in these spy activities.

I shall be pleased to furnished such additional information as you may desire from time to time, subject of course to the maintenance of adequate security.

I have discussed your letter with the Secretary of War, who has authorized me to answer it directly. I am sending a copy of the correspondence to him.

Sincerely,

L. R. GROVES,
Major General, United States Army.

The following material concerning Allan Nunn May is reprinted from The Report of the (Canadian) Royal Commission, Appointed under Order in Council P. C. 411 of February 5, 1946, to investigate the facts relating to and the circumstances surrounding the communication, by public officials and other persons in positions of trust of secret and confidential information to agents of a foreign power, June 27, 1946, Section VII, "Evaluation of Information and Material Handed Over," p. 617:

As to the question of atomic energy and the work done by nuclear physicists, we are able to say in the first place that on the evidence before us no one in Canada could have revealed how to make an atomic bomb. There was no one in Canada who had that information. In the second place there is no suggestion in the evidence that anyone who had any information on the subject made any disclosures except May. As to May, he did have certain information that would be of value to the Russians. He was in a position to get, where we do not know but possibly in Montreal, samples of Uranium 235 enriched and Uranium 233; he did get them and did deliver them to Lt. Angelov. These samples were considered so important by the Russians that upon their receipt, Motinov flew to Moscow with them. May also possessed considerable knowledge of the experimental plant at Chalk River, Ontario, which was described as "unique". In addition to May's work in Canada, he also did some work in the United States in collaboration with American scientists, but the evidence before us is that in such work also he could not properly have obtained the full story. How much of his information he handed over we are not able to say, but what he is known to have given, as shown by the documents and by his own written statement, we are told would be of considerable help to the Russians in their research work. May, in his written statement, did not particularise about the extent of the information he gave, but stated in effect that it was more than has since appeared (i. e. in the *Smyth* Report). He said that he gave his "contact" a "written report on atomic research as known to me. This information was mostly of a character which has since been published or is about to be published."

The following material concerning Allan Nunn May is also reprinted from The Report of the (Canadian) Royal Commission, June 27, 1946, pp. 447 through 457. Many of the names referred to in this extract denote individuals connected with the Canadian espionage ring either by their real names or their code names.

SECTION III. 13

ALLAN NUNN MAY

One of the many objectives of the Russian organization in Ottawa was the atomic bomb. The exhibits produced reveal how anxious the organization was to obtain as full information as possible about the work done by the nuclear physicists, in connection with the use of atomic energy.

As far back as March 28th, 1945, Lunan reported to Rogov :—

Badeau [Smith] informs me that most secret work at present is on nuclear physics (bombardment of radioactive substances to produce energy). This is more hush-hush than radar and is being carried on at the University of Montreal and at McMaster University at Hamilton. Badeau thinks that government purchasing of radium producing plant is connected with this research.

Lunan was here transmitting a report from Durnford Smith (Badeau) on the work of the National Research Council.

In mid-April of the same year one of the tasks given to Lunan and set out in the *"Organizational Directives"* for his group was :—

5. . . . Ask *Badeau whether* he could obtain *Uran No. 235*, let him be cautious. If he can, let him write in detail about the radium producing plant.

At about the same time, Motinov prepared a draft of a telegram for Zabotin to send to Moscow, which reads :—

To the Director,

The Professor reported that the Director of the National Chemical Research [Institute] Committee, Stacey, told him about the new plant under construction : Pilot Plant at Grand'Mere, in Province of Quebec. This plant will produce "Uranium". The engineering personnel is being obtained from McGill University and is already moving into the district of the new plant. As a result of experiments carried out with Uranium, it has been found that Uranium may be used for filling bombs, which is already in fact being done.

The Americans have developed wide research work, having invested in this business 660 million dollars.

"Grant"

This telegram was probably not sent. *"The Professor"* is Raymond Boyer. The location of the plant is wrongly given ; it was at Chalk River and not at Grand Mere. The mistake evidently occurred when Motinov later made his notes of what Rose had told him of Rose's conversation with Boyer, which also dealt with R. D. X.

At this time, according to another document, Angelov (*"Baxter"*) was given instructions to approach May (*Alek*) and to obtain from him a sample of Uran. 235, and information as to the location of the United States Atomic Bomb Plant. The same mistake as to the location of the plant appears in this exhibit.

There was some talk, too, that Smith might get into atomic research work. A report in Russian on one of the meetings of Lunan's Group, probably that of April 18th, says—

Badeau asks for permission to change to work on uranium. There is a possibility either by being invited or by applying himself, but he warned that they are very careful in the selection of workers and that they are under strict observation.

The same exhibit records Motinov's *"Conclusion"* :—

. . . 2. Not to recommend the transfer of Badeau to the production of uranium but to develop more widely the work in Research. In the future, for the purpose of more efficient direction, it is expedient to detach him from Back's group and to key him up as an independent contact man.

The matter was also taken up with Halperin (*Bacon*) because Lunan records, in a report dated 5th July, 1945 :—

Bacon . . . He is himself curious about the Chalk River Plant and the manufacture of Uranium. He claims that there is a great deal of talk and speculation on the subject but that nothing is known outside of the small and carefully guarded group completely in the know. He emphasized that he himself is as remote from this type of information as I am myself.

Evidently Lunan pressed Halperin to get Uranium–235 because another document records a report from him on Halperin (*Bacon*) as follows :—

Words in black brackets, which appeared as "faint type" in the printed report of the Royal Commission, indicate words crossed out in original documents.

Black's Group Mat. No. 1

Bacon.

[He] It has become very difficult to work with him, especially after my request for Ur 235 (Uran 235). He said that as far as he knows, it is absolutely impossible to get it. Thus for instance he declared that perhaps it (Uran) is not available in sufficient quantity. Bacon ex-

?

plained to me the theory of nuclear energy which is probably known to you. He refuses to put down in writing anything and does not want to give a photograph or information on himself. [I believe] I think that at present he has a fuller understanding of the essence of my requests and he has a particular dislike for them. With such a trend of thought as he has, [we can not obtain] it is impossible to get anything from him [except] with the exception of verbal descriptions, and I am not in a position to [unable to] understand everything fully where it concerns technical details.

I asked him what is taken into consideration in the construction of the very large plant (Chalk River, near Petawawa, Ontario), in the general opinion the principle of production of which is based on the physical properties of the nucleus; with regard to his expression of opinion that it is impossible to get Uran 235. He replied that he does not know. He believed that the project is still in the experimental stage.

In July, 1944, Dr. Cockcroft, who holds the chair of Jacksonian Professor of Natural Philosophy at Cambridge, England, and who is a scientist of international reputation, had been made director of Atomic Energy Project, Montreal and Chalk River, and worked in collaboration with Canadian scientists at the Montreal Laboratory of the National Research Council.

Dr. Allan Nunn May, a British temporary civil servant, formed part of the research group that came over to Canada, and was at the Montreal Laboratory as a group leader under Dr. Cockcroft. In the performance of his duties, May had access to a substantial amount of knowledge of the work that was being done in connection with the Atomic Energy Project. The evidence shows that before coming to Canada, he was an ardent but secret Communist and already known to the authorities at Moscow. Not long after his arrival here he was contacted on instructions from "*The Director*", and given the cover name "*Alek*" by the organization of Colonel Zabotin. In view of his background and the position he occupied, he was a logical person from whom the Russians could expect to obtain the available knowledge on atomic energy. By telegram dated July the 28th, 1945, "*The Director*" at Moscow sent a telegram to Colonel Zabotin with reference to Dr. Allan May ("*Alek*"), reading in part as follows:—

No. 10458
30.7.45

To Grant

Reference No. 218.

. . . Try to get from him before departure detailed information on the progress of the work on uranium. Discuss with him: does he think it expedient for our undertaking to stay on the spot; will he be able to do that or is it more useful for him and necessary to depart for London? [In the first half]

Director. 28.7.45

These instructions were promptly followed in Ottawa, for a few days later, on the 9th of August, 1945, the following telegram was sent to Moscow by Zabotin:—

241

To the Director,

Facts given by Alek: (1) The test of the atomic bomb was conducted in New Mexico, (with "49", "94–239"). The bomb dropped on Japan was made of uranium 235. It is known that the output of uranium 235 amounts to 400 grams daily at the magnetic separation plant at Clinton. The output of "49" is likely two times greater (some graphite units are [established] planned for 250 mega watts, i. e. 250 grams each day). The scientific research work in this field is scheduled to be published, but without the technical details. The Americans already have a published book on this subject.

Words in black brackets, which appeared as "faint type" in the printed report of the Royal Commission, indicate words crossed out in original documents.

(2) Alec handed over to us a platinum with 162 micrograms of uranium 233 in the form of oxide in a thin lamina. We have had no news about the mail.

9.7.45.   Grant.

On the same date, another telegram was forwarded by Zabotin giving information obtained from May on a man by the name of Norman Veall, upon whom we are also reporting (see Section IV, 1). This telegram disclosed that May advised against accepting any information about the atomic bomb from Veall.

243

To the Director,

Alek reported to us that he has met Normal Veal (he was at his home). Veal works in the laboratory of the Montreal branch of the Scientific Research Council. . . . He asked the opinion of Alek : Is it worth while for him (Veal) to hand over information on the atomic bomb. Alek expressed himself in the negative. Alek stated that Veal occupies a fairly low position and knows very little. . . .

9.8.45.   Grant.

A few days after May had handed over to the Russians information concerning the atomic bomb, and the above-mentioned quantity of uranium 233, Zabotin paid a social visit to a friend living in the vicinity of Chalk River. He then had the opportunity of seeing the plant from the river during a motor-boat cruise, and reported to "*The Director*" what he had seen. The latter, on the 14th of August, 1945, sent him a telegram which included the following :—

11438.
14.8.45.

To Grant
1. Your No. 231.

Wire what connections F——— has with the plant indicated by you, where is he working at present, and what are your mutual relations with him?

If possible, give a more detailed description of the exterior of the plant. . . .

Director.

May made two visits to the same plant: the first on the 16th August, 1945, and the second on the 3rd September. He also went on several occasions to the Chicago plant, doing experiments in collaboration with American scientists.

On August 22nd, 1945, "*The Director*" telegraphed Zabotin :—

Supplement to No. 11923

N 11931
22.8.45

To Grant

Take measures to organize acquisition of documentary materials on the atomic bomb!

The technical process, drawings, calculations.

Director,
22.8.45.

On the 31st August Zabotin, not having received any reply from Moscow as to the value of the information on the atomic bomb which he had sent, telegraphed to "*The Director*" as follows :—

275

To the Director

I beg you to inform me to what extent have Alek's materials on the question of uranium satisfied you and our scientists (his reports on production etc.)

This is necessary for us to know in order that we may be able to set forth a number of tasks on this question to other clients. Have you received all NN mail up to July of this year?

Grant

31.8.45

The evidence shows that May provided the Soviet espionage leaders with information on other subjects as well as on the atomic bomb. One of the documents is a telegram from Zabotin to Moscow, reading as follows:—

242

To the Director

On our task Alek has reported brief data concerning electronic shells. In particular these are being used by the American Navy against Japanese suicide-fliers. There is in the shell a small radio-transmitter with one electronic tube and it is fed by dry batteries. The body of the shell is the antenna. The bomb explodes in the proximity of an aeroplane from the action of the reflected waves from the aeroplane on the transmitter. The basic difficulties were: the preparation of a tube and batteries which could withstand the discharge of the shell and the determination of a rotation speed of the shell which would not require special adaptation in the preparation of the shell. The Americans have achieved this result, but apparently have not handed this over to the English. The Americans have used a plastic covering for the battery which withstands the force of pressure during the motion of the shell.

Grant.

9.7.45.

After his second visit to the Chalk River plant on September 3rd, 1945, Dr. May departed for England. The documents that have been produced reveal that Colonel Zabotin's organization was aware of this departure and that May was instructed to contact a person in London, England. This contact was being organized between Moscow, Londen and Ottawa.

The following telegrams were exchanged between Zabotin and *"The Director"* on this matter:—

No. 10458
30.7.45

To Grant
Reference No. 218.
[28.7.45]

Work out and telegraph arrangements for the meeting and the password of Alek with our man in London.

Director. 28.7.45

Grant
31.7.45

244

To the Director,

We have worked out the conditions of a meeting with Alek in London. Alek will work in King's College, Strand. It will be possible to find him there through the telephone book.

Meetings: October 7.17 27 on the street in front of the British Museum. The time, 11 o'clock in the evening. Identification sign:—A newspaper under the left arm. Password:—Best regards to Mikel (Maikl). He cannot remain in Canada. At the beginning of September he must fly to London. Before his departure he will go to the Uranimum Plant in the Petawawa district where he will be for about two weeks. He promised, if possible, to meet us before his departure. He said that he must come next year for a month to Canada. We handed over 500 dollars to him.

Grant.

11955
22.8.45

To Grant

Reference No. 244.

The arrangements worked out for the meeting are not satisfactory. I am informing you of new ones.

1. Place:
   In front of the British Museum in London, on Great Russell Street, at the opposite side of the street, about Museum Street, from the side

---

Words in black brackets, which appeared as "faint type" in the printed report of the Royal Commission, indicate words crossed out in original documents.

of Tottenham Court Road repeat Tottenham Court Road, Alek walks from Tottenham Court Road, the contact man from the opposite side—Southampton Row.

2. Time:

As indicated by you, however, it would be more expedient to carry out the meeting at 20 o'clock, if it should be convenient to Alek, as at 23 o'clock it is too dark. As for the time, agree about it with Alec and communicate the decision to me. In case the meeting should not take place in October, the time and day will be repeated in the following months.

3. Identification signs:

Alek will have under his left arm the newspaper "Times", the contact man will have in his left hand the magazine "Picture Post"

4. The Password:

The contact man: "What is the shortest way to the Strand?"
Alek: "Well, come along, I am going that way."
In the beginning of the business conversation Alek says: "Best regards from Mikel".
Report on transmitting the conditions to Alek.

18.8   Director.

                                    22.8.45
                                    Grant.

The evidence before us does not reveal whether the contact referred to in the above telegram was made.

In February, 1946, while our investigation was in progress, May was arrested in London on a charge of violating the Official Secrets Act. Before being arrested, Dr. May confessed his guilt. His written statement, signed by him, reads as follows:—

About a year ago whilst in Canada, I was contacted by an individual whose identity I decline to divulge. He called on me at my private apartment in Swail Avenue, Montreal. He apparently knew I was employed by the Montreal laboratory and he sought information from me concerning atomic research.

I gave and had given very careful consideration to correctness of making sure that development of atomic energy was not confined to U. S. A. I took the very painful decision that it was necessary to convey general information on atomic energy and make sure it was taken seriously. For this reason I decided to entertain proposition made to me by the individual who called on me.

After this preliminary meeting I met the individual on several subsequent occasions whilst in Canada. He made specific requests for information, which were just nonsense to me—I mean by this that they were difficult for me to comprehend. But he did request samples of uranium from me and information generally on atomic energy.

At one meeting I gave the man microscopic amounts of U.233 and U.235 (one of each). The U.235 was a slightly enriched sample and was in a small glass tube and consisted of about a milligram of oxide. The U.233 was about a tenth of a milligram and was a very thin deposit on a platinum oil and was wrapped in a piece of paper.

I also gave the man a written report on atomic research as known to me. This information was mostly of a character which has since been published or is about to be published.

The man also asked me for information about the U. S. electronically controlled A. A. shells. I knew very little about these and so could give only very little information.

He also asked me for introductions to people employed in the laboratory including a man named Veale but I advised him against contacting him.

The man gave me [200 ANM] some dollars (I forget how many) in a bottle of whiskey and I accepted these against my will.

Before I left Canada it was arranged that on my return to London I was to keep an appointment with somebody I did not know. I was given precise details as to making contact but I forget them now. I did not keep the appointment because I had decided that this clandestine procedure was no longer appropriate in view of the official release of information and the possibility of satisfactory international control of atomic energy.

The whole affair was extremely painful to me and I only embarked on it because I felt this was a contribution I could make to the safety of mankind. I certainly did not do it for gain.

As it will be seen, May clearly admits having done what has been revealed by the official documents from the Embassy, namely, the giving of uranium and a written report on atomic research as known to him. He denies having made the pre-arranged contact previously mentioned. The person who contacted him in Montreal and obtained the uranium and other information concerning the atomic bomb has been identified by Gouzenko as being Lieut. Angelov, one of the Secretaries of the Military Attache. It has also been established by the documents that the amount of money which May received was at least $700. plus two bottles of whisky.

After having elected to be tried by a jury in London, May, on the day set for his trial, pleaded guilty and was sentenced to ten years penal servitude.

After he had pleaded guilty and the United Kingdom Attorney-General had summarized the facts of the case, defending Counsel put in a plea for leniency. In passing sentence Mr. Justice Oliver said :—

Allar Nunn May, I have listened with some slight surprise to some of the things which your learned counsel has said he is entitled to put before me: the picture of you as a man of honour who had only done what you believed to be right. I do not take that view of you at all. How any man in your position could have had the crass conceit, let alone the wickedness, to arrogate to himself the decision of a matter of this sort, when you yourself had given your written undertaking not to do it, and knew it was one of the country's most precious secrets, when you yourself had drawn and were drawing pay for years to keep your own bargain with your country— that you could have done this is a dreadful thing. I think that you acted not as an honourable but a dishonourable man. I think you acted with degradation. Whether money was the object of what you did, in fact you did get money for what you did. It is a very bad case indeed. The sentence upon you is one of ten years' penal servitude.

We have no doubt of the importance of the information given by Dr. May on atomic energy; for that purpose we had the advantage of hearing Dr. Cockcroft whose collaboration has been most helpful in the determination of the extent and value of the secret data communicated. This is further dealt with in Section VII.

7. The Greenglass Case—Details

The particulars of this case are thoroughly developed in the testimony which David Greenglass himself gave at the trial of Julius and Ethel Rosenberg and Morton Sobell on March 9 and March 12, 1951. The following is a verbatim account of what was said, as taken down by the court reporter. The remarks of counsel for the Government and the defense have been omitted except where this would impair understanding of Greenglass' statements. Also included is the testimony of Dr. Walter S. Koski, an expert witness for the prosecution.

From stenographer's minutes of Case 134–245, *United States of America* v. *Julius Rosenberg et al.* Before Hon. Irving R. Kaufman, district judge, United States District Court, Southern District of New York, March 9, 1951.

DAVID GREENGLASS, called as a witness in behalf of the Government, being first duly sworn, testified as follows:

＊　　　＊　　　＊　　　＊　　　＊　　　＊　　　＊

Direct examination by Mr. COHN:

Q. Mr. Greenglass, will you try to keep your voice up so the Court and jury can get the benefit of your testimony. Are you the David Greenglass who is named as a defendant in the indictment here on trial?—A. I am.

Q. That indictment charging conspiracy to commit espionage?—A. Yes.

Q. Have you entered a plea to that indictment?—A. I have.

Q. What is that plea?—A. Guilty.

Q. Are you now in the custody of the United States Marshal?—A. I am.

Q. Now, prior to the time you were remanded to the custody of the United States Marshal, what was your home address?—A. 265 Rivington Street.

Q. That is here in Manhattan?—A. Yes.

Q. How old are you?—A. 29.

Q. When were you born?—A. March 3, 1922.

Q. Are your parents alive?—A. My father is dead. My mother is alive.

Q. Do you have any brothers and sisters?—A. I have two brothers and one sister.

Q. Your sister is the defendant Mrs. Ethel Greenglass Rosenberg; is that correct?—A. That is true.

Q. And another defendant, Julius Rosenberg, is your brother-in-law?—A. That is true.

Q. Is Mrs. Rosenberg older or younger than you are?—A. Older.

Q. What are the names of your brothers?—A. One brother is Samuel. One is Bernard.

Q. Are you yourself married?—A. I am.

Q. What was your wife's maiden name?—A. Ruth Printz.

Q. How do you spell that?—A. P-r-i-n-t-z.

Q. When were you married?—A. November 29, 1942.

Q. Do you have any children?—A. I have two.

Q. How old are they?—A. One is nine months old and one is four years old.

Q. Boys or girl?—A. One is a girl. One is a boy.

Q. Where were you educated, Mr. Greenglass?—A. I was educated in New York.

Q. Would you tell us briefly the schools which you attended here in New York?—A. I went to P. S. 4, P. S. 97, Haaren Aviation School, Brooklyn Polytechnic, and Pratt Institute.

Q. What field have you pursued since your graduation from public school?—A. I am a machinist.

Q. Have you studied the work of a machinist and related problems while you were at Aviation School?—A. Yes.

NOTE.—Counsel for the U. S. Government: Irving Saypol, United States attorney; Miles J. Lane, James Kilsheimer, and Roy Cohn, assistant United States attorneys. Other counsel whose names appear in transcript material represent defendants.

Q. And also at Pratt Institute; is that correct?—A. Yes, sir.

Q. Is that correct?—A. Yes, sir.

Q. After you left school and prior to 1943 did you have any practical experience as a machinist?—A. I did.

Q. Here in New York?—A. In New York.

Q. Now, in 1943 did you enter the Army of the United States?—A. I did.

Q. As a private?—A. Private.

Q. When in 1943 did you go into the Army?—A. April 1943.

Q. After that did you have basic training?—A. Yes.

Q. Where was that?—A. Aberdeen, Maryland.

Q. Were you thereafter assigned to work as a machinist while in the Army?—A. I was.

Q. After that did you go to ordnance school?—A. I went to ordnance school; yes.

Q. What did you do out there?—A. It was a shop——

Mr. E. H. BLOCH. Out where?

Q. The ordnance school: where was the ordnance school, Mr. Greenglass?—A. In Aberdeen, Maryland.

Q. I think you said it was a shop?—A. It was a shop.

Q. And did you pursue your trade as a machinist in that shop?—A. I did.

Q. How long were you at Aberdeen?—A. Until July.

Q. That is July of 1943?—A. Yes.

Q. Now, am I correct in stating that during the next year, July 1943 to July 1944, you were stationed at various posts, Army posts, throughout the United States?—A. I was.

Q. You were stationed at a number of them?—A. A number of them.

Q. Different parts of the country?—A. Yes.

Q. Now, in July of 1944, did you receive a new assignment?—A. I did.

Q. To what location? At what location?—A. To Oak Ridge, Tennessee, the Manhattan Project.

Q. The Manhattan Project District?

The COURT. When was that?

The WITNESS. It was July 1944.

Q. July of 1944. You were assigned to the Manhattan District Project of the United States Army, is that correct?—A. That is right.

Q. Did you at that time know what the Manhattan District Project of the United States Army was?—A. I did not.

Q. You know now it was the project in charge of construction of the atomic bomb, is that correct?—A. I do.

Q. Now, when you were out at Oak Ridge, Tennessee, in July of 1944, how long did you stay out there?—A. About two weeks.

Q. Now, during that period were you given any security lectures?—A. I was.

Q. Did they concern the new duties you were to undertake?—A. Yes, they did.

Q. Were you told anything about the nature of those duties and the nature of the work at Manhattan Project?—A. I was.

Q. What were you told?—A. I was told that it was a secret project.

Q. Were you told at that time what was going on at that project, what was being constructed?—A. No.

Q. You were told nothing about that, is that correct?—A. Nothing at all.

Q. Was the Espionage Act mentioned to you in connection with revealing any information as to what was going on in the Manhattan Project?—A. It was.

Q. After your two weeks' orientation at Oak Ridge, Tennessee, were you then assigned to report to some other place in the United States?—A. I was.

Q. Where was that?—A. Los Alamos, New Mexico.

Q. How did you go out there?—A. Train all the way.

Q. About when did you report at Los Alamos?—A. August 1944.

Q. When you reported at Los Alamos were you given certain instructions concerning the duties you were to pursue out there?—A. I was interviewed for a job.

Q. Did there come a time when you were told that you would work as a machinist in the shop?—A. That is right.

Q. Were you told at that time the nature of the work being done at Manhattan Projects?—A. No.

Q. Was the fact that it was secret reaffirmed to you?—A. It was.

Q. Were you told just how much you were to know about what was going on at Manhattan Project?—A. I was told I was to know as much as was necessary to do my job.

Q. And nothing more?—A. Nothing more.

Q. Now, would you tell us at this point when it was that you learned for the first time that the Manhattan Project District was the district of the United States Army concerned with the construction of the atomic bomb?—A. When my wife came to visit me in November 1944, she told me that Julius had told her——

Mr. E. H. BLOCH. I object to any conversation between this witness' wife and himself outside the presence of the defendant Julius Rosenberg.

The COURT. She is named as a coconspirator.

Mr. E. H. BLOCH. I respectfully except.

The COURT. Objection overruled.

Q. Will you tell us again? I think you said the time was November 1944?—A. Right.

Q. What did your wife tell you?—A. She told me that Julius had said that I was working on the atomic bomb.

Q. And that was the first you knew of it?—A. That was the first I knew of it.

Q. You had never been told that by anybody in an official capacity of the United States Government?—A. No, sir.

Q. Now, going back to August of——

The COURT. Will you just slow up the slightest bit because I am trying to make some notes.

Mr. COHN. I will be glad to. I am trying to get over some of the preliminary points.

Q. In August of 1944, Mr. Greenglass, when you took up your duties at Los Alamos, will you tell us——

The COURT. Excuse me. When was that conversation with your wife? November?

The WITNESS. At the end of November 1944.

The COURT. Very well.

Q. About several months after you first went to Los Alamos; is that right?—A. It was.

Q. During that first few months you did not know just what was being done at Los Alamos?—A. That's right.

Q. Now, I think you said you were assigned to work as a machinist?—A. I was.

Q. And where, physically, was your work done?—A. It was at a shop called the "E" building shop or the "student shop."

Q. Where was that located?—A. In "E" building, in the tech. area, at Los Alamos.

Q. By "the tech. area," you mean the technical area?—A. Technical area at Los Alamos.

Q. Out at Los Alamos, this shop was located in one of the buildings out at Los Alamos; is that right?—A. That's right.

Q. You were assigned to work there as a machinist?—A. I was.

Q. Now, you said the "E" shop; did this letter "E" have any significance?—A. It was the building I was in.

Q. That was the building you were in?—A. Yes.

Q. Now, were you a member of a group out there? Was this building and was the machine shop under the jurisdiction of a particular group of the Manhattan District Project, at Los Alamos?—A. It was.

Q. What was the name of that group?—A. It was the "E" group.

Q. The "E" group?—A. Right.

Q. Now, did the "E" group have a head or a leader?—A. It did.

Q. What was he called?—A. His name was Kistiakowski.

Q. Is that Dr. George B. Kistiakowski of Harvard University?—A. That's right.

Q. And do you know what his professional standing is, in what field he is known?—A. Yes, I do.

Q. What is that field?—A. Thermodynamics man.

The COURT. Speak up.

Mr. E. H. BLOCH. I didn't get that, I am sorry.

The WITNESS. He is a thermodynamics man.

Q. Thermodynamics?—A. Physical chemistry.

Q. In general terms, what was group "E" concerned with?—A. With high explosives.

Q. High explosives?—A. Yes.

Q. Did you have any other superiors in addition to Professor Kistiakowski?—A. I did.

Q. Will you tell us?—A. The foreman of the shop was a man by the name of De Mars, a civilian. I don't know how to spell his name.

Mr. COHN. I don't either, your Honor, but we will check it.

Q. You say he was the foreman of the shop?—A. That's right.

Q. Did you have any superiors between the foreman of the shop and Dr. Kistiakowski, who was the leader of "E" group?—A. I did.

Q. Who was that?—A. His name was Fitzpatrick.

Q. Was he a civilian?—A. He was a GI.

Q. What was his rank?—A. Well, later on he became a Master Sergeant. I don't recall his rank at the time.

Q. And what was his title?—A. He was in charge of procurement and the machine shop of "E" group.

Q. He was in charge of procurement and this machine shop, this shop "E"?—A. For the "E" group.

Q. In other words, the structure was, out in Los Alamos, the "E" group, headed by Dr. Kistiakowski and concerned with high explosives; under Dr. Kistiakowski there was Sergeant Fitzpatrick——A. That's right.

Q. In charge of and concerned with procurement, and insofar as the shop itself, it had a foreman and that foreman was Mr. De Mars, at the beginning; I think you said?—A. That's right.

Q. And you were one of the machinists?—A. I was one of the machinists.

Q. About how many machinists would you say were assigned to that shop?—A. Oh, there were about—the greatest amount was about 10 machinists.

Q. Would the number vary from time to time?—A. Yes; it would vary.

The COURT. May I suggest, Mr. Cohn, that you stand back a little bit.

Mr. COHN. All right.

Q. Now, did there come a time when Mr. De Mars was transferred, gave up his duties as foreman?—A. He did.

Q. Did you get another foreman out at the shop?—A. I did.

Q. What was his name?—A. Bob Holland.

Q. Holland, H-o-l-l-a-n-d?—A. Right.

Q. Now, after Mr. Holland's assignment, was there any change in your duties?—A. I became the assistant foreman.

Q. You became the assistant foreman?—A. Yes.

Q. Did there come a time when Mr. Holland left?—A. There was.

Q. About when was that?—A. Oh, the end of '45, beginning of '46.

Q. After Mr. Holland left, was there any further change in your duties?—A. I became the foreman of the shop.

Q. Did you continue to hold that position until you were discharged from the Army of the United States?—A. I did.

Q. When were you so discharged?—A. In the last day of February 1946.

Q. Were you honorably discharged?—A. I was.

The COURT. Then for what period of time were you foreman?

The WITNESS. I would say a period of about two, two and a half months.

Q. Just prior to your discharge from the Army; is that right?—A. That's right.

Q. Will you tell us, were you a noncommissioned officer when you were discharged from the Army?—A. I was.

Q. What rank?—A. T/4, sergeant.

Q. T/4, sergeant.—A. Yes.

Q. Sergeant T/4?—A. Yes.

Q. Now, going back to the time when you undertook your duties as a machinist at this "E" shop out at Los Alamos, would you tell the Court and jury in general terms just what your duties were, what you did over the period of time you were working in the machine shop, as a general proposition?—A. Well, the shop itself took jobs from various scientists and made apparatus whenever they needed it; and there were two methods of jobs coming through the shop. One was to—when a scientist needed a piece of apparatus, he just sent it through procurement and it was sent to either one of the three shops in the technical area.

Q. There were three shops; is that right?—A. There were three shops.

Q. What were the names of the other two?—A. "V" and "C" shop, which were bigger than mine; both were bigger than mine.

Q. Yours was the smallest shop; is that right?—A. Ours was the smallest shop.

Q. Go ahead.—A. That was one way; and they would be distributed according to how much work each shop had. The other way was go directly to Fitz and say, "How about getting this job done?" And usually it was put through, or the sketch or piece of paper or the scientist talking to one of us machinists to do it.

Q. In other words, it was your job to machine this particular——A. Apparatus.

Q. Apparatus or product that the scientist required in connection with his experimentation on atomic energy; is that correct?—A. That is correct.

Q. Now, did the physical location of your "E" shop remain the same during your entire stay at Los Alamos?—A. No, it didn't.

Q. When was there a change?—A. Oh, it was in the fall 1944, we had a building built and the whole procurement section moved into that building. It was called the "Theta" building.

Q. That building, and the shop under which you undertook your duties was known as "Theta" shop?—A. Theta.

The COURT. How do you spell that "Theta"?

The WITNESS. T-h-e-t-a.

Mr. COHN. It is a Greek letter.

The WITNESS. A Greek letter.

Q. Was there any change in your duties when you went over to Theta shop?—A. They remained the same.

Q. You were doing the same thing, but the physical location had changed; you had been in the "E" shop before and you were now in the Theta shop?—A. That's right.

Q. Did you continue to work in the Theta shop, in the various capacities you have described, until the time that you left Los Alamos?—A. I did.

Q. Did you continue to do work such as that which you have described to us?—A. That is correct.

Q. Now, you have told us about the security talks you had at Oak Ridge and about what was told you concerning the secret nature of your work when you got out to Los Alamos. In addition to these oral instructions, were you given any written material containing security regulations and telling you just what you were at liberty to disclose and what you should not disclose?—A. I was given such a book.

Mr. COHN. May this be marked for identification, your Honor?

(Marked "Government's Exhibit 1" for identification.)

Mr. E. H. BLOCH. May we look at it?

Mr. COHN. As soon as I offer it in evidence, Mr. Bloch, certainly.

Q. Would you just look at this, look through it for a minute, Mr. Greenglass [handing to witness], have you examined Government's Exhibit 1 for identification?—A. I did.

Q. Do you recognize that?—A. It is a photostat of the booklet that I received at Los Alamos.

Mr. COHN. I offer it in evidence, your Honor.

Mr. A. BLOCH. Objected to on the ground it is incompetent, irrelevant, and immaterial, not binding on my defendant.

The COURT. Overruled.

Mr. A. BLOCH. Exception.

Mr. E. H. BLOCH. I suppose when Mr. Bloch said his defendant, that means all defendants?

The COURT. That is correct.

Mr. E. H. BLOCH. Could I see it?

Mr. COHN. Yes [handing].

Mr. E. H. BLOCH. I will try to be as quick as I can.

Mr. COHN. It is all right.

(Government's Exhibit 1 previously marked for identification received in evidence.)

The COURT. Are you going to call certain portions to the attention of the jury?

Mr. COHN. I am, your Honor. I might read just a few brief portions to the jury, pass it around so that they can examine the whole thing, and if any of the gentlemen of the defense feel I have omitted anything I should have read, I assume they will be at liberty to do that now that this has been received as Government's Exhibit 1, your Honor.

The COURT. All right.

Mr. COHN. I may say to the jury, the exhibit itself is marked "Restricted." The word on the beginning of the first page is "Security." The first two paragraphs read as follows:

"This handbook has been designed to provide members of the technical area staff and their families with a concise summary of existing security regulations.

It should be understood that to obey these regulations is a minimum requirement. There is a further obligation on the part of everyone to maintain a constant and intelligent interest in the prevention and reporting of all incidents whose occurrence endangers the security of the project. It is a basic policy of the project that everyone working here should know whatever is required for doing his job well. It is therefore of greatest importance for each person to understand that he is in a position of trust with regard to all such information and also with regard to information which he may accidentally gain about other confidential matters."

There is further descriptive material. On page 2 there is a section entitled "Communication."

"(A) There must be no conversation outside the technical area, or in the presence of unauthorized persons, and no information in personal letters, conveying any of the following kinds of information:

"1. The purpose of the project.

"2. The general problems being worked on.

"3. Technical data connected with 1 or 2 above.

"4. The scheduling or general progress of the work.

"5. Any over-all account of the personnel employed on the project.

"6. The procurement or presence here of essential materials and installations.

"By 'unauthorized persons' are meant persons whom you do not know to have the permission of their group or divisional leaders or the director to receive the information in question.

"(B) There must be no conversation outside the post, or in the presence of unauthorized persons, and no information in personal letters, conveying any of the following kinds of information:

"1. The professions or former connections of persons working in the technical area.

"2. The name of the contractor under whom the project is being run.

"3. Affiliation of this project with other war projects.

"4. The size of the project or post, or other significant features such as water supplies, fire-protection installations, etc.

"5. The general kinds of work going on in the technical area. We are engineers; the technical area should be called only 'the technical area.'

"By 'unauthorized persons' are meant persons who do not live in or have access to the post, or who, living here, have no reason to receive the particular class of information.

"6. Your address, P. O. Box 1663, Sante Fe, New Mexico, may be given to family, friends, and in private business dealings. Do not use Los Alamos stationery in private correspondence."

Then there are further restrictions concerning the receipt of mail; travel— the employees are instructed not to establish or maintain social relations with anyone living in neighboring communities; not to have friends visit them out there; are told not to fill out any questionnaires, licenses, applications or anything else without first consulting the Personnel Office as to the propriety of the detailed information requested by that application; and to report any people without the proper badge——

By Mr. COHN:

Q. By the way, were badges worn out there?—A. They were.

Q. Having different significance?—A. They were.

Q. How did they go, by color?—A. By color.

Q. What did a color represent?—A. A white badge was authorized to go to the seminars and be let in on all the information that was available on the bomb.

* * * * * * *

Mr. A. BLOCH. Will you read the last answer?

(Answer read.)

Q. That was a white badge?—A. That was a white badge.

Q. Were there any other colors?—A. There was a red badge which allowed the bearer to get all the information necessary to be able to do his job; and then there was a blue badge which allowed—well, it allowed the bearer to go into the tech area to do various jobs like steamfitting or ditch-digging, but not to be around any of the equipment or to see any of the experiments.

Q. Now, I assume that is what this regulation refers to when it says that you are to report any person wearing the wrong badge to the authorities?—A. That is right.

Q. When observed in a certain area?—A. That is right.

Q. Did you yourself have a badge?—A. I did.

Q. Now, specifically, you told us that Dr. Kistiakowski was out at Los Alamos and was in fact the leader of Group E?—A. Right.

Q. And that his reputation is in the field of physical chemistry?—A. That is right.

Q. Thermodynamics?—A. That is right.

Q. While out at Los Alamos did you come to learn the identity of any other scientists who were present and working on atomic energy?—A. That is correct.

Q. Would you name one or two of those?

Mr. E. H. Bloch. Is it contended that this testimony will connect up any of the defendants?

Mr. Cohn. Quite definitely, your Honor.

The Witness. I did get to know a number of scientists and some of world fame, for instance, Dr. Oppenheimer, whom we knew as the head of the project.

Q. J. Robert Oppenheimer?—A. That is right, and there was Neil Bohr, whom I first knew as Baker.

Q. What do you mean by that?—A. It was a pseudonym to keep his identity secret.

Q. You mean that Dr. Bohr was known at Los Alamos by an assumed name, that of Baker?—A. That is right, Mr. Baker.

Q. And you knew at first that there was a man named Mr. Baker, a scientist?—A. That is right.

Q. Was there a period of time during which you yourself did not know who Mr. Baker actually was?—A. That is correct.

Q. And did there come a time when you found out who he was?—A. That is right.

Q. And who is he?—A. Mr. Baker was Neils Bohr. He is a nuclear physicist.

Q. Considered one of the outstanding in the world, is that correct?—A. That is correct.

Q. Do you recall whether the fact that Dr. Bohr was out in Los Alamos was secret information?

Mr. E. H. Bloch. When was this? Will you fix the time, please?

Q. Will you tell us the best you remember when you first knew that Mr. Baker, a man named Mr. Baker was out there?—A. It was about September or October of 1944.

Q. You knew him only as Baker, is that right?—A. That is right.

Q. Was it shortly thereafter you found out who he really was?—A. That is right.

Q. And you were told he was Dr. Bohr, is that correct?—A. That is right. In passing one of my colleagues said, "That's Baker and he is Neils Bohr."

Q. You knew that the information as to who Dr. Bohr out there was was a secret?—A. I did.

Q. As a matter of fact, I think that this very security pamphlet states that the identity of scientists out there and their former occupation was not to be discussed by any unauthorized person, is that right?—A. That is right.

Q. I assume as a practical matter that one's former occupation in a particular field of science would be a clue to the particular work he might be doing?—A. That is correct.

Q. Is that the reason for this regulation?

Mr. E. H. Bloch. If he knows.

The Witness. That is the reason for it.

Q. In addition to Mr. Baker whom you came to know as Neils Bohr and Dr. Oppenheimer, may I ask you specifically, did you know that Dr. Harold Urey was connected with the Manhattan project?—A. I did.

Q. About what point after your arrival at Los Alamos did you learn that fact?—A. Oh, it must have been about December or so.

The Court. When did you learn about Dr. Oppenheimer? I do not think you told us about that.

The Witness. That was almost at the beginning of the time I was there.

Mr. E. H. Bloch. Your Honor, I will object and am objecting to whether or not this particular witness knew some of the most renowned scientists at Los Alamos unless this particular information is related to the issues in this case so far as it bears upon the guilt or innocence of the defendants.

Mr. Cohn. I would be glad to state to your Honor that the name of each scientist which has been spoken by Mr. Greenglass from this stand will be directly related to the defendants in this case and specifically to Mr. Bloch's client.

The Court. Very well.

Mr. Cohn. I make that representation.

The Court. Very well.

Q. And there were other scientists there; is that correct?—A. That is correct.

Q. Whose identities you had learned?—A. Yes.

Q. Now, was one of the scientists who was present at Los Alamos and whose name and presence you came to know, Dr. Walter Koski?—A. That is correct.

Mr. COHN. I believe Dr. Koski is here in court.

Would you rise, Dr. Koski?

(A man rises in courtroom.)

Q. Do you recognize Dr. Koski here in court?—A. I do.

Q. Did you do any work at any time in connection with apparatus that Dr. Koski required in the course of his experimentation on atomic energy?—A. I did.

Q. Did you specifically work on the machining of a flat type lens mold and other molds which Dr. Koski required in the course of his experimentation on atomic energy?—A. I did.

Mr. E. H. BLOCH. May I just make a suggestion. I am going to suggest to the Court and the Court indirectly to Mr. Cohn that when on subjects which have been referred to in previous documents that his questions not be leading and suggestive and that he try to avoid leading and suggestive questions.

Q. You say you yourself——

The COURT. Just a moment. What was the last question and answer?

(Question and answer read.)

(Question read.)

Q. Now did there come a time when the first atomic explosion took place?—A. Yes.

Q. When was that?—A. July 1945.

Q. Where?—A. Alamogordo, New Mexico.

Q. In the course of your employment at Los Alamos did you hear discussion concerning this atomic explosion?—A. I did.

The COURT. Was that after the explosion or did you hear about the anticipated explosion?

The WITNESS. I heard of an explosion to take place at Alamogordo.

Q. Was that before?—A. I heard that before. Afterwards I heard of the atomic explosion that took place at Alamogordo.

Q. Now am I correct in stating that during the entire period of your stay in Los Alamos, 1944 to the time you were discharged in 1946, you worked in the machine shop and in the Theta shop on apparatus and equipment in connection with experimentation on atomic energy?—A. I did.

Q. Was that work pursued in the manner you have described here from sketches supplied and verbal descriptions by the particular scientists out there who required the apparatus?—A. I did.

Q. I think you have told us, Mr. Greenglass, that your sister Ethel was a number of years older than you are; is that correct?—A. She is.

Q. Do you remember in what year she was married to the defendant, Julius Rosenberg?

The COURT. How much older is she?

The WITNESS. Six years older.

Q. Do you remember the year in which she was married to the defendant, Julius Rosenberg?—A. 1939.

Q. Had you come to know Julius Rosenberg before your sister married him?—A. I did.

Q. Was he around your house?—A. Yes; he was.

Q. And you were 17 years old at the time they were married; is that correct?—A. That is correct.

Q. Now did you have any discussion with Ethel and Julius concerning the relative merits of our form of government and that of the Soviet Union?

Mr. A. BLOCH. Objected to as incompetent, irrelevant, and immaterial, not pertinent to the issues raised by the indictment and the plea.

Mr. E. H. BLOCH. And upon the further ground that this will obviously lead to matters which may only tend to confuse the jury and inject inflammatory matter which will make it difficult or almost impossible for the jury to confine themselves to the real issues in the case.

*   *   *   *   *   *   *

The COURT. All right. Objection overruled.

Mr. E. H. BLOCH. Of course, I am joining in the objection already made.

(Record read.)

Mr. E. H BLOCH. I object to the question on the further ground it is leading and suggestive.

The Court. Overruled.

Mr. E. H. Bloch. Exception.

By Mr. Cohn:

Q. You may now answer.—A. I did have such discussion.

Q. Over what period of time, roughly?—A. From about 1935 to about 1946 or 1946.

Q. To 1945 or 1946; is that right?—A. Yes.

Q. Were those discussions numerous?—A. At the beginning, yes.

Q. And at any time in the course of those discussions did the stated position of either Ethel or Julius Rosenberg change? Did their views change?

Mr. E. H. Bloch. I object to it.

Mr. A. Bloch. I object.

Mr. E. H. Bloch. I object to the form of the question.

The Court. You mean we haven't had the views yet and he is asking about a change?

Mr. E. H. Bloch. Yes.

The Court. You better ask about the views, first.

Q. Which system of government did they tell you they preferred?

Mr. E. H. Bloch. May I again object and ask some clarification? I understand from this witness' answers that he said that he had discussions with these defendants from 1935 to 1944 or 1945.

The Court. 1945 or 1946.

Mr. E. H. Bloch. 1945 or 1946. I am going to ask your Honor to compel Mr. Cohn to detail the time of each discussion so that appropriate objection may be made for the record.

The Court. I thought you didn't want it, but if that is what Mr. Bloch wants, all right. If you can remember each time you had such a discussion you can tell us, and if you can't remember the exact date or the exact month but you remember the year, you can tell us and then tell us what the discussion was.

Mr. Cohn. I think, your Honor, of course, if Mr. Bloch thought I didn't go far enough or he thought anything further was required for his purposes, he could clarify it on cross-examination.

The Court. Yes.

Mr. Cohn. But if your Honor wants me to——

The Court. Let us get as much as we can this way.

Mr. Cohn. May I still try to keep it brief?

The Court. Yes. Try to get from him what he can remember at this time. Then Mr. Bloch can detail it more for him.

By Mr. Cohn:

Q. I think you said these discussions with your sister began in 1935.—A. I did.

Q. When did they begin, so far as the defendant Julius Rosenberg was concerned?—A. About 1937.

Q. All right. Can you remember any specific occasions on which they had these discussions with you?—A. Do you mean the early period or later?

Q. Start with the early period. Mr. Bloch wants everything. Let me ask you this: In the early period how frequently would they express their views regarding the relative merits of the two countries?—A. I would say two or three times a week.

Q. Two or three times a week.

Mr. Bloch. Your Honor, I submit on the basis of that I should be required to take up the time of the Court and jury——

The Court. No. He has answered that it was two or three times a week now. Now let us find out whether the conversations ran along the same line on each occasion.

Mr. Cohn. That is the question to which Mr. Bloch objects.

Q. Did the conversations run along the same line over a period of years?

Mr. E. H. Bloch. I object to the form of the question. It is a question for the jury to decide.

The Court. I will overrule it.

Mr. E. H. Bloch. Exception.

The Court. I am going to give you a chance to go into what you want to within the limits that I shall describe if there should be limits.

Mr. E. H. Bloch. I understand your Honor's orientation, but I still feel that proper questions must be asked on direct. If they are improper it is not incumbent upon me to cross-examine.

The Court. That is right; if they are improper, I will be the one to rule that they are improper, and I have ruled that they are not improper.

The WITNESS. Well, roughly, they did. The conversations on the merits of socialism over capitalism I think in the beginning were more vehement.

Mr. E. H. BLOCH. Mr. Greenglass, please, it is difficult to hear you.

The WITNESS. In the beginning they were more vehement.

Mr. E. H. BLOCH. Well now, I object to that.

The COURT. Yes, I will sustain that.

Q. Talking about Socialism over capitalism, did they specifically talk about Socialism as it existed in the Soviet Union and capitalism as it existed here?—A. They did.

Q. Which did they like better? Did they tell you?

Mr. E. H. BLOCH. I object to the question as leading and suggestive.

The COURT. I will sustain the objection on that ground, which they liked better. But you tell us whether or not on any occasion they told you that they preferred one over another.

The WITNESS. They preferred Socialism to capitalism.

The COURT. What type of Socialism?

The WITNESS. Russian Socialism.

Q. Now, you say in the early period these conversations were to your knowledge two or three times a week?—A. Yes.

Q. That is your best estimate?—A. Yes.

Q. Where did they take place?—A. At my mother's home, 64 Sheriff Street.

Q. What do you describe as the early period? Mr. Bloch wants to know. It started in 1935.—A. Well, I would say before 1939.

Mr. E. H. BLOCH. Then, if the Court please, I move to strike out the testimony upon the ground that in addition to the grounds already urged, it is too remote to the charges and the issues in this case.

The COURT. You say it started then and continued, didn't it?

Mr. E. H. BLOCH. I know, but I am now talking about these particular discussions which he claims were had between 1935 and 1939. I object to any testimony about any ideas that these defendants are alleged to have had.

The COURT. Did I not understand that these conversations continued up to, as you put it, 1945 or 1946? What is this limitation of 1939 that Mr. Bloch speaks of?

Mr. COHN. Mr. Bloch has been asking me to fix the number of times, where the occasions were and all that. Mr. Greenglass spoke, your Honor, about an early period at which these discussions took place. I assume by the year 1939— that was before Mr. and Mrs. Rosenberg were married; they took place in the home of his mother. I suppose after that they took place some place else and after that some place else.

The COURT. But these conversations continued, did they?

The WITNESS. Yes, they did.

The COURT. Along the same lines?

The WITNESS. They did.

By Mr. COHN:

Q. Up until 1945 or 1946?—A. That is right.

The COURT. We will take a recess.

(Short recess.)

(Jury in box.)

By Mr. COHN:

Q. Mr. Greenglass, when you went out to Los Alamos, was your wife out there with you?—A. No, she wasn't.

Q. I think you told us she went out there in August of 1944; is that right?—A. That's right.

Q. When after August of 1944 did you see your wife?—A. She came to visit me on our second wedding anniversary. It was November 29, 1944.

Q. It was in November, November 29, 1944?—A. That's right.

Q. For how long a period of time was she out in Los Alamos?—A. I got a three-day pass plus a two-day week end, which made five days.

Q. Where did she stay? Was she out at Los Alamos?—A. No, she stayed at Albuquerque.

Q. Where, at an apartment, hotel?—A. In a hotel.

Q. In a hotel?—A. That's right.

Q. You say you got a three-day pass and you worked it in with the week end?—A. That's right.

Q. You had five days off; is that right?—A. That's right.

Q. You joined your wife at the hotel in Albuquerque?—A. Albuquerque.

Q. She remained there for the entire five days?—A. She did.

Q. Now, was there any time during those five days when you had a conversation with your wife concerning the atom bomb?—A. I did.

Q. When during that five-day period was that conversation had?—A. In the latter half of the furlough. We went for a walk out on Route 66, past the city, Albuquerque City limits, and not yet to the Rio Grande River, and my wife started the conversation.

Mr. E. H. BLOCH. If the Court please, before the witness gets into the conversation, I want to record our objection to any conversation had between this witness and his wife outside the presence of the defendants.

The COURT. Very well.

Mr. E. H. BLOCH. I assume I am speaking for all the defendants.

Mr. A. BLOCH. I understand this is a general objection to the introduction of this kind of testimony and you are taking it subject to connection?

The COURT. No, I am taking his testimony concerning his conversation with his wife on the ground that he and his wife are co-conspirators together with the defendants.

Mr. A. BLOCH. If they are co-conspirators.

The COURT. Alleged co-conspirators. Of course, if they are merely having a conversation that isn't in furtherance of the objective and does not in any way mention the defendants, why, of course there would be no materiality to it. That would be the objection, no materiality.

Mr. E. H. BLOCH. Well, if the Court please, isn't there implicit in the Court's mind that there still would have to be that connection whereby the Government will prove to the satisfaction of the Court and the jury that there was a conspiracy between this witness, his wife and the defendants Rosenberg?

The COURT. Of course.

Mr. E. H. BLOCH. That is implicit.

The COURT. Of course.

Mr. E. H. BLOCH. So in that sense it is being taken subject to connection and subject to motion to strike in the event the Government fails to do that.

Mr. COHN. May we have the last question, Mr. Reporter, please?

(Question read by reporter.)

Q. Will you tell us, Mr. Greenglass, what your wife said and what you said.— A. My wife said that while she was still in New York Julius Rosenberg invited her to dinner at their house at 10 Monroe Street. She came to dinner and later on there was a conversation between the three present, my wife, my sister, and my brother-in-law.

It went something like this: Ethel started the conversation by stating to Ruth that she must have noticed that she, Ethel, was no longer involved in Communist Party activities——

Mr. E. H. BLOCH. Now, if the Court please, this is just what I was afraid of, and I move to strike it out, any reference to Communist——

Mr. COHN. I object to it being struck out, your Honor, on the ground that it is directly relevant to the charge in this indictment which will emerge as this conversation unfolds.

The COURT. I will overrule the objection.

Mr. E. H. BLOCH. I respectfully except.

The COURT. The mere fact that the word "Communism" is mentioned does not taint all of the testimony and make it inadmissible if it is otherwise relevant.

Mr. E. H. BLOCH. But apart from the lack of casual connection between Communist affiliations and sympathies with the crime in question, the introduction of this testimony also introduces an element of proof of another separate and distinct crime.

The COURT. Well, you have already stated your objection. You stated it yesterday, and you stated it, I believe, the day before, too.

Mr. E. H. BLOCH. I think that is so, your Honor.

The COURT. And I have your objection and I have made my ruling.

Mr. COHN. May I have the last part of the answer?

(Record read.)

Q. Go ahead, Mr. Greenglass.—A. That they don't buy the Daily Worker any more or attend meetings, club meetings. And the reason for this is that Julius has finally gotten to a point where he is doing what he wanted to do all along, which was that he was giving information to the Soviet Union.

And he then went on to tell Ruth that I was working on the atomic bomb project at Los Alamos, and that they would want me to give information to the Russians. My wife objected to this, but Ethel said——

Mr. E. H. BLOCH. I object to the characterization.

The COURT. Is this what your wife told you?

Mr. COHN. Mr. Greenglass is relating what his wife said to him. I assume that he is doing his best to recall the words that were spoken.

Mr. E. H. BLOCH. If she used the word "object" of course I will withdraw my objection.

The COURT. Did your wife use the word "object"?

The WITNESS. She told me that she didn't think it was a good idea.

Mr. E. H. BLOCH. All right.

The COURT. Very well.

The WITNESS. And that she didn't want to tell me about it.

The COURT. Proceed.

The WITNESS (continuing:). But they told her that I would want to know about it and I would want to help, and that at least—the least she could do was tell me about it. So that was the conversation. At first—she asked me what I thought about that—at first, I was frightened and worried about it and I told her——

Mr. E. H. BLOCH. I object to his reactions or his state of mind.

The COURT. Strike out his reactions.

Q. What did you tell your wife?—A. I told my wife that I wouldn't do it. And she had also told me that in the conversation Julius and Ethel had told her that Russia was an ally and as such deserved this information, and that she was not getting the information that was coming to her. So later on that night after this conversation I thought about it and the following morning I told my wife that I would give the information.

Q. Does that complete the conversation to the best of your memory that took place between you and your wife?—A. That's right. Then when I had told her what the conversation was—I mean, I told her I would do it, she asked me for specific things that Julius had asked her to find out from me.

Q. You mean specific information about the Manhattan Project?—A. That's right.

Q. Would you tell us as you recall it what your wife asked you?—A. She asked me to tell her about the general lay-out of the Los Alamos Atomic Project, the buildings, number of people and stuff like that; also scientists that worked there, and that was the first information I gave her.

Q. You say she asked you for that information, is that right?—A. She asked me for that information. When I gave it to her, she memorized the information.

Mr. E. H. BLOCH. I object to that.

Q. Did you have any conversations——

Mr. E. H. BLOCH. I move to strike that out.

The COURT. There was no answer.

Mr. E. H. BLOCH. No, what I am objecting to is that she memorized that and he is not in a position to know it.

The COURT. Strike it out.

Q. Did you have any conversations with your wife concerning whether——

The COURT. Wait a minute. You gave her the answers to all of these questions?

The WITNESS. I gave her the answers to all of these questions; yes.

By Mr. COHN:

Q. Do you know whether she wrote this information down or not?—A. She did not write the information down.

Q. Did you have any conversation with her as to whether she was going to write it down or not?—A. She told me that she was instructed not to write it down, but to memorize it.

The COURT. Instructed by whom?

The WITNESS. Instructed by Julius.

Q. In giving to your wife the names of the scientists working at Los Alamos on that occasion, can you now recall any of the names which you furnished to her?—A. I gave her Oppenheimer's name. I gave her Bohr's name, and Kistiakowski's name.

Q. Did you tell her about this Bohr-Baker situation?—A. I did.

Q. You say you gave her a general description of the lay-out at Los Alamos, is that right?—A. That's right.

Q. How about the number of people there, the personnel, did you give any estimate of figures on that?—A. I gave her an estimate of how many people there were in the technical area.

Mr. A. BLOCH. May I ask to have the last answer repeated.

The COURT. Repeat it, please.

(Answer read.)

Q. Of course, the repeating of this specific information is forbidden in that security book which is in evidence as Government's Exhibit 1, is that correct?

Mr. E. H. BLOCH. I think the exhibit speaks for itself, your Honor.

Mr. COHN. I will withdraw it, your Honor.

The COURT. I was about to overrule the objection because of its unimportance.

Mr. COHN. It is unimportant and that is why I withdrew the question.

The COURT. It speaks for itself and the answer would merely have been cumulative, so it makes no difference.

Mr. E. H. BLOCH. I agree.

Q. After you furnished this information to your wife, did your wife return to New York?—A. My wife returned to New York and I had told her that I would be in New York in January on furlough, so she left for New York, knowing that I was going to be there.

Mr. COHN. Raise your voice a little because Mr. Bloch has some trouble hearing you.

Q. Did you actually have a furlough in January?—A. I arrived home January 1st, 1945.

Q. January 1st?—A. 1945, yes.

Q. How long was your furlough?—A. It was a 15-day furlough with travel time.

Q. How long was that as a practical matter?—A. About 21 days or 22 days.

Q. When you say you arrived home, where were you then residing, where were you and your wife then living?—A. 266 Stanton Street, in Manhattan.

Q. Here in Manhattan?—A. Right.

Q. After your arrival in New York did there come a time when you saw the defendant Julius Rosenberg?—A. Yes, he came to me one morning and asked me to give him information, specifically anything of value on the atomic bomb, whatever I knew about it.

Q. Now, where did this conversation take place?—A. In my home at 266 Stanton Street.

Q. Did you say this was in the morning?—A. This was in the morning and he told me to write up this information at night, late at night, and he would be back the following morning to pick it up.

Q. About how long after you had arrived in New York did this conversation take place?—A. A few days after I arrived.

Q. And did he outline to you in any further detail the information he wanted?—A. He asked me what I was doing out there and I told him I was working on lenses, H. E. lens molds.

Q. That is the lens molds in connection with Dr. Kistiakowski's work that you told us about?—A. That is right.

Q. What else?—A. And he told me to write it up, to write up anything that I knew about the atomic bomb.

Q. Anything else?—A. He gave me a description of the atom bomb.

Q. Did you do any writing at that time?—A. I wrote up the information he wanted that evening. It included sketches on the lens molds and how they were used in experiments.

Q. Anything else?—A. Plus a description of it.

Q Anything else?—A. Plus a list of scientists who were on the project.

Q. Do you recall the names of any of these scientists?—A. Yes, I gave him the same ones I had given him originally, plus, I gave him a scientist, Baker. I also gave him a scientist by the name of—well, there was one Hans Baker.

Q. Do you know what his field was?—A. Yes, his field was theoretical physics.

Q. Did you furnish that information?—A. I gave that information, too.

Q. And you say there were some other scientists whose names you do not recall?—A. I don't recall at this moment.

Q. Was this information turned over to Rosenberg?—A. It was, the following morning.

Q. Where?—A. At my home.

Q. At your home?—A. Yes.

Q. Up at 266 Stanton Street?—A. That's right.

Q. Now, you turned that information over to the defendant Rosenberg the following morning in your home, is that right?—A. Yes.

*   *   *   *   *   *   *

From Stenographer's Minutes of Case 134–245, *United States of America* vs. *Julius Rosenberg, et al.* Before Hon. Irving R. Kaufman, District Judge, United States District Court, Southern District of New York, March 12, 1951.

* * * * * *

Mr. COHN. We would like Mr. Greenglass back.

DAVID GREENGLASS resumed the stand.

The COURT. Now, Mr. Greenglass, will you please remember to speak up?

The WITNESS. I will.

Direct examination continued by Mr. COHN:

Q. Now, Mr. Greenglass, I think that on Friday afternoon before we adjourned, we were at the point where Rosenberg had returned to your apartment to get this information on the atom bomb that he had asked you to write down; is that correct?—A. That is correct.

Q. Will you tell us again—first of all, did you in fact furnish him with written information concerning the atom bomb?—A. I did.

Q. Will you tell us just what information you furnished him with on that day?—A. I gave him a list of scientists who worked on the project. I gave him some sketches of flat type lens molds, and I gave him some possible recruits.

Q. What kind of recruits?—A. For Soviet espionage.

Mr. E. H. BLOCH. I move to strike out the latter part of his answer.

The COURT. I will strike that out and permit you to tell us what Mr. Rosenberg said to you about recruiting scientists or recruiting anybody to help. What were his words, in substance?

The WITNESS. He said he wanted a list of people who seemed sympathetic with communism and would help furnish information to the Russians.

The COURT. Very well.

Q. And you furnished him with such a list; is that correct?—A. I did.

Q. Now I want to come specifically to these sketches you told us about of this lens. Exactly do you remember how many sketches you gave him?—A. I gave him a number of sketches, showing various types of lens molds.

Q. Was this that lens mold in connection with Dr. Caskey, that you told us about on Friday afternoon, which had been constructed at the shop, the Los Alamos shop in which you were working?—A. That was the same lens mold.

Q. Now, did you give Rosenberg a sketch of the lens mold; did you tell him how the lens mold was used?

Mr. E. H. BLOCH. If the Court please, I am going to ask Mr. Cohn not to be leading at this point.

Q. Tell us exactly what you gave Rosenberg with reference to the lens mold.— A. I gave him a sketch of the lens mold. I marked them "A, B, C," the parts of the mold, and I defined what these markings meant.

Q. Where were these definitions contained, on the same sheet of paper?—A. On a separate sheet of paper.

Q. The sketch was on one sheet and the description on another sheet?—A. That is right.

Mr. COHN. May this be marked for identification, Your Honor?

(Marked Government's Exhibit 2 for identification.)

Q. Now, Mr. Greenglass, have you, at our request, prepared a copy of the sketch of the lens mold which you furnished to Rosenberg on that day in January?— A. I did.

Q. Would you examine Government's Exhibit 2 for identification [handing] and tell me if that is the sketch which you prepared.—A. That is the sketch that I prepared.

Mr. COHN. We offer it in evidence, your Honor.

Mr. E. H. BLOCH. Before I make any objection, may I have a voir dire question here?

The COURT. Go ahead.

Mr. E. H. BLOCH. When did you prepare this?

The WITNESS. During this trial, yesterday.

Mr. E. H. BLOCH. I object to its introduction upon the ground it is incompetent, irrelevant, and immaterial. The witness is here. He testified orally to things. This is not a proper way of corroborating the witness. In fact, it is improper, I submit, to corroborate in this way.

NOTE.—Counsel for the U. S. Government: Irving Saypol, United States attorney; Miles J. Lane, James Kilsheimer, and Roy Cohn, assistant United States attorneys. Other counsel whose names appear in transcript material represent defendants.

Mr. COHN. Well, if your Honor wants to hear me on that, I think the jury is certainly entitled to see what the witness has testified he gave to the defendant in this case, what information concerning the atom bomb and things in connection with it he gave to the defendant in this case.

The COURT. In other words, you put this in the same category, as I understand it, of chart evidence. After the witness testifies to something a chart may be produced for the purpose of enlightening the jury or making it easier for the jury to understand. You are not introducing this as the exhibit that was turned over.

Mr. COHN. Oh, no; not at all, your Honor. In fact, I will ask Mr. Greenglass——

By Mr. COHN:

Q. When did you last see the very sketch which you turned over to Rosenberg?—A. In January 1945.

Q. When you handed it to Rosenberg?—A. That is right.

Q. And you have not seen it since then?—A. No.

The COURT. Objection overruled.

Mr. E. H. BLOCH. Before your Honor rules, may I ask one more question along your Honor's line of thinking, if I may?

The COURT. Go ahead.

Mr. E. H. BLOCH. After looking at this Government's Exhibit 2 for identification, are you saying that that paper that you have in your hand represents a true copy of the sketch that you turned over to Rosenberg?

The WITNESS. To the best of my recollection at this time; yes.

Mr. E. H. BLOCH. Well, then, if your Honor please, I renew my objection, because I submit that this is not analogous to introducing a chart in evidence. A chart is introduced for the purpose of elucidating the jury on matters which may be complex, and it is a sort of over-all picture by which the jury may be enabled to follow certain details. Here, this exhibit is being introduced because it purports, according to this witness, to be a true copy of what he allegedly turned over to Rosenberg. Now, I submit that is a violation of the rule against corroborating the witness by extrinsic evidence while he is on the stand, and I believe it is improper.

Mr. SAYPOL. May I address myself to the question? I submit, if the Court please, that counsel misunderstands the objective in utilizing this exhibit. It is based entirely on the secondary evidence rule. The actual sketch, obviously, is not available, as the witness has testified. Certainly there may be made available for the use of the jury, in conjunction with the witness's testimony, a recently prepared replica which, as he has testified, to the best of his recollection, is a replica of that which he furnished to the defendant.

Mr. E. H. BLOCH. As far as the best evidence rule is concerned, your Honor, I could see the cogency of Mr. Saypol's argument if it would be the contention of the prosecution that this document, which they now attempt to introduce in evidence, was made at or contemporaneously with or prior to the time.

The COURT. What you are saying does not go to the basic question of whether or not a foundation has been laid for its introduction. What you are saying goes to the weight to be given to the document.

Mr. E. H. BLOCH. I think it goes to both, your Honor. I think it goes to the fact that no proper foundation has been laid under the present——

The COURT. I will receive it. Objection overruled.

Mr. E. H. BLOCH. I respectfully except.

(Marked "Government's Exhibit 2" in evidence.)

By Mr. COHN:

Q. Now, Mr. Greenglass, while it is being marked, I might ask you——

The COURT. Just a moment. Let it be marked.

Q. Now, would you address yourself to Government's Exhibit 2 in evidence, Mr. Greenglass; does that——

Mr. E. H. BLOCH. I am sorry, Mr. Cohn, but now I would like to look at it a little more carefully so I may be enabled to follow the witness intelligently.

Mr. COHN. Certainly, Mr. Bloch [handing].

Mr. E. H. BLOCH. Thank you very much.

All right.

Q. Addressing yourself to Government's Exhibit 2 in evidence, Mr. Greenglass, does that exhibit contain certain letters, "A," "B," "C"?—A. They do.

Q. Now, what do those letters have reference to? Do they have reference to this other paper?—A. Yes, they have reference to another paper, where I put down the meaning of these letters.

Q. Would you tell us now, as best as you remember it, exactly what descriptive language was contained on this piece of paper you furnished Rosenberg along with this sketch?

Mr. E. H. BLOCH. That is objected to, your Honor, on the same grounds that I objected to the introduction of this document.

The COURT. Overruled.

Mr. E. H. BLOCH. I respectfully except.

The WITNESS. "A" refers to the curve of the lens; "B" is the frame; "C" shows approximately how wide it is.

The COURT. All right, now you had better give us that slowly so we can all understand it.

"A" refers to what?

The WITNESS. The curve of the lens, the outside curve; "B" to the frame; and "C" to the width. It is a four-leaf clover design like; it looks something similar.

Mr. E. H. BLOCH. We can't hear the witness, your Honor, I am sorry.

The WITNESS. It has four curves on it, and these—it is hollow in the center and it was used to pour "H. E." into it.

Q. What do you mean by "H. E."?—A. High explosive. It then took on the shape, the H. E. took on the shape of the mold and the mold was removed and you had a high-explosive lens.

Mr. COHN. Your Honor, may I pass it to the jury?

The COURT. Yes.

(Government's Exhibit 2 in evidence passed to the jury.)

Q. I think you have already told us that this lens, mold, along with other things constructed in your shop, were used in connection with experimentation on the atomic bomb; is that correct?—A. They were.

Q. By the way, did you have any conversation with Rosenberg concerning the writing on the descriptive material?—A. I did. My wife——

Mr. E. H. BLOCH. Will you fix the time, please?

Q. Will you tell us just when this conversation took place, in relation to the time you turned over the material?—A. It took place in the morning after I had written this information out. Julius came to the house and received this information, and my wife, in passing remark that the handwriting would be bad and would need interpretation, and Julius said there was nothing to worry about as Ethel would type it up, retype the information.

The COURT. Excuse me a moment. May I have that answer reread?

(Last question and answer read.)

Q. Did you have any further conversation with Rosenberg on the occasion when you turned over this material?—A. Not at—he asked me to come to dinner, my wife and myself, for an evening a few days later—I can't remember—a day or two later.

Q. At his home?—A. Yes; at his home.

Q. Did you accept the dinner invitation?—A. I did.

Q. Did there come a time when you and your wife did in fact go to Rosenberg's home in response to the dinner invitation?—A. We did.

Q. About how soon after this meeting at which you turned over the material?—A. It was a day or two later.

Q. Now, where did Rosenberg live at that time?—A. 10 Monroe Street, in Knickerbocker Village.

Q. In Knickerbocker Village?—A. Yes.

Q. Do you remember what time you arrived at his apartment?—A. I would say it was about 7 o'clock or so.

Q. Now, I would like you to tell the Court and the jury exactly what happened from the time you entered the apartment on that night, until the time you left? By that I mean, tell us who was there, tell us what was said and by whom?

The COURT. What was the date, did you say?

Mr. COHN. I believe the date was fixed, your Honor, as two or three days or a day or two—two or three days after the meeting in Greenglass' apartment, at which he turned over the information to Rosenberg.

The COURT. Very well.

The WITNESS. When I got to the apartment with my wife, there was Julius and Ethel Rosenberg and a woman by the name of Ann Sidorovich.

Mr. E. H. BLOCH. What was that name?

Q. Just stop there for a moment. What did you say that name was?—A. Ann Sidorovich.

Mr. COHN. May we have this marked for identification, please?

(Marked "Government's Exhibit 3" for identification.)

Q. Now, had you ever met Ann Sidorovich before?—A. I had never met her before; no.

Q. Did you know any members of her family?—A. I knew her husband.

Q. What was his name?—A. Mike Sidorovich.

Q. How long a period of time did you know him?—A. I knew him for some years.

Q. I would like you to examine now Government's Exhibit 3 for identification and tell me if you recognize the people on that picture?—A. This is Mike and Ann Sidorovich.

Mr. Cohn. We offer it in evidence, your Honor. May it be received?

(Government's Exhibit 3 previously marked for identification received in evidence.)

Mr. Cohn. May I exhibit that to the jury, your Honor?

The Court. Yes.

(Government's Exhibit 3 shown to the jury.)

Q. Now, keep your voice up, Mr. Greenglass, and tell us—you have told us who was present, Mr. and Mrs. Rosenberg and this woman, Ann Sidorovich; now, would you tell us exactly what happened on that evening, exactly what was said and by whom?—A. Well, the early part of the evening we just sat around and spoke socially with Ann and the Rosenbergs, and then Ann Sidorovich left. It was at this point that Julius said that this is the woman who he thinks would come out to see us, who will come out to see us at Albuquerque, to receive information from myself.

Q. What kind of information?—A. On the atomic bomb. And she would probably be the one to come out to see us. We then ate supper and after supper there was more conversation, and during supper—and during this conversation there was a tentative plan brought forth, to the effect that my wife would come out to Albuquerque to stay with me, and when this woman Ann or somebody would come out to see us, they would go to Denver, and in a motion-picture theater they would meet and exchange purses, my wife's purse having this information from Los Alamos, and of course, that is the way the information would be transmitted.

Q. Now, was anything said about the reason for Ann Sidorovich being present at the Rosenberg's home on that particular night when you were there?—A. Yes; they wanted us to meet this Ann Sidorovich, so that we would know what she looked like; and that brought up a point, what if she does not come?

Q. You mean, there was a possibility that somebody else would come?—A. That's right. So Julius said to my wife, "Well, I give you something so that you will be able to identify the person that does come."

Q. In other words, if Ann Sidorovich would come, she was up in the apartment that night; you were up in the apartment that night; she knew what you looked like; you knew what she looked like; but if somebody else would come, this would be mutual identification; is that right?

Mr. E. H. Bloch. Mr. Cohn, please don't repeat the answer.

Mr. Cohn. If I do so, your Honor, it is for the purpose of clarity. Strange names are coming in. However, I won't do it.

Mr. E. H. Bloch. You know why I don't want you to do it, because sometimes reemphasis——

Mr. Cohn. I will settle it by saying that I won't do it, your Honor.

May we have the last from the witness?

(Last answer read.)

Q. All right, go ahead from there.—A. Well, Rosenberg and my wife and Ethel went into the kitchen and I was in the living room; and then a little while later, after they had been there about 5 minutes or so, they came out and my wife had in her hand a Jello box side.

Mr. E. H. Bloch. Side?

Mr. Cohn. Side.

Mr. E. H. Bloch. S-i-d-e?

Mr. Cohn. That's right.

Q. You mean the side of an ordinary Jello box——A. That's right.

Q. About what size Jello box, the small size?—A. The kind you buy in your home.

Q. Right.—A. And it had been cut, and Julius had the other part to it, and when he came in with it, I said, "Oh, that is very clever," because I noticed how it fit, and he said, "The simplest things are the cleverest."

Q. Now, let me see if I understand that. Your wife had one side; is that correct?—A. That's right.

Q. Who kept the other side?—A. Julius had the other side.

Q. Was there any conversation as to what would be done with these two sides?—A. Well, my wife was to keep the side she had, and she was to use it for identification with the person who would come out to see us, and at this point we discarded the idea, discarded the idea of going to Denver.

Q. Now, I want to stay with the Jello box for a minute here. Am I correct in assuming that the last time you saw the other part—your wife had one part of this side of the Jello box?—A. That's right.

Q. And the last you saw of it on that night was in Rosenberg's hand; is that correct?—A. That's right.

Mr. COHN. May this be marked for identification, please?

(Marked "Government's Exhibit 4" for identification.)

Mr. COHN. Your Honor, at this point I would like—this will be quite important—to have the witness, as best he remembers it, take this Jello box and cut the correct side into two parts, just as he remembers it was cut on that night, in January of 1945, and I would like to ask him to indicate to the Court and jury which side he kept and which side Rosenberg kept. May I do that?

The COURT. All right.

Q. Will you take Government's Exhibit 4 for identification and this pair of scissors, and address yourself to the appropriate side and cut it into two pieces [handing to witness]? Cut it into two pieces resembling the two pieces you saw that night in Rosenberg's apartment?

(Witness cuts exhibit.)

Q. The side that was cut was one of the thin sides; is that correct?—A. That's right; this is the side I had [exhibiting].

Q. That was the side you had?—A. That's right.

The COURT. Mark that for identification.

Mr. COHN. Yes; and may we have this marked for identification as "Government's Exhibit 4–A"?

(Marked "Government's Exhibit 4–A" for identification.)

Q. Where did you last see this other side on that night?—A. In Julius' hand.

Mr. COHN. May we have the other side marked as 4–B for identification, your Honor?

(Marked "Government's Exhibit 4–B" for identification.)

Mr. E. H. BLOCH. Just to clarify myself, if I may, 4–A——

The COURT. 4–A is——

Mr. E. H. BLOCH. Is the side which the witness kept.

The COURT. That is right.

Mr. E. H. BLOCH. And 4–B is the side that Rosenberg is alleged to have retained.

The COURT. It was, which was last seen in his hand.

Mr. E. H. BLOCH. Pardon me, which was last seen in his hand.

Mr. COHN. Right, on that night in January 1945.

We offer them in evidence, your Honor, 4–A and 4–B.

(Government's Exhibits 4–A and 4–B previously marked for identification received in evidence.)

Mr. COHN. May I exhibit them to the jury, your Honor?

The COURT. Yes.

(Government's Exhibits 4–A and 4–B shown to the jury.)

By Mr. COHN:

Q. Now, I think Mr. Greenglass, we were at the point where you were continuing with further conversation that was had after this Jello box incident had been effected; would you continue and tell us just what was said and by whom?—A. Well, it was at this point that the plan to meet in Denver was discarded, and I suggested the Safeway store, meeting the person we were to meet at a Safeway store, outside, outside of the Safeway store, in Albuquerque, and this was thought to be a pretty good idea. The exact date was left in abeyance, since my wife would follow me out to Albuquerque later and that could be set in that time, I mean, between when I left to go back to Los Alamos and when she came out.

Q. The date was left in abeyance and would be set before your wife joined?—A. That is correct.

Mr. E. H. BLOCH. If the Court please, I am not interfering with this witness' testimony, but very frankly, I am a little confused as to who said what and who said other things?

The COURT. All right.

Mr. E. H. BLOCH. And I haven't interrupted. He is putting in "since" and giving some of us difficulty in understanding him.

By the COURT:

Q. You said that the idea of Denver was discarded; was that at that same meeting at Julius Rosenberg's home?—A. That's right.

Q. Now, at this point, that is the point where the Jello box was cut up, did I understand that this Ann Sidorovich had already left?—A. Oh, she had left.

Q. You and your wife and Mrs. Rosenberg and Julius Rosenberg were present; is that right?—A. That's right.

Q. You are in the living room and they go into the kitchen on this Jello box situation?—A. That's right.

Q. You say her husband was not present?—A. Her husband was not present.

Q. That is right. Now, at what point—before you left that evening you had decided to abandon the idea of meeting in a movie house in Denver?—A. That is right.

Q. Then who suggested the Safeway Store?—A. I did.

Q. And who said it was a good idea?—A. Julius said it was a good idea.

The COURT. Very well. Let me ask you one other question: Did you say what part of the Safeway Store you would meet at?

The WITNESS. In front of it, not in the store.

The COURT. In front of it.

The WITNESS. That's right.

By Mr. COHN:

Q. Now I think the last thing you said was the date, the actual date of this meeting was left up in the air until such time as your wife joined you, and when she came out she would know the further details?—A. That is correct.

Q. Now, was there any further conversation between you and your wife and the Rosenbergs on that evening?—A. Well, the Rosenbergs told my wife that she wouldn't have to worry about money because it would be taken care of—I mean, she would be able to get out there and live out there, if she wasn't able to work, the money would be forthcoming.

By the COURT:

Q. Was that in your presence?—A. In my presence.

Q. Both of them said that?—A. Julius, and Ethel backed it up. Earlier in the evening, during these conversations, my wife had remarked to Ethel that she had looked kind of tired and she said she was tired because she——

By Mr. COHN:

Q. Who said this?—A. My wife had remarked to Ethel that she looked tired.

Q. Ethel looked tired?—A. And Ethel remarked that she was tired between the child and staying up late at night, keeping—typing over notes that Julius had brought her—this was on espionage.

Mr. E. H. BLOCH. I move to strike out the last.

The COURT. Did she say that?

The WITNESS. She said "in this work." She also stated that she didn't mind it so long as Julius was doing what he wanted to do.

Q. During this evening, was any reference made by either of the Rosenbergs, to the material which you had turned over to Julius a couple of days before?—A. Well, we discussed the lenses—we generally talked shop about what I had done at Los Alamos, and we discussed lenses all during this evening, and, you know, whatever was going on at Los Alamos, scientists——

Q. Was anything said about——A. Scientists, of that nature; things of that nature; and he said that he would like to meet somebody who would talk to me more about lenses.

Q. Did he tell you who this person he wanted you to meet was?—A. He said it was a Russian he wanted me to meet.

Q. Did he give you any further identification on that night?—A. No.

Q. Now, after the conclusion of this evening, did you return home?—A. At the end of this evening I returned home.

Q. About what time do you think you left Rosenberg's apartment?—A. Oh, it was twelve or maybe possibly later.

Q. Did you and your wife have any further conversation about anything that had transpired when you returned home?—A. Well, she showed me the piece, the Jello side, the Jello box side, and she put it in her wallet.

Q. In her wallet?—A. That's right.

Mr. E. H. Bloch. If the Court please, I am objecting to any conversation between this witness and his wife outside of the presence of the defendant Rosenberg and the other defendants, and likewise any acts that may have taken place, in which the Rosenbergs weren't present. I just want to reserve objection for the record.

The Court. Very well. You know the ruling; you know the reason for my ruling.

Mr. Cohn. Of course, your Honor, I offer them as statements by coconspirators in furtherance of the conspiracy.

Q. Now, did anything further come of Julius's statement that he wanted you to discuss this lens with the Russian?—A. Yes.

Q. Tell us.—A. A few nights later—well, an appointment was made for me to meet a Russian on First Avenue, between 42nd and 59th Streets—it was in that area.

Q. Who made the appointment?—A. Julius made the appointment.

Q. When was it in relation to the dinner meeting in January?—A. It was a few days after. I took my father-in-law's car and drove up there. It was about eleven-thirty at night. I remember coming up the street. It was quite dark and there was a lighted window. I passed that in parking—it was a saloon—I parked up the block from it, and in a little while Julius came around the corner, looked into the car, saw who I was; said, "I will be right back"; brought back a man; introduced the man to me by first name, that I don't recall at this time, and the man got into the car with me. Julius stayed right there and we drove around——

Q. Let me see if I understand it. When you say "Julius stayed right there," was Julius in the car or not?—A. He was not in the car.

Q. He merely effected the introduction?—A. He just introduced me to him.

Q. And remained on the street?—A. And remained on the street.

Q. Where did you drive?—A. Well, we drove all over that area. He just told me to keep driving and he asked questions about lenses.

Q Did he ask you specifically about this high explosive lens?—A. He did. He asked about high explosive lenses and he wanted to know pertinent information, type of H. E. used.

Mr. E. H. Bloch. I move to strike out "pertinent" as a conclusion.

The Court. Strike out "pertinent." Tell us what he wanted to know.

The Witness. He wanted to know the formula of the curve on the lens; he wanted to know the H. E. used, and means of detonation; and I drove around——

The Court. And what, means of detonation?

The Witness. That's right; and I drove around, and being very busy with my driving, I didn't pay too much attention to what he was saying, but the things he wanted to know, I had no direct knowledge of and I couldn't give a positive answer.

Q. Now, about how long did this drive with the Russian last?—A. About twenty minutes or so.

Q. Where did it terminate?—A. At the same place that it originated.

Q. Did you see Julius any more on that night?—A. Yes; he came back—I mean, he was around there, and the Russian got out and they went off together, and I drove back home.

Q. Did Julius give you any instructions?

Mr. E. H. Bloch. Now, if the Court please, just a second. I move to strike out the characterization of the man that he met, especially in the last answer, as "Russian."

The Court. That is denied. The testimony, as I understand it, is that Julius said he wanted to introduce him to a Russian.

Mr. E. H. Bloch. Yes; but whatever the defendant Rosenberg may have said doesn't substitute for the fact, which I contend the prosecution must prove, namely, that this man was a Russian or a Russian subject or citizen.

The Court. We will understand that when he said "a Russian" we will accept that the Russian that he is characterizing as such was the man whom Mr. Rosenberg had characterized as "a Russian."

By Mr. Cohn:

Q. Did Julius give you any instructions as to where you should go and what you should do after you concluded this drive with the man whom you described as "a Russian?"—A. He said, "Go home now. I will stay with him." He was going to have something to eat with him.

Q. Did you in fact return home?—A. I went home.

Q. Did you tell your wife where you had been?—A. Yes; I told my wife where I had been.

Q. Now, I think you told us you arrived in New York on this furlough on January 1, 1945; is that correct?—A. That is correct.

Q. About when did you leave New York and return to Los Alamos?—A. About the 20th.

Q. Did your wife go with you?—A. She did not go with me.

Q. Did there come a time when she joined you at Albuquerque?—A. She did.

Q. Will you tell us when that was?—A. That was in the springtime, it was about March or April—early—late March, early April, I think.

Q. Of 1945?—A. Of 1945.

Q. Where did your wife live when she got out to Albuquerque?—A. Well, at first she stayed in a fellow GI's apartment, a man by the name of Delman. He and his wife had gone east on furlough. Then she stayed at a fellow by the name of Spindel's apartment. Then we had our own place.

Q. Now, when you say you had your own place, was it a house, an apartment, or what?—A. It was an apartment at 209 North High Street.

Q. 209 North High Street. That was in Albuquerque?—A. In Albuquerque.

Q. Were you able to go home every night?—A. No; I wasn't.

Q. When did you go to the apartment?—A. Well, usually Saturday night. I would start down and get there sometime Saturday evening.

Q. When would you return to Los Alamos?—A. Sunday—I mean Monday, early in the morning.

Q. Were you in this apartment—was your wife in this apartment, were you in there over the week ends during the month of May and during the month of June, in 1945?—A. I was.

Q. Now, Mr. Greenglass, did Ann Sidorovich ever come out to see you?—A. No; she didn't.

Q. Did somebody else come out to see you?—A. Yes.

Q. Was it a man or woman?—A. It was a man.

Q. And when was this visit?—A. First Sunday in June 1945.

Q. Did you at that time know the name of this man?—A. I did not.

Q. Do you now know his name?—A. Yes; I do.

Q. What is it?—A. Harry Gold.

Mr. E. H. BLOCH. Your Honor, again I would like to state for the record that I am objecting to anything that happened out in New Mexico, outside the presence of the defendant Rosenberg, and more specifically, upon the ground that there is nothing in this witness' previous testimony to lay the foundation for the introduction of this evidence, reflected by the last question or by the last two questions.

The COURT. That last objection is very vague, but there is just no question in my mind as to the competence of this evidence, so I will overrule your objection.

Mr. E. H. BLOCH. I respectfully except.

(Government's Exhibit 5 marked for identification.)

Q. Do you recognize this picture, Mr. Greenglass [showing]?—A. Yes; I do.

Q. Who is that?—A. That is Harry Gold.

Mr. COHN. We offer it in evidence, your Honor.

(Government's Exhibit 5 for identification received in evidence.)

The COURT. Excuse me. Where did he come to see you, in Albuquerque?

The WITNESS. In Albuquerque.

Mr. COHN. Would your Honor want to take the morning recess at this point?

The COURT. Yes. We will take a recess at this point.

(Short recess.)

(Jury in box.)

Mr. E. H. BLOCH. If the Court please, may I ask whether the prosecution would have any objection to offering for identification the remaining portions of the Jello box, from which the witness cut the sides?

Mr. COHN. It has already been done, your Honor.

The COURT. It has already been done.

Mr. E. H. BLOCH. It is marked?

Mr. COHN. Exhibit 4 for identification and the two parts have been received in evidence.

Mr. E. H. BLOCH. 4-A and 4-B. I was a little confused about that. Thank you very much.

Mr. COHN. Did you want to examine it?

Mr. E. H. BLOCH. No; thank you very much.

Q. When did you say it was that Harry Gold came to your house, Mr. Green-glass?—A. It was the third Sunday in June 1945.

Q. What time of day?—A. It was in the morning.

Q. Who was home?—A. I and my wife were home.

Q. Would you tell us exactly what happened from the first minute you saw Gold?

Mr. E. H. BLOCH. My objection still stands, your Honor.

The COURT. Yes; overruled.

Mr. E. H. BLOCH. Exception.

The WITNESS. There was a knock on the door and I opened it. We had just completed eating breakfast, and there was a man standing in the hallway who asked if I were Mr. Greenglass, and I said "Yes." He stepped through the door and he said, "Julius sent me," and I said, "Oh," and I walked to my wife's purse, took out the wallet and took out the matched part of the Jello box.

Q. That was Government's Exhibit 4–A, is that correct?—A. The Jello box.

Q. The piece you retained that night?—A. Yes.

Q. After you produced that did Gold do anything?—A. He produced his piece and we checked them and they fitted, and the identification was made.

Q. In other words, he had——A. He had the other part of the box.

Q. And you had last seen that in Rosenberg's apartment that night in January 1945?—A. That is right.

Q. Now, after mutual identification was effected, did you have any conversation with Harry Gold?—A. Yes. I offered him something to eat and he said he had already eaten. He just wanted to know if I had any information, and I said, "I have some but I will have to write it up. If you come back in the afternoon I will give it to you." I started to tell him the story about one of the people I put into the report, and he——

Q. Who was one of the people you put into the report?—A. A fellow by the name of Bederson, and he cut me short.

Q. What kind of person was he? Why did you put him in the report?—A. Well, I considered him good material for recruiting into espionage work.

Mr. E. H. BLOCH. I move to strike out the part of the answer with respect particularly to the words "espionage work" as reflecting only the operation of this witness' mind.

The COURT. No; I will overrule it. The witness has already testified that Mr. Rosenberg had asked him on a previous occasion to send such names of anybody whom he considered to be a good recruit, and I am overruling the objection.

Mr. E. H. BLOCH. Then, if the Court please, may I ask Mr. Cohn to clarify what report this witness is referring to?

The COURT. Yes.

Mr. COHN. I will be glad to do that, your Honor.

The COURT. Yes.

Q. In which report had you mentioned the name which you discussed with Gold?—A. I mentioned it in that particular report that I gave him that day.

Q. The report you gave Gold later that day?—A. Yes.

Q. You discussed the name before you embodied it in the report?—A. That is right.

Q. Tell us just what was said by you and Gold.

Mr. E. H. BLOCH. When was this? Morning or afternoon?

Q. Mr. Bloch wants to know when was it?—A. This particular time was the morning. He cut me short on the business with Bederson. He said he didn't want to know about it and he left and I got to work on the report.

Q. Where did you work on the report?—A. Right in the living room, my combination living room and bedroom there.

Q. Tell us exactly what you did.—A. I got out some 8-by-10 ruled white line paper, and I drew some sketches of a lens mold and how they are set up in an experiment, and I gave descriptive material that gives a description of this experiment.

Q. Was this another step in the same experiment on atomic energy concerning which you had given a sketch to Rosenberg?—A. That is right, and I also gave him a list of possible recruits for espionage.

Mr. E. H. BLOCH. I move to strike out the last two words, "recruits for espionage."

The COURT. Overruled.

Mr. E. H. BLOCH. I respectfully except. Of course, I don't like to be popping up and down, your Honor. I want to make it clear that I am objecting to this entire line of testimony with respect to this incident between the witness and Gold in New Mexico as not binding upon the defendant.

The COURT. Overruled.

The WITNESS. I gave this list of names and also sketches and descriptive material.

Q. What time of day was it that you gave this material to Harry Gold?—A. It was later in the afternoon. He came back about 2:30 or 3 o'clock and picked it up.

Q. Did all these sketches and descriptive material concern experimentation on the atomic bomb?—A. That is right.

Q. Tell us exactly what happened when he came back at 2:30?—A. Well, when he came back to the house he came in and I gave him the report in an envelope and he gave me an envelope which I felt and realized there was money in it and I put it in my pocket.

Q. Did you examine the money at that point?—A. No; I didn't.

Q. Did you have any discussion with Gold about the money?—A. Yes; I did. He said, "Will it be enough?" And I said, "Well, it will be plenty for the present." And he said, "You need it," and we went into a side discussion about the fact that my wife had a miscarriage earlier in the spring, and he said, "Well, I will see what I can do about getting some more money for you."

Q. Was there any further discussion with Gold?—A. Well, he wanted to leave immediately and I said, "Wait, and we will go down with you," and he waited a little while. We went down, and we went around by a back road and we dropped him in front of the USO. We went into the USO, and he went on his way. As soon as he had gone down the street my wife and myself looked around and we came out again and back to the apartment and counted the money.

Q. How much was it?—A. We found it to be $500.

Q. $500?—A. Yes.

Q. What did you do with the money?—A. I gave it to my wife.

Q. Going back to these sketches which you gave to Harry Gold, do you remember just what sketches you gave to Harry Gold concerning a high-explosives lens mold on that occasion?—A. I gave sketches relating to the experiment set up; one showing a flat—the face of the flat-type lens mold.

Q. Face view?—A. Face view of the flat-type lens mold.

Q. Have you prepared, at our request, a sketch of this face view?—A. I have.

Mr. COHN. Let this be marked for identification.

(Marked "Government's Exhibit 6" for identification.)

Q. Would you examine Government's Exhibit 6 for identification? By the way, you prepared that on June 15, 1950; is that correct?—A. I did.

The COURT. Well, 6 for identification, I take it, is a replica of——

Mr. COHN. Well, 6 for identification was prepared on June 15, 1950, your Honor, last year.

The COURT. Oh, I see.

Q. In answer to his Honor's question is this to the best of your recollection an accurate replica of the face view which you gave Harry Gold in June 1945?—A. That is right.

Mr. COHN. I will now offer it in evidence, your Honor.

Mr. E. H. BLOCH. Before ruling, may I have one question on the voir dire?

The COURT. Go ahead.

By Mr. BLOCH:

Q. When you made this sketch on June 15, 1950, and I am referring now to Government's Exhibit 6 for identification, did you rely solely on your memory?—A. I did.

*   *   *   *   *   *   *

The COURT. Is this to your present knowledge an exact replica of the sketch which you turned over even to the extent of the comments on the side?

The WITNESS. It is.

Mr. COHN. Your Honor, Mr. Saypol reminds me that I did not show the jury the picture of Mr. Gold [handing to jury].

By Mr. COHN:

Q. Will you address yourself to Government's Exhibit 6 in evidence and tell the jury what that represents?—A. I showed a high-explosive lens mold. I showed the way it would look with this high explosive in it with the detonators on, and I showed the steel tube in the middle which would be exploded by this lens mold.

Q. Now, did you prepare on that Sunday in June 1945 and give to Harry Gold on that same day any other sketches concerning this high-explosive lens mold for atomic energy?—A. I showed him a schematic view of the lens-mold set up in an experiment.

Q. Now, have you similarly prepared for us a replica as you remember it of the sketch which you gave Harry Gold on that day?—A. I did.

By the COURT:

Q. What do you call this sketch, a schematic view of it?—A. Yes. Well, none of those are to scale. So they are all schematic.

Q. What is the difference between 7 for identification, now being marked, and 6?—A. Well, this shows an experiment.

Q. Actually, the mold being used in an experiment?—A. That is right. The set-up.

(Marked "Government's Exhibit 7" for identification.)

By Mr. COHN:

Q. Is this the sketch Mr. Greenglass [handing to witness]?—A. Yes; it is.

Mr. COHN. We offer it, your Honor.

Mr. E. H BLOCH. May I ask a question?

Mr. COHN. Surely.

Mr. E. H. BLOCH. To protect the record.

By Mr. E. H. BLOCH:

Q. Mr. Greenglass, in connection with Government's Exhibit 7 for identification, can you tell us when you prepared this?—A. I prepared it during this trial.

Q. When specifically, if you remember?—A. Yesterday.

Q. And did you rely solely upon your memory in preparing this?—A. I did.

Mr. E. H. BLOCH. Now, if the Court please, I make the same objection upon the same grounds heretofore urged.

The COURT. Same ruling.

Mr. E. H. BLOCH. (Continuing). To the introduction of Exhibits 2 and 6.

The COURT. Same ruling.

Mr. E. H. BLOCH. I respectfully except.

(Marked "Government's Exhibit 7".)

By the COURT:

Q. Now, the comments on the bottom of No. 7: Were they the same comments you had on the sketch?—A. No. That is just to identify it.

The COURT. Oh, well. No, no comments.

By Mr. COHN:

Q. In other words, that is to describe this, but your recollection is you did not put such description on the same piece of paper when you gave it to Gold?—A. No.

Mr. COHN. We have no objection at all if that description is cut off, your Honor.

Mr. E. H. BLOCH. I can't even answer that, Mr. Cohn, because I am objecting to the introduction of the document.

Mr. COHN. You don't even want to see it cut off.

Q. Now, you had better look at this in altered form, Mr. Greenglass. Is this the sketch?—A. That is the sketch.

Q. Does this sketch, Government's Exhibit 7 in evidence similarly have letters such as the first one, A, B, C, and D, and so forth?—A. Yes.

Q. What do those letters refer to?—A. They refer to the parts of this sketch.

Q. Were letters such as that on the sketch which you gave to Gold?—A. That is right.

Q. Did those letters refer to descriptive material?—A. They did.

Q. Where was this descriptive material?—A. On a separate sheet of paper.

Q. Did you give that descriptive material to Gold?—A. I did.

Q. Will you tell us the language you used on the separate piece of paper in describing this exhibit to Harry Gold?

Mr. E. H. BLOCH. I am sorry. I would like it to be clarified; has this exhibit already been marked in evidence?

The COURT. Yes.

*   *   *   *   *   *   *

By Mr. COHN:

Q. Now, would you tell us just what you wrote on this other sheet of paper to describe this exhibit and the letters contained thereon?

Mr. E. H. BLOCH. That is objected to, your Honor.

The COURT. Overruled.

Mr. E. H. BLOCH. I respectfully except.

The WITNESS. "A" is the light source which projects a light through this tube "E", which shows a camera set up to take a picture of this light source.   Around the tube it is a cross-section of the high explosive lens "C" and a detonator "B" showing where it is detonated, and the course is that when the lens is detonated it collapses the tube, implodes the tube, and the camera through the lens "F" and the film "D" shows a picture of the implosion.

Q. By the way, Mr. Greenglass, I think you have already told us you knew at all times that all of these sketches and descriptive material were secret?—A. I did.

By the COURT:

Q. Were there constant experiments going on?—A. Constant.

Q. And the sketches in 6 and 7 were what were considered an advance on these sketches marked as Government's Exhibit 2?—A. Yes.

Q. Well, let us eliminate the word "advance"; they were just another step?—A. That is right.

Mr. COHN. In line with your Honor's explanation, we have now arrived at the point where we have the secured permission of the Court to interrupt the testimony of Mr. Greenglass and put someone on the stand, concerning these other matters.

Mr. SAYPOL. Your Honor's remarks were quite pertinent.

Mr. H. E. BLOCH. I object to Mr. Saypol's statement that your Honor's remarks were quite pertinent.   I think the Court itself spontaneously realized that there was a question in your Honor's mind which has not been proved by any evidence——

The COURT. No, I didn't realize anything of the kind.   Please don't comment on what I have said.

Mr. E. H. BLOCH. Well, I am objecting to Mr. Saypol's statement.

The COURT. We will strike Mr. Saypol's statement and strike yours, too.

(Witness Greenglass temporarily excused.)

WALTER S. KOSKI, called as a witness on behalf of the Government, being first duly sworn, testified as follows:

Direct examination by Mr. SAYPOL:

Q. Dr. Koski, what is your profession?—A. Physical chemistry.

Q. You have heard us suggest to witnesses that they speak up loudly and to keep you in the same category we will ask you to. You say you are a physical chemist?   Is that what you said?—A. I am.

Q. Are you engaged in that capacity now?—A. I am.

Q. Where are you so engaged?—A. Johns Hopkins University

Q. Exactly in what capacity are you so engaged at Johns Hopkins?—A. I am associate professor of physical chemistry.

Q. Collaterally, do you have any other association in your profession?—A. I am consultant at the Brookhaven National Laboratories.

Q. Consultant in what?—A. I am corroborating in a program which has as its objective to measure certain properties of radioactive nuclei.

Q. Nuclear chemistry?—A. Nuclear chemistry or nuclear physics.

Q. What activity so far as is related to your field is conducted at Brookhaven?—A. Nuclear chemistry.

Q. Is that something related to some sort of measurements?—A. It relates to the measurement of certain properties of radio-active nuclei.

Q. What has been your education?—A. I have a Ph. D. in physical chemistry.

Q. Is that from Johns Hopkins, too?—A. It is.

Q. When?—A. June 1942.

Q. What was your employment from 1942 to 1944?—A. I was a research chemist at the Hercules Powder Co.

Q. In 1944 did you become associated with the United States Government?—A. I did.

Q. In what capacity?—A. As an engineer at the Los Alamos Scientific Laboratories.

Q. How long did you continue your work there?—A. Up to about September 1947.

Q. That is about the time that you became associate professor of chemistry at Johns Hopkins?—A. That was.

Q That was also the time when you took on this retainer as consultant at Brookhaven Laboratories?—A. It was.

Q. That is the Brookhaven National Laboratories, to be exact?—A. Correct.

Q. Referring now to this period between 1944 and 1947 when you were at Los Alamos, can you tell me generally what instructions were issued to you, if any, concerning the character of the work that was being done there, what your position was to be in respect to publicization?

Mr. A. BLOCH. That is objected to on the ground that it is not binding on the defendant and it is hearsay.

The COURT. Overruled.

Mr. A. BLOCH. And therefore incompetent, irrelevant, and immaterial.

The COURT. Overruled.

Mr. A. BLOCH. Exception.

The WITNESS. We were informed that all work done at Los Alamos was of a highly classified nature.

Q. When you say "classified" do you mean that it was restricted or secret?—A. Secret.

Q. Was that knowledge imparted to you in the form of instructions on one or more occasions?—A. It was imparted to us verbally and by written material.

Q. Is that the atmosphere that prevailed in connection with all of the work that was conducted there?

Mr. E. H. BLOCH. I object to the word "atmosphere."

The COURT. All right. Strike it out.

Q. Prior to your arrival at Los Alamos in 1944 did you have knowledge of the work that was going on there?—A. I did not.

Q. Did there come a time when you learned the nature of the activities?—A. There did.

Q. (Continuing). That were being conducted there?—A. There did.

Q. Just reverting for a moment, Doctor, remember, we were discussing secrecy and restriction at Los Alamos: Have you ever seen that before [showing paper to witness]?—A. Yes, I have.

Q. Will you tell us the circumstances under which you saw it and when?—A. This was—this is a restricted document that was sent to all people coming into the laboratory.

Q. Did you read it at the time?—A. I did.

Q. Did you familiarize yourself with the contents?—A. I did.

Q. Did you observe as well as you could the instructions that were contained in Government's Exhibit 1?—A. I did.

Q. Those related to what?—A. Related to security and the secrecy of all technical information.

Q. Going forward now at the point where you were interrupted, you say there came a time when you learned after arrival what the nature of the work was that was being done at Los Alamos?—A. Correct.

Q. What was the knowledge that you acquired as to the nature of the work?—A. The objective of the laboratory was to construct a nuclear weapon or atomic bomb.

Q. At this point will you tell us whether you performed any particular phase of that work, research, I take it, incidental to the development or incidental to the project?—A. I did.

Q. What did your work involve?—A. My work was associated with implosion research connected with the atomic bomb.

Q. So that we as laymen may understand, when you say implosion research, does that have something to do with explosives?—A. The distinction between explosion and implosion is in an explosion the shock waves, the detonation wave, the high-pressure region is continually going out and dissipating itself. In an implosion the waves are converging and the energy is concentrating itself.

Q. I take it, concentrating itself toward a common center?—A. Toward a common center.

Q. In other words, in explosion it blows out; in implosion it blows in?—A. Yes.

Q. Is implosion one of the physical reactions incident to the over-all action in the atomic bomb?—A. It is.

Q. So, as I understand you, your precise job was to make experimental studies relating to this phenomena of implosion?—A. It was.

Q. Mr. Koski, in the performance of that work did you have occasion to use what has been called here a lens, a device called a lens?—A. I did.

Q. What is the lens as you knew it in connection with your experiments?—A. A high-explosive lens is a combination of explosives having different velocities and having the appropriate shape so when detonated at a particular point, it will produce a converging detonation wave.

Q. Well, once again, so that we as laymen might understand, I take it our common conception of a lens is a piece of glass used to focus light, is that right?—A. Yes, that is right.

Q. What is the distinction between a glass lens and the type of lens you were working on?—A. Well, a glass lens essentially focuses light. An explosive lens focuses a detonation wave or a high-pressure force coming in.

Q. What are the physical steps which are involved and which were involved in the production of a lens of the type you have described?—A. The procedure in general was to first make a design of this lens. Then I would go down to the Theta shop which was one of the shops which constructed such material for us.

Q. I take it the design for the mold would be prepared by you or under your supervision?—A. Yes.

Q. Then by the same token, the design or the sketch, we may call it that, may we not?—A. Yes.

Q. Would then be taken by you or somebody under your supervision, probably you, to the Theta shop for mechanical work incidental to its manufacture—A. Correct.

Q. And then the mold having been manufactured in the Theta shop—that was a machine shop?—A. That was a machine shop.

Q. What would you do with the mold in relation to the explosive for the component part of the lens?—A. This mold was taken out to our laboratory, at a remote site. There this mold was used to cast the high explosive necessary in this lens.

Q. You say to test the high explosive?

The COURT. Cast.

The WITNESS. Cast.

Q. That is, to shape the explosive?—A. That is right.

Q. In the course of the conduct of those experiments did you have occasion to utilize different and successively changing designs of lenses?—A. We did.

Q. In other words, as you developed a lens and tested it and experimented with it, the results that you obtained would be utilized by you in the development, in the design of other lenses which would make up for any observed defect in the preceding lenses?—A. They were.

Q. In this work about that time, that is, around 1945, starting the latter part of 1944 into 1945 up to the middle of 1945, did you work particularly on what is known as a flat-type lens?—A. I did.

Q. Was this flat-type lens and your related experiments, were they involved in the development of the atomic bomb?—A. They were.

Q. Now, in the course of your work when you required a lens of your own intended design or your idea, will you describe for us the procedure which you would follow and which you did follow to the end that you should ultimately have a mold for the lens?—A. I went down to the Theta shop and there discussed with the people in charge of the shop——

Q. Do you remember their names?—A. Mr. Fitzpatrick and Mr. Marshman. They were sergeants at the time. I told them what we needed, gave them rough sketches and verbally explained whatever information they needed to construct this mold for us.

Q. About that time did you—do you have a recollection of having seen the defendant Greenglass in the Theta shop?—A. I have seen Mr. Greenglass in the Theta shop.

Q. Considering the nature of the work that you had with high explosives, what was the physical location of your laboratories and your experimental area in relation to, say, the Theta shop or the balance of the project?—A. We had offices and small laboratories in the same area that the Theta shop was located in. Our actual experimental work, however, was done at a remote site.

Q. Were there reasons for conducting your work at a remote site?—A. The reasons were that we were handling large amounts of high explosives and they were detonated, and there were very heavy shocks.

Q. Now you have told us in the course of your experimentation several different models of the flat-type lens were prepared under your instructions, is that right?—A. That is correct.

Q. Now, once again will you explain why that was necessary?—A. Would you repeat that question?

Q. I think you have told us already that it was necessary to have different models, that is, as you progressed, and as you observed the results of experiments, and you varied the design of the lens itself; that is the form in which the explosive was contained in the lens?—A. Correct.

Q. Was that work at Los Alamos, your experiments, classified as secret?—A. They were.

Q.—Did that apply to all technical work that was being conducted at Los Alamos?—A. It did.

Q. I show you Government's Exhibit 1—by the way, just withdrawing that: You have been in attendance here and you have heard the witness Greenglass testimony, the defendant Greenglass' testimony, have you not?—A. I have.

Q. I show you Government's Exhibit 2, rather. Will you examine that, please? Do you recognize that exhibit as a substantially accurate representation—as a substantially accurate replica of a sketch that you made at or about the time which you have testified to at Los Alamos in connection with your experimentation?—A. I do.

Q. Is that a reasonably accurate portrayal of a sketch of a type of lens, mold or lens that you required in the course of your experimental work at the time?—A. It is.

Q. Would you recognize it as a reasonably accurate replica of the one you submitted to the Theta machine shop?—A. Yes.

Q. For processing?—A. Yes.

Q. In the manner in which you have testified?—A. I do.

Q. I show you Government's Exhibit 6, as to which you have heard Mr. Greenglass testify, and I ask you whether your answers are the same in respect to that exhibit after you have examined it?—A. They are.

Q. Do you recall that in the course of your experimentation at or about that time in 1945 you obtained from the Theta shop molds of the design indicated by those exhibits?

Mr. E. H. BLOCH. Now if the Court please, I am going to ask that——

The COURT. I can't hear you.

Mr. E. H. BLOCH. I am going to ask that this be made specific. I think Mr. Saypol referred to the year 1945. I want to draw your attention to what Mr. Greenglass testified as to his position from the time he came to Los Alamos to work to the time he left.

Mr. SAYPOL. Will you suffer an interruption?

Q. Do you recognize those as depictions——

The COURT. I can't hear you, Mr. Saypol.

Mr. SAYPOL. I am sorry.

Q. Do you recognize those exhibits, that is, 2 and 6, as accurate replicas of sketches submitted by you in 1944 and 1945 to the Theta shop as the result of which molds, lens molds were supplied to you for your experimentation?

Mr. E. H. BLOCH. Now, if the Court please, I have no objection to the substance of this question but I ask that the time be more definitely fixed.

By the COURT:

Q. If you can remember the approximate month. If you can remember the day, so much the better. If you can remember the approximate months of those years when those respective sketches were submitted to the Theta shop, let us have it.—A. I cannot.

The COURT. Very well.

Mr. E. H. BLOCH. Then I object to the question as too general.

The COURT. Overruled.

Mr. E. H. BLOCH. I respectfully except.

By the COURT:

Q. You do remember that they were some time during the years 1944 and 1945?—A. They were.

   *   *   *   *   *   *   *

By the COURT:

Q Do you remember whether it was the latter part of 1944?—A. It was approximately from the middle of 1944 until about the middle of 1945.

By Mr. SAYPOL:

Q. You have listened, you said, to the testimony of the defendant Greenglass in relation to Exhibits 2 and 6. Can you tell us whether his testimony is a reasonably accurate description of the devices portrayed in Exhibits 2 and 6 and the functions they had in connection with your experiments?

Mr. E. H. BLOCH. I object to the form of the question.

The COURT. Overrruled.

   *   *   *   *   *   *   *

By Mr. SAYPOL:

Q. Now, in respect to Government's Exhibit 7, will you examine that, please, Dr. Koski? Having examined it, having heard Greenglass's testimony as to what it depicts, will you tell us whether it is familiar to you?—A. It is.

Q. What does it portray to you?—A. It is essentially—it is a sketch, a rough sketch of our experimental set-up for studying cylindrical implosion.

Q. Did you hear Mr. Greenglass testify as to the description, written description of that experiment that he delivered to one Harry Gold in June 1945?—A. I did.

Q. Is Government's Exhibit 7 and the details of the information as testified to by Mr. Greenglass that he said he imparted to Gold in June 1945 a reasonably accurate—are they reasonably accurate descriptions of the experiments and their details as you knew them at the time?

Mr. A. BLOCH. Objected to upon the ground that it is an attempt to characterize the testimony of another witness; not calling for fact.

(Question read.)

The COURT. I will strike from that question, "as testified to by Mr. Greenglass." Now do you understand the question?

The WITNESS. I do.

The COURT. Can you answer it?

The WITNESS. They are.

Q. That is the experiment that you yourself were conducting in conjunction with the development of the atomic bomb?—A. They are.

Q. In your special field as you knew it at the time, 1944 and 1945, did you have knowledge that the experiments which you were conducting and the effects as they were observed by you could have been of advantage to a foreign nation?

Mr. E. H. BLOCH. Objected to upon the ground that this witness has not been qualified as a political expert; merely as a scientific expert. I object to the question as calling for a conclusion.

The COURT. I will overrule it.

Mr. E. H. BLOCH. I respectfully except.

The WITNESS. I wonder if you would repeat the question.

(Last question read.)

The WITNESS. I did.

Q. And would that knowledge have been of advantage to a foreign nation?

Mr. E. H. BLOCH. The same objection, your Honor.

The COURT. Overruled.

Q. This question follows my previous question and your answer: In that field in which you were engaged do you know whether anywhere else there had been similar prior experimentation?

Mr. E. H. BLOCH. Same objection, your Honor.

The COURT. Overruled.

Mr. E. H. BLOCH. Exception.

The WITNESS. To the best of my knowledge and all of my colleagues who were involved in this field, there was no information in textbooks or technical journals on this particular subject.

Q. In other words, you were engaged in a new and an original field?—A. Correct.

Q. And up to that point and continuing right up until this trial has the information relating to the lens mold and the lens and the experimentation to which you have testified continued to be secret information?—A. It still is.

Q. Except as divulged at this trial?—A. Correct.

The COURT. As far as you know, only for the purposes of this trial?

The WITNESS. Correct.

Mr. SAYPOL. Will your Honor allow a statement for the record in that respect? The Atomic Energy Committee has declassified this information under the Atomic Energy Act and has made the ruling as authorized by Congress that subsequent to the trial it is to be reclassified.

The COURT. Counsel doesn't take issue with that statement.

Mr. E. H. BLOCH. No, not at all. I read about it in the newspapers before Mr. Saypol stated it.

Mr. SAYPOL. May I have just a moment, if the Court please?

(Mr. Saypol confers with associates.)

Mr. SAYPOL. You may examine.

Mr. E. H. BLOCH. Will you bear with me for three or four moments, your Honor, since I am not a scientist, I don't want to query about matters which might appear asinine.

Cross-examination by Mr. E. H. BLOCH:

Q. Dr. Koski, did you turn over any of the sketches requested in Government's Exhibits 2, 6, and 7 to the defendant Greenglass?—A. I did not.

By the COURT:

Q. Was the defendant Greenglass in a position where by reason of his employment in the Theta shop he could see the sketches which you turned over?—A. He was.

By Mr. BLOCH:

Q. Mr. Greenglass was a plain ordinary machinist, was he not?

Mr. SAYPOL. I object to characterizations.

The COURT. I will permit the characterization.

The WITNESS. Correct.

Q. Now, you heard Mr. Greenglass testify about the E shop, did you not?—A. Yes.

Q. Then I believe he testified that there were two other shops similar to the E shop in that technical area and finally there came a time when there was a new building which was called the Theta building and all the shops moved in there, is that correct?—A. That is not correct.

Mr. SAYPOL. Well——

The COURT. Well, what is not correct?

The WITNESS. The Theta shop was a separate shop. All of the shops didn't move into this building.

Mr. SAYPOL. I want to know what is incorrect.

By the COURT:

Q. You are not characterizing that Greenglass had testified to that and therefore was incorrect?—A. No.

Q. You are characterizing that the statement of counsel as formulated in his question is incorrect.

Mr. E. H. BLOCH. Well let us clarify it for everybody's sake then.

By Mr. E. H. BLOCH:

Q. Was there an E shop?—A. There was.

Q. And did that E shop at some time move into another building?—A It did.

Q. What was that other or new building called?—A. That was the location of the Theta shop.

Q. Now, was the Theta shop in existence and used for work at the project while the E shop was being used for work?—A. There might have been some overlapping but I am not sure.

Q. Now, were there other shops besides the E shop—I believe he characterized them as the E. C. shops; you correct me if I am wrong—that also moved into the new building, or the Theta shop at the time that you started to use the Theta shop, is that correct?—A. No. The Theta, E and C shop never were in the same building.

Q. Now, when the personnel of the E shop moved into the Theta building were the same number of machinists used for the work which you supervised?—A. I do not recall the details about the machinists. I usually contacted their superiors.

Q. In fact, you very seldom had any conversations with any machinists, is that right?—A. Rarely, but not completely—on occasions we did have.

Q. It was very rare?—A. It was rare.

Q. Now, did you know when the defendant Greenglass became an assistant foreman?—A. I did not.

Q. Did you know when he became a foreman?—A. I did not.

Q. Now, just two more questions, Doctor. Do these exhibits——

The COURT. What are the numbers?

Mr. E. H. BLOCH. I am going to mention them.

Q. (Continuing.) 2, 6, and 7, purport to be a complete picture of these lenses in the scientific sense?

The COURT. Do you understand, Doctor, what he means by a complete picture?

The WITNESS. I am not clear as to what you mean.

Mr. E. H. BLOCH. Well, maybe I am a little too vague.

Mr. SAYPOL. To preserve accuracy, I think the testimony that 2 and 6 are sketches of molds, and 7 is a description of an experiment.

The COURT. That is right.

Mr. SAYPOL. Am I correct?

Mr. E. H. BLOCH. That is correct.

The WITNESS. That is correct.

Mr. SAYPOL. So counsel's question to the extent that it refers to 7 should be corrected.

Q. Well, let us satisfy everybody. I will tell you what I am driving at, Dr. Koski : is it not a fact that a scientist would not consider Government's Exhibits 2, 6, and 7, whether or not two of them relate to a lens and one of them relates to some kind of cylindrical apparatus, until the scientists knew the dimensions of the lens or the cylindrical apparatus?—A. This is a rough sketch and of course is not quantitative but it does illustrate the important principle involved.

Q. It does omit, however, the dimensions?—A. It does omit dimensions.

Q. It omits, for instance, the diameter, does it not?—A. Correct.

Q. Now is it not a fact that——

The COURT. You say it does, however, set forth the important principle involved, is that correct?

The WITNESS. Correct.

The COURT. Can you tell us what that principle is?

The WITNESS. The principle is the use of a combination of high explosives of appropriate shape to produce a symmetrical converging detonation wave.

Q. Now, weren't the dimensions of these lens molds very vital or at least very important with respect to their utility in terms of success in your experiments?—A. The physical over-all dimensions that you mention are not important. It is the relative dimensions that are.

Q. Now the relative dimensions are not disclosed, are they, by these exhibits?—A. They are not.

Mr. E. H. BLOCH. That is all.

Redirect examination by Mr. SAYPOL :

Q. The important factor from the experimental point of view is the design, is it not?—A. Correct.

Q. That was original, novel at the time, was it not?—A. It was.

Q. Can you tell us, Doctor, whether a scientific expert in the field you were engaged in could glean enough information from the exhibits in evidence so as to learn the nature and the object of the experiment that was involved in the sketches in evidence?—A. From these sketches and from Mr. Greenglass' descriptions, this gives one sufficient information, one who is familiar with the field, to indicate what the principle and the idea is here.

Q. And would I be exaggerating if I were to say, colloquially, that one expert, interested in finding out what was going on at Los Alamos, could get enough from those exhibits in evidence which you have before you to constitute a tip-off as to what was going on at Los Alamos?

*   *   *   *   *   *   *

(Last question read as follows :

"Q. And would I be exaggerating if I were to say colloquially that one expert, interested in finding out what was going on at Los Alamos, could get enough from those exhibits in evidence which you have before you to reveal what was going on at Los Alamos?")—A. One could.

Mr. E. H. BLOCH. Of course, my objection still goes.

Q. Rather than using the preliminary——

Mr. E. H. BLOCH. I am sorry.

Mr. SAYPOL. Let me finish the question.

Mr. E. H. BLOCH. I want to preserve the record. Go ahead.

Q. Rather than using my former question, as to suggesting that it would be an exaggeration, is it not a fact that one expert could ascertain at that time if shown Exhibits 2, 6, and 7, the nature and the object of the activity that was under way at Los Alamos in relation to the production of an atomic bomb?—A. He could.

Mr. SAYPOL. That is all.

Mr. E. H. BLOCH. Will your Honor bear with me just a moment? No further questions.

Mr. SAYPOL. May I address one further question? It is technical and it has been suggested to me.

By Mr. SAYPOL :

Q. There was a question put to you by counsel regarding the fact that the exhibits do not show the dimensions. Then there was some statement as to relative dimensions. Distinguishing between relative dimensions and design, is it not the fact that design of the component was the primary fact of importance in these sketches?—A. It was.

Q. So that the sketches, particularly 2 and 6, do show relative dimensions in that they show the relations of each of the factors in the lens, one to the other?—A. They do.

Mr. SAYPOL. That is all.

Recross-examination by Mr. E. H. BLOCH :

Q. Well, Doctor, when you gave instructions to I believe you said it was Sergeant Fitzpatrick—was that the name?—A. Yes.

Q. And the other gentleman?—A. Marshman.

Q. Did you detail with any specificness the measurements of the lens or of the component parts of the lens that you wanted constructed?—A. I gave specific instructions. I gave rough sketches, and then while this lens mold was in progress we had to send down one of our men to sketch out, to precisely draw the shape of this lens on the metal from which it was being cut.

Q. When you say precisely draw, are you saying now that precision work was necessary in the construction of this mold lens?—A. The shape of this lens is an important factor.

Q. So aside from the shape—I am trying to direct your mind, Doctor, to the precision, quality of the work that was entailed and necessary in the construction of the lens.—A. It had to be a precision job.

Mr. E. H. BLOCH. That is all.

By the COURT :

Q. While there might have been some other details that might also have been of some use to a foreign nation which were not contained on Exhibits 2, 6, and 7, the substance of your testimony as I understand it was that there was sufficient on Exhibits 2, 6, and 7 to reveal to an expert which was going on at Los Alamos?—A. Yes, your Honor.

Mr. SAYPOL. That is all.

Mr. E. H. BLOCH. That is all.

The COURT. It is about twelve minutes of one now. In view of the fact that I turned over Court's Exhibits 1 to 5, we will take a rather extended luncheon recess. So you may now have an hour and forty-five minutes to examine them and we will return here at two-thirty, ladies and gentlemen.

(Recess until 2: 30 p. m.)

AFTERNOON SESSION

(Jury not present.)

Mr. E. H. BLOCH. May we approach the bench?

* * * * * * *

(The jury returned to the courtroom at 2 : 40 p. m.)

The COURT. Do you want the witness Greenglass?

Mr. SAYPOL. Yes, we have sent for him.

Mr. COHN. Yes.

DAVID GREENGLASS resumed the stand.

Direct examination continued by Mr. COHN :

Q. Mr. Greenglass, one thing I forgot to ask you about this morning in connection with the meeting up at Rosenberg's apartment, when you and your wife went there for dinner after Ann Sidorovich had left the apartment. Did you have a conversation with Mr. and Mrs. Rosenberg?—A. Yes, I did.

Q. Will you tell us what they said to you at that point?—A. At that point——

Mr. E. H. BLOCH. If the Court please, I submit the question has already been asked and already been answered. I have no objection if Mr. Cohn wants to direct the witness to some specific item which you feel——

Mr. COHN. That is precisely what I am doing.

The COURT. All right.

The WITNESS. Well——

Mr. E. H. BLOCH. Let us not have a rehash of the testimony.

The WITNESS. Well, at this point Mr. and Mrs. Rosenberg told me they were very happy to have me come in with them on this espionage work and that now that I was in it there would be no worry about any money they gave to me, it was not a loan, it was money given to me because I was in this work and that it was not a loan.

Q. Did they say anything about the source of that money?—A. They said that it came from the Russians who wanted me to have it.

Q. Now you have told us about the visit of Harry Gold to you in June about the material that you turned over to him. When after that was the next occasion when you saw Julius Rosenberg?—A. It was on my furlough in September 1945.

Q. Where—you got a furlough in September of 1945.—A. That is right.

Q. Where did you go on that furlough?—A. I went home, but I no longer had the apartment at 266 Stanton Street, so we stayed in an apartment where I had been living before I was married, which was in the building that my mother lives in, 64 Sheriff Street.

Q. In other words, you came from New Mexico to New York for the furlough?—A. That is right.

Q. Did your wife come with you?—A. She did.

Q. Now how long after you arrived in New York——

The COURT. Which furlough is this?

Mr. COHN. September 1945, your Honor. The other one was January 1945.

The COURT. That is right.

Q. You had not been in New York from January 1945 until September 1945; is that right?—A. I had not; no.

Q. And this meeting with Harry Gold took place out at New Mexico?—A. That is right.

Q. Now, in September 1945, after you returned to New York, when was it that you first saw Julius Rosenberg?—A. It was the morning after I came to New York.

Q. Now, would you tell us what happened? Where did you see him?—A. He came up to the apartment and he got me out of bed and we went into another room so my wife could dress.

Q. Did you have a conversation in that other room?—A. I believe we did.

Q. What did he say to you?—A. He said to me that he wanted to know what I had for him.

Q. Did you tell him what you had for him?—A. Yes. I told him "I think I have a pretty good"—"a pretty good description of the atom bomb."

Q. The atom bomb itself?—A. That's right.

Q. Now at this point, Mr. Greenglass, I want to take you back to your testimony on Friday afternoon. Did I understand you to say—well, I will be a little more specific. I am going to take you back to meeting in January 1945 when you had a conversation with Rosenberg at your apartment. Did I understand your testimony to be that Rosenberg had given you a description of the atom bomb?—A. He did.

Q. He gave you a description of the atom bomb?—A. That is right.

Q. Will you tell us the conversation you had with him at the time when he gave you this description of the atom bomb?—A. Well, he said to me he would have to give me an idea of what the bomb was about so that I would be able to know what I am looking for. He then gave me a description of what I later found out to be was the bomb that was dropped at Hiroshima.

Mr. E. H. BLOCH. Now if the Court please, I do not like to interrupt the witness. I move that everything after what he subsequently found out——

The COURT. Yes, Mr. Bloch.

Mr. E. H. BLOCH. Be deleted from the answer.

The COURT. Strike it out.

Mr. E. H. BLOCH. And I also request at this time in connection with this very specific inquiry that the witness be cautioned against using the word "description" but let us have specifically what the description was.

Mr. COHN. I intend to come to that directly, your Honor. I can't do everything at one time.

Mr. E. H. BLOCH. I am sorry, I did not mean by way of criticism.

Mr. COHN. I am sure of that, Mr. Bloch.

Your Honor, I object to striking the remainder of that answer. I think it is important, and I don't see why the witness is not able to give us knowledge that he gained in the course of his official duties at Los Alamos.

The COURT. You haven't brought out what he subsequently learned, and I don't know from whom.

Q. All right, tell us under what circumstances you subsequently learned that this bomb was the type atom bomb dropped on Hiroshima?

Mr. E. H. BLOCH. I object to that as not binding on the defendant.

The COURT. Now, of course, I take it you are bringing into force his knowledge of a particular project and from what he learned there so that he could apply what was told by Rosenberg to him to the knowledge that he learned and concluded that that was the bomb on Hiroshima.

Mr. COHN. Yes, your Honor.

The COURT. Very well, I will overrule that.

*   *   *   *   *   *

Mr. E. H. BLOCH. At any rate, my objection stands.

Q. You say in the course of your work?—A. In the course of my work at Los Alamos I came in contact with various people who worked in different parts of the project and also I worked directly on certain apparatus that went into the bomb, and I met people who talked of the bombs and how they operated.

Q. And on the basis of that knowledge and information?—A. I gave the sketches and these reports.

Q. Right; and was it on the basis of that same knowledge and information that you learned that this bomb which Rosenberg had described to you was the type of atom bomb that was dropped on Hiroshima?—A. That is right. They——

Q. All right. Could you tell us, as you remember, in exactly what words Rosenberg described this type atom bomb to you?—A. He said there was fissionable material at one end of a tube and at the other end of the tube there was a sliding member that was also of fissionable material and when they brought these two together under great pressure, that would be—a nuclear reaction would take place. That is the type of bomb that he described.

Q. Was that the first time you had ever heard a description of that type atom bomb?—A. That is right.

Q. Or of any type atom bomb, is that right?—A. That is right.

Q. Now did Rosenberg tell you at that time why he was describing this type atom bomb to you?—A. He was describing it to me so that I should know what to look for, what I could——

The COURT. He told you that?

The WITNESS. That is right.

Q. After he gave you that description, the Hiroshima type, did you, in ensuing months, gather information concerning the atom bomb?—A. I did.

Q. Will you tell us just how you went about that?—A. I would usually have access to other points in the project and also I was friendly with a number of people in various parts of the project and whenever a conversation would take place on something I didn't know about I would listen very avidly and question——

Mr. E. H. BLOCH. I move to strike out the word "avidly."

The COURT. Overruled.

Mr. E. H. BLOCH. Exception.

Mr. COHN. May we have the last few words?

(Record read.)

The WITNESS (continuing). And question the speakers as to clarify what they had said. I would do this surreptitiously so that they wouldn't——

Mr. E. H. BLOCH. I move to strike that out.

The COURT. You would do it, I take it, so that they wouldn't know it.

The WITNESS. I would do it so they wouldn't know.

The COURT. Strike out "surreptitiously." Go ahead.

Q. Now, in addition to that fact, you yourself were working on various things used in connection with the experimentation used on the atom bomb?—A. That is correct.

Q. Is that correct?—Yes, sir.

Q. Such as this high explosive lens?—A. High explosive lens molds were made in my shop and I got—as a matter of fact, there were molds used on the atom bomb.

Q. Was it on the basis of this knowledge which you had accumulated over those months that you told Rosenberg you thought you had a pretty good description of the atom bomb itself?—A. I did.

Q. Did you at a later time give to Rosenberg a description of the atom bomb itself?—A. I did.

Q. Now, was this atom bomb which you described to him the same type atom bomb he had described to you in January?—A. It was not.

Q. Would you explain that to us?—A. One type of bomb, the one that he described to me, was dropped at Hiroshima, and it was the only type bomb of that nature that was made. The one I got most of my knowledge on, got the knowledge—the information on, was of a different nature. It was a type that worked on an implosion effect.

Q. It was a different type atomic bomb?—A. That is right.

Mr. E. H. BLOCH. If the Court please, I move to strike out the answer upon the ground that this witness has not been qualified as an expert.

The COURT. Overruled.

Mr. E. H. BLOCH. I except.

Q. Was this type atom bomb a type which was manufactured at Los Alamos to your knowledge, after the Hiroshima bomb was no longer in process of manufacture?—A. That is right.

Q. Did you give Rosenberg the description at that time?—A. No. It was later in the afternoon.

Q. All right. Now, I believe we are at the point where Rosenberg—you told Rosenberg you had a pretty good description of the atom bomb. What did he say to you at that point?—A. He said he would like to have it immediately, as soon as I possibly could get it written up he would like to get it.

Q. He wanted it written up?—A. Yes.

Q Now, would you tell us what you did?—A. Oh, besides that, during this conversation he gave me $200 and he told me to come over to his house. I then went to see my—well, he then left and I was there alone with my wife.

Q Did you have any discussion with your wife?—A. My wife didn't want to give the rest of the information to Julius, but I overruled her on that. I told her that——

Mr. E. H. BLOCH. I object to this, not only because it was not in the presence of the Rosenbergs, but because the witness is stating conclusions.

Q. Tell us what you said to her.—A. I have said that "I have gone this far and I will do the rest of it, too."

Q. How about the money, what did you do with the $200?—A. I gave that to my wife.

Q. What happened after this conversation between you and your wife?—A. We went down and had——

By the COURT:

Q. Before you get to that point, when did you turn over Exhibits 2 and 6?

Mr. COHN. 2 and 6, your Honor, are the first two exhibits on the high-explosive lens.

The COURT. They are replicas of it. When were they turned over?

By Mr. COHN:

Q. When did you give the first sketch, the first lens mold sketch?—A. That was in January 1945.

The COURT. What about the second one, 6?

Mr. COHN. Do we have the exhibit here?

Q. This is Exhibit 2. His Honor's question was, When did you turn that over?—A. That was in January 1945.

Q. To whom did you give it?—A. I gave that to Julius Rosenberg.

Q. Now, Exhibit 6?—A. I gave that to Harry Gold.

Q. In June of 1945 at Albuquerque?—A. In June 1945, at Albuquerque.

Mr. COHN. Does that clarify it, your Honor?

The COURT. How about 7?

Q. Exhibit 7. Am I correct in stating you gave Exhibit 7 to Gold at the same time you gave him Exhibit 6?—A. I gave that too; that is right.

The COURT. Very well.

Q. Now tell us what you did after you had this discussion with your wife?—A. Well, we went down—it was late in the morning—we had a combination breakfast and lunch, and I came back up again and I wrote out all the information and drew up some sketches and descriptive material.

Q. Did you draw up a sketch of the atom bomb itself?—A. I did.

Q. Did you prepare descriptive material to explain the sketch of the atom bomb?—A. I did.

Q. Was there any other material that you wrote up on that occasion?—A. I gave some scientists' names, and I also gave some possible recruits for espionage.

Q. Now about how many pages would you say it took to write down all of these matters?—A. I would say about 12 pages or so.

Q. About what time did you complete preparing this report?—A. It must have been about 2 in the afternoon.

Q. Now, tell us what you did after you prepared these 12 pages of written material, including the sketch of the atom bomb and a description of the sketch?—

A. My wife and myself got into my father-in-law's car and we drove around to Julius's house. We went up to the house and I gave Julius the information which——

Q. Gave him all of this written information?—A. That is right.

Q. Including this sketch?—A. That is right.

Mr. COHN. May we have this marked for identification, your Honor?

(Marked Government's Exhibit 8 for identification.)

Q. Have you prepared for us, Mr. Greenglass, a replica of the sketch—I believe it is a cross-section sketch of the atom bomb—a replica of the sketch you gave to Rosenberg on that day?—A. I did.

Q. I show you Government's Exhibit 8 for identification, Mr. Greenglass, and ask you to examine it and tell us whether or not that is a replica of the sketch, cross section of the atomic bomb?—A. It is.

Q. And how does that compare to the sketch you gave to Rosenberg in September 1945?—A. About the same thing. Maybe a little difference in size; that is all.

Q. Except for the size?—A. Yes.

Q. It is the same?—A. Yes.

Q. By the way, who was present when you handed the written material including this sketch over to Rosenberg?—A. My wife, my sister, Julius, and myself.

Q. By your sister you mean Mrs. Rosenberg?—A. That is right.

Mr. COHN. We offer this in evidence, your Honor.

Mr. E. H. BLOCH. I object to it on the same ground urged with respect to Government's Exhibits 2, 6, and 7, and I now ask the Court to impound this exhibit so that it remains secret to the Court, the jury, and counsel.

Mr. SAYPOL. That is a rather strange request coming from the defendants.

Mr. E. H. BLOCH. Not a strange request coming from me at the present.

Mr. SAYPOL. We have discussed that with the Court, as counsel knows, and I think nothing else need be said. If I had said it or my colleague, Mr. Cohn, had said in, there might have been some criticism.

The COURT. As a matter of fact, there might have been some question on appeal. I welcome the suggestion coming from the defense because it removes the question completely.

Mr. SAYPOL. And I am happy to say that we join him.

The COURT. All right. It shall be impounded. Let me see it. Do you have any objection to the descriptive words on the bottom wherein it is stated, cross section A-bomb, not to scale?

Mr. E. H. BLOCH. I haven't seen the exhibit itself, your Honor.

The COURT. Show it to counsel.

(Handed to Mr. E. H. Bloch.)

Mr. E. H. BLOCH. No, I have no objection to that.

Mr. COHN. May the exhibit be received in evidence, your Honor?

The COURT. Yes.

(Government's Exhibit 8 for identification received in evidence.)

The COURT. It will be sealed after it is shown to the jury.

Mr. COHN. Yes, your Honor. I would like to interrogate the witness on the basis of it for a moment. Mr. Saypol calls my attention to the fact that all defense counsel have not joined in this request that this document be impounded. I wonder if the defendant Sobell's counsel cares to join?

The COURT. I thought it was understood that where one counsel spoke and the other one didn't object to what he said, by his silence he acquiesced in what the other counsel was saying.

Mr. KUNTZ. I thought your Honor made that a rule throughout this trial.

By Mr. COHN:

Q. Now, Mr. Greenglass, address yourself to that sketch and tell us, if you will, just what you wrote as best you remember of the descriptive material you gave to Rosenberg in September 1945, the descriptive material in that sketch.—A. Well, I had this sketch marked A, B, C, D, E, F, and those referred to various parts of the bomb.

Q. Now tell us exactly what you wrote in this descriptive material.

Mr. E. H. BLOCH. Before you answer the question, may we come up to the bench, your Honor?

*　　　*　　　*　　　*　　　*　　　*　　　*

(The following proceedings were resumed in the presence and hearing of the jury:)

The COURT. Now, there is a matter of some concern to me personally that the witness is about to testify to, and the concern I have is as to the method that this testimony should be handled.

Now, Mr. Saypol, Mr. Cohn was about to take detailed proof on certain descriptive matters concerning the atom bomb which the witness contends was turned over to the defendant, Julius Rosenberg; that while it might not be in the best interests of the country, was yet a matter that is necessary in the trial of a case and under our democratic form of government.

Mr. Bloch, I understand that you are willing to concede the testimony concerning that particular phase of it, is that correct?

Mr. E. H. BLOCH. I was willing to do this, your Honor—I want to restate it very clearly. I thought that in the interest of national security any testimony that this witness may give of a descriptive nature concerning the last Government exhibit might reveal matters which should not be revealed to the public.

The COURT. Therefore?

Mr. E. H. BLOCH. And therefore I felt that his testimony on this aspect should be revealed solely to the Court, to the jury, and to the counsel, and not to the public generally.

The COURT. Well, now, Mr. Saypol, do you wish to say something?

Mr. SAYPOL. Yes. I feel free to address myself to the subject in the light of the fact that the situation as it exists is not of my creation but that of one of counsel for the defendants. The character of the proof has been offered, this witness and the preceding one, has been the subject of very grave consideration by my colleagues, myself, by agencies of the Government, including the Department of Justice, the Atomic Energy Commission, and the Joint Congressional Committee on Atomic Energy.

We are cognizant that there had to be balanced on the one hand the disclosure of the type of information that has come out and is about to come out in order to supply the requirements of the Constitutional Rights of defendants to full confrontation. That subject has been expended upon by our courts. That weighed against the national security. The matter is of such gravity that the Atomic Energy Commission held hearings, at which I was represented, as did the Joint Congressional Committee, and representatives of the Atomic Energy Commission have been in attendance here at the trial, as your Honor knows, have been in constant consultation with me and my staff on the subject.

At least one of the counsel for the defendants made the offer to preserve the confidential character of this information. I think I stated before that solely for the purposes of this trial, the Atomic Energy Commission had released— had authorized the release of this information so that the Court and the jury might have it. If all counsel for the defendants had joined in Mr. Bloch's suggestion, it would have been ideal. In the presence of a conflict amongst the defendants as the prosecutor, my view is that of my colleagues, where I say frankly that the decision is not one that I would freely care to make myself, although I am not unequipped to do so, nor am I hesitant, but it has unanimity amongst us. Since there is no concurrence among counsel for the defendants, it is my view that we should go forward with the proof as it has proceeded, unless the Court of its own volition, bearing in mind, as I know it will, the Constitutional factors as they relate to the defendants, itself chooses to make an appropriate direction.

The COURT. Ladies and gentlemen, in as plain and simple language as I can possibly put it to you, under our form of government we do not have what has been characterized as "star-chamber proceedings," where a defendant is not permitted to hear the testimony against him or only a portion of testimony is given and certain portions are withheld. When the defendant is put on trial, under our form of government, I am happy to say, he is entitled to full confrontation, and that means confrontation of all the evidence which the Government contends to prove the guilt of the defendant or defendants.

Now, there are some courses open to the Court and I am about to pursue one of these courses reluctantly, but necessarily so. I am going to ask spectators in the courtroom to please leave the courtroom during the course of the taking of these proceedings on the balance of this testimony.

When the court reconvened the room had been cleared of spectators except that representatives of the press were present. The court reporter transcribed what then was said but copies of this portion of the transcript have thus far been circulated only among counsel for both sides. However, next day's New York Times (March 12, 1951), carried an account of what happened and the following is an extract:

Reading from the sketch, Greenglass described 30–6 high-explosive lenses, each of which carried two detonators. He explained that two detonators were used to make sure the lenses fired if one detonator proved defective. He introduced the word "implosion" to describe an explosion focused inward instead of outward. He said 72 condensers were used to fire the detonator.

Greenglass then described a "barium plastic sphere" which he said acted to protect the high explosive from the plutonium constituting the bomb core. Inside the barium sphere, he said, a plutonium sphere was placed.

Inside the plutonium sphere, he said, a beryllium sphere provided a source of neutrons to discharge into the plutonium. At this stage, he said, the plutonium was "highly sensitive" because of the pressure concentrated against it. As the neutrons discharged, he added, "nuclear fission takes place."

On completion of the description, Judge Kaufman ordered the court stenographer not to transcribe that testimony. He said the stenographer would read it from his notes to any defense lawyer, but that the court did not want the testimony made permanent in writing.

Greenglass added, under Mr. Cohn's questioning, that the bomb switch was set off by a "barometric pressure device" and that the bomb itself was dropped by parachute. The latter statement went unchallenged.

Mr. E. H. BLOCH. If the Court please, would this be an appropriate time to take a recess?

The COURT. We will take a short recess.

(Short recess.)

*   *   *   *   *   *   *

(The following took place in the presence of the jury:)

By Mr. COHN:

Q. Now will you tell us just what happened, Mr. Greenglass, after you handed this sketch and the descriptive material concerning the atomic bomb to Rosenberg? What did he do? What did the others there do?—A. Well he stepped into another room and he read it and he came out and he said, "This is very good. We ought to have this typed up immediately." And my wife said, "We will probably have to correct the grammar involved," because I was more interested in writing down the technical phases of it than I was in correcting the grammar. So they pulled—they had a bridge table and they brought it into the living room, plus a typewriter.

Q. What kind of typewriter?—A. A portable.

Q. Then what?—A. And they set that up and each sentence was read over and typed down in correct grammatical fashion.

Q. Who did the typing, Mr. Greenglass?—A. Ethel did the typing and Ruth and Julius and Ethel did the correction of the grammar. While this was going on, sometimes there would be stretches where you could do—there wasn't too much changing to be made, and at this time Julius told me that he had stolen the proximity fuse when he was working at Emerson Radio.

Q. He gave you that information on that occasion, is that right?—A. Yes.

Mr. E. H. BLOCH. If the Court please, I move to strike that out, the reference to stealing the proximity fuse, upon the ground that it is not related to the charge here. It imputes to the defendants the commission of a separate crime. I think it injects inflammatory and prejudicial material into this trial, and I ask that it be stricken.

By the COURT:

Q. Did he tell you what he did with that proximity fuse?—A. He told me that he took it out in his brief case. That is the same brief case he brought his lunch in with, and he gave it to Russia.

The COURT. All right. Strike out "stealing" and we will let this latter part stand.

Mr. E. H. BLOCH. I respectfully except to your Honor's ruling.

*   *   *   *   *   *   *

Q. Did you know that Julius had been working at the Emerson Electric Company?—A. Yes, I did know it.

Q. Did you know what type of work he had been doing?—A. Yes, he was an engineer and inspector out there.

Q. Do you know whether or not he had any connection with Government work?—A. He worked for the Signal Corps, actually.

Q. He was actually in the employ of the Signal Corps?—A. That's right.

Q. Now, about how long a period of time would you say was consumed in this process of the typing by Mrs. Rosenberg of the information that you had furnished to Mr. Rosenberg?—A. Oh, I would say it took most of the afternoon, about 5 o'clock, I guess, when we got done with it.

Q. Did you have any further conversation before you left the apartment on that occasion?—A. Well, that——

Q. Go ahead. I was going to ask you specifically about one thing. Did you have any conversation with Rosenberg about how long you were going to remain in the United States Army at Los Alamos?—A. Oh, I just stated that as soon as I could possibly get out, I was going to go out of the Army, get a discharge.

Q. What did he say?—A. He said he'd want me to stay up in Los Alamos if I could get a job up there as a civilian, stay there as a civilian.

Q. Did he tell you why he wanted you to stay there?—A. Well, he said that he wanted me to stay there so I could continue to give information.

Q. What did you say?—A. I said I would like to leave the place, I would like to come home.

Q. Now, about how long did this furlough last in September? Was it a long furlough or a short one?—A. It was fairly long. The dates on this occasion— I don't know exactly when I got back to Los Alamos.

Q. By the way, you turned—I think you have told us on several occasions that you turned over this sketch and descriptive material to Rosenberg, is that right?—A. I said that before.

Q. And that it was typed by Mrs. Rosenberg?—A. That's right.

Q. Do you know what happened to the original notes after the typing was completed?—A. The original notes were taken and burnt in the frying pan and then flushed down the drain.

Q. Who did that?—A. Julius did that.

Q. Pardon me?—A. Julius did.

Q. Now, did you return to Los Alamos in September?—A. I did return to Los Alamos in September; yes.

Q. Did there come a time when you obtained a discharge from the Army of the United States?—A. I did.

Q. When was that?—A. In February 1946, last day in February.

Q. Where were you discharged?—A. El Paso, Texas, Fort Bliss.

Q. Did you go from Los Alamos to El Paso?—A. That's right.

Q. Did you receive an honorable discharge?—A. I did.

Q. After you were discharged from the Army at the end of February 1946, where did you go?—A. I went back to New York City.

Q. Where did you take up residence?—A. First at my mother's home, with an aunt, at 64 Sheriff Street, in the same building that my mother lives, and then at 265 Rivington Street.

Q. And you have resided at 265 Rivington Street up to the time of your confinement on this charge?—A. That's right.

Q. Now, after your discharge from the Army and after you returned to New York, did you go into business?—A. I went into business with Julius Rosenberg and my brother and a man by the name of Goldstein.

Q. Which brother of yours do you refer to?—A. Bernard.

Q. Bernard Greenglass; is that correct?—A. That's right.

Q. How long did you remain in business with Rosenberg and your brother?— A. Till August 1949.

Q. In that three-year period, from 1946——

The COURT. What business was that?

The WITNESS. It was two businesses, G. & R. Engineering.

By Mr. COHN:

Q. Is that Greenglass and Rosenberg or Goldstein?—A. Greenglass, Goldstein and Rosenberg.

Q. The "G" covered two partners whose names began with "G" and the "R" is for Rosenberg?—A. That's right; and then the Pitt Machine Products Corporation.

The COURT. Until when did you say that was in business—until when?

The WITNESS. I was in business until August 1949.

Q. Now, what type of businesses were these, just in general terms?—A. Machine shops.

Q. Machine shops?—A. That's right.

Q. You continued to work as a machinist; is that correct?—A. That's right.

Q. Now, during those three years from February or March 1946, until August of 1949, did you see Rosenberg at business from time to time?—A. We did—I did.

Q. Did you have any conversations with him?—A. I did.

Q. Am I correct in stating you saw him very frequently in business?—A. Yes; every day almost.

Q. Did any of these conversations relate to espionage activities?—A. They did.

Q. Would you tell us those conversations which you recall? Try to tell us as best you can when——

Mr. A. BLOCH. I object to it unless the time is fixed.

Mr. COHN. I am just trying to get him to state it.

Q. Try to tell us as best you can, if you can remember, when or around when each conversation took place.—A. Well, in '46 or '47 Julius Rosenberg made an offer to me to have the Russians pay for part of my schooling and the GI Bill of Rights to pay for the other part, and that I should go to college for the purpose of cultivating the friendships of people that I had known at Los Alamos and also to acquire new friendships with people who were in the field of research that are in those colleges, like physics and nuclear energy.

Mr. E. H. BLOCH. I am sorry, may I inquire if the witness is now stating what Rosenberg said to him.

The COURT. So I understood.

Mr. E. H. BLOCH. Are these the words?

The WITNESS. Approximately the words.

Mr. E. H. BLOCH. All right.

Q. Did he mention any particular institutions which he desired to have you attend?—A. Well, he would have wanted me to go to Chicago, University of Chicago, because there were people there that I had known at Los Alamos and it was a well-known institution and it was doing a lot of good work in the field of nuclear physics.

Q. Did he mention any other institutions?—A. M. I. T., and then later on when N. Y. U. had a nuclear engineering course he wanted me to take that.

Q. Did he give you the name of any scientists with whom he desired you to build up friendships?—A. No; he told me that at Chicago University there were some people that I had gone to school with, I mean, I had been at Los Alamos with, and that I should cultivate their friendships.

Q. Did he specify how much of this money would be furnished by the Russians?—A. He specified that the GI Bill of Rights would pay for my schooling and they would give a certain amount of money for living of the student, and he said the Russians would pay additional money so I could live more comfortably.

Q. Now, did you ever agree to go to any of these schools?—A. I said I would try, but I never bothered.

Q. You never, in fact, did go; is that right?—A. That's right.

Q. Now, did Rosenberg tell you anything about activities of this kind in which he had engaged?—A. Well, he had told me that he had people going to schools in various places.

The COURT. Will you fix the time when he told you this.

The WITNESS. It was during this period of 1946 to 1949.

The COURT. All right.

A. (Continuing.) He told me that he had people going to school in various up-State institutions. He never made mention of the institutions, but he said that he was paying students to go to school.

Q. Did he tell you anything else concerning his activities along these lines?—A. He told me that he had people giving him information in up-State New York and in Ohio.

Q. Did he tell you why they were giving him that information?—A. They were giving information to give to the Russians.

Q. Did he mention any particular place in up-State New York from which he was getting information?—A. He mentioned the fact that he was getting information from General Electric at Schenectady.

Q. General Electric in Schenectady?—A. That's right.

Q. I think you told us he also mentioned he was getting some information from someone in Cleveland; is that right?—A. That's right.

Mr. E. H. BLOCH. I am sorry. I think he said Ohio.

Mr. COHN. I think he did.

Q. Well, I will ask you, as a matter of fact, did he say anything to you about getting information from someone in Cleveland?—A. Well, once in the course of the conversation in which we were talking about turret lathes, and where they are manufactured, he told me that he had seen the huge Warner-Swasey turret lathe plant at Cleveland, Ohio, and when later I asked him how come he was in Cleveland, Ohio, he stated to me that he was out there to see one of his contacts.

Q. He did not tell you this contact worked at Warner-Swasey?—A. No, he did not.

Q. He said he had been out to see one of his contacts in Cleveland and while there to see the contact he had gone to this plant?—A. That is right.

Q. Did he mention anything to you about a man named Joel Barr?—A. There was a Joel Barr that used to come around to the place and work on certain of his own work and this Joel Barr subsequently went to Europe.

Q. Did Rosenberg ever——

Mr. E. H. Bloch. I move to strike out the latter part of his answer unless this witness can positively tell this Court and jury that he knows about that fact.

The Court. Or unless he was told it by Rosenberg.

Mr. E. H. Bloch. Or, of course, unless it was part—that is not the way he stated it though, your Honor.

Q. Did you ever have any conversation with Rosenberg about Joel Barr leaving this country?—A. Rosenberg once told me that Joel Barr was leaving this country to study music in Belgium.

Q. That was in the 1946 to 1949 period?—A. It was in the end of the year 1947.

Q. End of the year of 1947?—A. That is right.

Q. Now, don't tell us what the conversation was at this time, but tell us, as a matter of fact, did you, in the year 1950, have another conversation with Rosenberg about Joel Barr's trip to Europe?—A. That is right.

Q. Did Rosenberg mention to you any Government projects concerning which he had obtained information from any of his contacts?—A. Well, once in the presence of my brother, he mentioned a sky platform project.

Q. A sky platform project?—A. Yes.

Q. Which brother do you refer to?—A. Bernard Greenglass.

Q. He was present when Rosenberg mentioned that?—A. That is right.

Q. Did you have any conversation with Rosenberg about the sky platform project?—A. Yes; I had a conversation with him later. I asked him in privacy.

Mr. A. Bloch. Can you fix the time?

Q. Can you tell us about when this conversation occurred, the period of years, or months, however you can do it?—A. I would say this was '47, late '47. He told me he had gotten this information about the sky platform from one of the boys, as he put it.

Q. Did he tell you just what information had been given to him by one of the boys concerning the sky platform project? Did he describe it to you at all?—A. Yes; he did. He described it in front of my brother, too.

Q. How did he describe it?—A. He said that it was some large vessel which would be suspended at a point of no gravity between the moon and the earth and as a satellite it would spin around the earth.

Q. Did he tell you from what part of the country that information had been obtained, where the contact was?—A. I don't recall that.

Mr. E. H. Bloch. I did not get that last answer.

(Answer read.)

Q. Did Rosenberg tell you anything about his own dealings with the Russians?—A. Yes; he did.

Q. What did he tell you?—A. He told me that he—if he wanted to get in touch with the Russians he had a means of communicating with them in a motion-picture theatre, an alcove where he would put microfilm or messages and Russians would pick it up. If he wanted to see them in person he would put a message in there and by prearrangement they would meet at some lonely spot in Long Island.

Q. Did he mention anything else along those lines?—A. Well, he——

Q. Let me ask you this, did he mention any other projects, Government projects concerning which he had obtained information?—A. He once stated to me in the presence of a worker of ours that they had solved the problem of atomic energy for airplanes, and later on I asked him if this was true, and he said that he had gotten the mathematics on it, the mathematics was solved on this.

Q. Did he say from where he had gotten this?—A. He said he got it from one of his contacts.

Mr. E. H. Bloch. If the Court please, that last answer, I wonder whether the witness could clarify who was meant by him when he said "they."

The WITNESS. "They" meaning scientists in this country.

Q. Now, what did you do in August 1949 when you terminated your business association with Rosenberg?—A. I got a job.

Q. Where did you obtain that work?—A. I got a job at Arma Engineering Corporation in research and development, model shop.

Q. Did you continue to see Rosenberg and your sister from time to time socially?—A. I did.

Q. Mr. Greenglass, do you remember the month of February 1950, last year?—A. I do.

Q. Did you see Rosenberg in your apartment on the day in February 1950?—A. I did.

Q. Now before I ask you for the conversation on that date——

Mr. COHN. Your Honor, I might say I have one more topic left which I do not think I can complete this afternoon; I think I can complete it fairly early in the morning. I do have one or two things which I omitted in the course of my examination today. I wonder if I can go back and go over them before recess?

The COURT. Go ahead.

Q. You told us on Friday afternoon, Mr. Greenglass, about the atomic explosion that took place at Alamogordo, New Mexico, is that correct?—A. That is correct.

Q. In July of 1945?—A. That is right.

Q. Did you ever furnish any information concerning that atomic explosion to Rosenberg or to Gold?—A. Yes; I furnished information to Gold. I stated to Gold——

Mr. E. H. BLOCH. Could we have the time fixed, please?

The WITNESS. June 1945.

Q. You say you stated to Gold. Did you state it verbally or was it part of the written report you gave Gold?—A. Part of the written report.

Q. Will you tell us what you put in that report concerning this explosion?—A. I had told him that the explosion at Alamogordo was to be an equivalent amount of H. E., as they thought the atom—the nuclear fission would amount to; in other words, I had thought at the time that it was going to be an H. E. explosion at Alamogordo.

Q. Did you put that information in this report?

The COURT. By "H. E." you mean heavy explosive?

The WITNESS. High explosive.

Q. Did you put that information in this report?—A. That is right.

The COURT. That was before the explosion had taken place?

The WITNESS. That is right.

The COURT. How long before the explosion?

The WITNESS. About a month before—it was a little more than a month before.

Q. Now, did Rosenberg ever say anything to you about any reward that he had received from the Russians for the work that he had been doing?—A. He stated that he had gotten a watch as a reward.

Q. Did he show you that watch?—A. He did.

Q. Did he tell you that he had received that watch?—A. I don't recall that.

Q. Did he mention anything else that he or his wife received from the Russians as a reward?

Mr. E. H. BLOCH. Now, Mr. Cohn, I was wondering whether you would fix the time of this last watch incident.

Mr. COHN. I will try to do so.

Q. Can you remember when Rosenberg told you about the watch?—A. I believe it was in January 1945.

Q. During your furlough in January 1945?—A. Yes.

Q. Now, did he ever mention anything else that he or his wife had received as a reward from the Russians?—A. His wife received also a watch, a woman's watch, and I don't believe it was at the same time.

Q. Your recollection is that she received that at a different time?—A. Later, at a later date.

Q. When were you told about a watch that Mrs. Rosenberg had received, do you remember that?—A. I don't recall when that was but I do recall that my wife told me of it.

Q. You got that information from your wife, is that right?—A. That is right.

Q. Now, was there anything else that they received which they told you about?—A. I believe they told me they received a console table from the Russians.

Q. A console table?—A. That is right.

Q. When did they tell you about that?—A. That was after I had gotten out of the Army.

Q. Did you ever see that table?—A. I did.

Q. At their home?—A. I did.

Mr. COHN. I think this would be a very good stopping point, your Honor.

From stenographer's minutes of Case 134–245, *United States of America* v. *Julius Rosenberg et al.* Before Hon. Irving R. Kaufman, district judge, United States District Court, Southern District of New York, March 13, 1951.

DAVID GREENGLASS resumed the stand.

Direct examination continued by Mr. COHN:

Q. Mr. Greenglass, I think yesterday afternoon you told us that Rosenberg told you that he had received a watch from the Russians; is that correct?—A. That is correct.

Q. Now, did he tell you he received anything along with that watch?—A. He said he received a citation.

Q. Did he describe the citation at all?—A. He said it had certain privileges with it in case he ever went to Russia.

Q. Mr. Bloch, Sr., would like you to keep your voice up.—A. I will.

Q. Now, I asked you yesterday afternoon if you remembered a visit you received from Rosenberg in February of 1950. Do you remember such a visit?—A. I do.

Q. How do you fix the date of that visit?—A. Well, it was a few days after Fuchs was taken in England.

Q. A few days after the news of Dr. Fuchs' arrest in England appeared in the papers; is that right?—A. That is right.

Q. Where did this conversation with Rosenberg occur?—A. Partly in my home and partly on the street and in a park.

Q. Did it begin at your home?—A. Yes; it did.

Q. Will you tell us just what happened when he arrived there and what happened after that; what he said and what you said?—A. He came up to my apartment and awakened me. It was about in the middle of the morning. I slept late because I work at night. He said that he would like me to go for a walk with him and we went down the street, down Sheriff Street, toward the Hamilton Fish Park, and we walked around the park, and in the park, and during this walk he spoke to me about Fuchs. He told me, he said, "You remember the man who came to see you in Albuquerque? Well, Fuchs was also one of his contacts"; and this man who came to see me in Albuquerque would undoubtedly be arrested soon, and if so, would lead to me.

The COURT. You mean this is what Rosenberg told you?

The WITNESS. That is right.

And Rosenberg said to me that I would have to leave the country; think it over, and we will make plans to go. Well, I told him that I would need money to pay my debts back so I would be able to leave with a clear head, and Rosenberg said that he didn't think it was necessary to worry about it. But I insisted on it, so he said he would get the money for me from the Russians. He then went on to say—I protested further——

Mr. E. H. BLOCH. I move to strike out the word "protested."

The COURT. Tell us what you said.

The WITNESS. I said, "I wouldn't be able to"—"I didn't think it was wise to go right to the Consulate here and ask for a passport," and he said, "Oh, they let other people out who are more important than you are," and I said, "Is that so?" And he said, "Yes. Well, they let Barr out, Joel Barr, and he was a member of our espionage ring."

Q. Was this the Joel Barr you told us yesterday that Rosenberg had told you had gone to Europe at the end of 1947?—A. That is right.

Q. Go ahead.—A. Well, the conversation continued for a little while, and he said, "You'll just have to leave, and I want to"—oh, I also said to him, "Why doesn't this other guy—fellow leave, the one who came to me in Albuquerque?" And he said, "Well, that's something else again," and I went back home after that.

NOTE.—Counsel for the U. S. Government: Irving Saypol, United States attorney; Miles J. Lane, James Kilsheimer, and Roy Cohn, assistant United States attorneys. Other counsel whose names appear in transcript material represent defendants.

Q. Now, did you have any further conversations with Rosenberg at later times?—A. Yes.

Q. About leaving the country?—A. Yes; I did.

Q. Will you tell us when the next conversation took place?—A. Well, my wife was in the hospital; she had been badly burned in an accident, and it was about the middle of April, it was just about after—before she came out of the hospital, Julius came to see me, and he said I would have to leave the country, and—well, that was about the gist of the conversation.

Q. He told you again that you were to leave; is that right?—A. Yes; and that he wanted me to go. Then again there was——

Q. Had he given you any money up to this point?—A. No; no money was given to me up to this point.

Q. When was the next conversation?—A. The next conversation was after my wife had gotten out of the hospital about May or just before—it was probably a little before May, and he came up to my apartment in order to get some stocks from me, some shares that I had for a business enterprise I was in with him, and he at this time told me that I would have to leave the country as soon as possible, he would get the information for me to leave. Then——

The COURT. Was there any discussion of the country you were to go to?

The WITNESS, It came—he said I would have to go via Mexico, but he didn't give me the complete information as to that until a little later.

The COURT. All right.

Q. Now when was the next conversation with Rosenberg on this subject?—A. Well, it was after my wife came out of the hospital after giving birth to our youngest child.

Q. About when was that?—A. It was May—it was May 22nd or 23rd, something like that.

Q. In that vicinity?—A. In that vicinity.

Q. Yes?—A. It was the next day after she came out of the hospital that Rosenberg came to see me.

Q. Up at your apartment?—A. At my apartment.

Q. Tell us what he said and what you said.—A. And he came into the apartment and he had a Herald Tribune in his hand with a picture of Harry Gold on it and he said, "This is the gentleman who came to see you in Albuquerque."

Q. Is that the copy of the Herald Tribune carrying the case of Harry Gold's arrest for espionage?—A. That is right.

Q. What did he say?—A. He said, "This is the man who saw you in Albuquerque."

I looked at it and I said I couldn't tell from that picture, and he said, "Don't worry, I am telling you this is the man and you will have to go out—you will have to leave the country," and he gave me a thousand dollars then and said he would give me $6,000 more. We then went for a walk.

Q. Where did you walk, do you remember?—A. We walked down to Delancey Street and down the Drive.

Q. Did you meet anybody you knew during the course of this walk with Rosenberg?—A. Not on this walk.

Q. Go ahead.—A. We walked down Delancey Street to the Drive, the East River Drive and walked along the Drive, during this time he told me what was necessary to—how I was to leave the country.

Q. Tell us exactly.

Mr. A. BLOCH. I move to strike that out. It is not a conversation. It is simply a conclusion of the witness how he was to leave the country.

The COURT. Overruled.

Mr. A. BLOCH. Exception.

Q. Will you tell us exactly what Rosenberg said to you on that subject?—A. Well, he said that I would have to get a tourist card——

Q. For what country, did he tell you?—A. To go to Mexico.

Q. In other words, the first place you were to go to was Mexico?—A. That is right. First I was to go to the border area and at the border area get a tourist card. In other words, not to get the tourist card at some Mexican Consulate in this city but to wait till we get to the border.

Q. Yes?—A. He told me that in order to get the tourist card you have to have a letter or you have to be inoculated again at the border—a letter from the doctor saying you were inoculated.

Q. For what, did he tell you?—A. For smallpox.

Q. Did he tell you how he found that out?—A. He said he went to see a doctor and a doctor told him about it and I said I would attend to that. He then told me I would have to have passport pictures made up.

Q. Passport pictures?—A. Of myself, my wife and my family, and also he gave me a certain form letter to memorize and sign "I. Jackson" at the end of the letter. This letter was to be used when I get to Mexico City. I was to write to the Secretary to the Ambassador of the Soviet Union and state in that letter—I don't recall completely right now but something to the effect about the position of the Soviet Union in the U. N.

Q. Something favorable or unfavorable?—A. It was favorable.

Q. Yes. How were you to sign the letter?—A. I was to sign the letter "I. Jackson."

Q. "I. Jackson."—A. Then I was to wait three days at some place—first, of course, to get a place to stay, some place away from the center of town. Then I was to go with a guide to the city in my hand——

Q. A guide?—A. A guide.

Q. I see. A. (Continuing.) To the city in my hand, with my middle finger in the—between the pages of the guide—go to a place called Plaz de la Colon and look at the statue of Columbus there—and this would be about 5 o'clock in the afternoon, three days after I had sent the letter.

Q. In other words, you would write a letter and three days after that at 5:00 in the afternoon you were to be in front of the statue of Columbus in Mexico City with this travel guide in your hand, is that right?—A. That is correct.

Q. All right.—A. I was then to wait until some man was to come up close to me and then I would say, "That is a magnificent statue," and that I was from Oklahoma and I hadn't seen a statue like it before, and this man was to say, "Oh, there are much more beautiful statues in Paris." That was to be our identification.

Then he was to give me my passports and additional money so that I could go on with my trip. I was then supposed to continue on probably via Vera Cruz——

Q. Vera Cruz?—A. Vera Cruz.

Q. A seaport in Mexico?—A. That is right.

Mr. E. H. BLOCH. If the Court please, the witness said "probably." Now I want to make certain that he is speaking of his own knowledge.

The COURT. Yes, he used the word "probably."

A. (Continuing.) Well, I was to continue on to Vera Cruz and then to Sweden or Switzerland, one or the other. I would—in Sweden I was to go to the statue of Linnaeus in Stockholm and repeat—after sending a letter to the Ambassador of the Soviet Union—to the Secretary of the Ambassador of the Soviet Union, with the same type of letter and also "I. Jackson" as the signature again.

I would then go three days later to the statue of Linnaeus and with a guide in my hand, with my finger in the place, and a man would come up to the statue about the same time in the evening, about 5 o'clock, and I would repeat that it was a beautiful statue, a magnificent statue—something to that effect, and the man would say, "There are much more beautiful ones in Paris," and that was to make our contact.

Then he was to give me my means of transportation to Czechoslovakia, and that is where I was to go.

Q. Was that to be your permanent place?—A. Supposedly that was where I was supposed to go; so far as what went after that I didn't know.

Q. Did Rosenberg tell you what you were to do when you arrived in Czechoslovakia?—A. Yes, he did.

Q. What did he tell you?—A. He told me to write to the Ambassador of the Soviet Union and say that I was here.

Q. Were you to sign "I. Jackson" this time?—A. My full name was to be signed, "I. Jackson."

Q. All right. Now, did you write down these instructions or did Rosenberg write them down?—A. Nobody wrote them down. I was told to memorize them at this time and I did memorize them.

Q. Did you have any further conversation with Rosenberg on that day?—A. Well, that was the end of the conversation on that day except that—he said that he probably—that he had to leave the country himself and he was making plans for it, and I said, "Why you?" He said that he was a friend—that he knew Jacob Golos, this man Golos, and probably Bentley knew him.

Q. And that he himself was going to leave the country, is that right?—A. And that he himself was going to leave the country.

Q. Now was there any further conversation with Rosenberg on that occasion?—A. Only that he asked me to memorize it and get everything attended to—the passport photographs.

Q. What did he tell you about the passport photos; he wanted you to get those yourself, is that right?—A. That is right.

Q. Did he tell you exactly what he wanted to know with regard to the passport photos?—A. Yes; he wanted five copies, five pictures, each with myself, by myself, my wife, and then my wife and the children and then myself with the children, and then I think all of us together, the family altogether.

Q. In other words, these five different poses, he wanted five copies of each one?—A. That is right.

Q. What were you to do with those passport photos after you had obtained them?—A. I was to give it to him later on.

Q. Did he say anything to you about a lawyer?—A. Earlier, up in the apartment, he said that he had gotten a lawyer himself, that I should get one but I never did get one.

Q. Now——A. He said, in case you are picked up before you leave the country.

Q. Now, was there any further conversation with Rosenberg on that day, or did that complete it?—A. That completed the conversation on that day.

Q. All right, what happened next with relation to this subject?—A. Well, the following Sunday was the first time the baby had ever come down.

Q. This was the baby that was born a short time before?—A. Yes.

The COURT. The baby had come down where?

The WITNESS. In the street.

The COURT. I see.

The WITNESS. Before that it was in the house.

The WITNESS (continuing). So we took the whole family and went over to a photo shop on Clinton and Delancey Street; it is near the Apollo Theatre.

Q. Near the Apollo Theatre?—A. Yes.

Mr. E. H. BLOCH. I just want to clarify this. Is it the contention that the Rosenbergs were along?

Mr. COHN. No, I don't think he said so.

Mr. E. H. BLOCH. I wanted to clarify it.

Q. The Rosenbergs weren't with you, were they?—A. No.

Q. Just you, with your wife and the two children, the people who were supposed to be in these passport photos; is that right?—A. That is right.

Q. Go ahead.—A. We went over to the shop and we had these pictures taken; and later that evening I picked the pictures up after they were done; and it was during the week—it was Memorial Day, and I remember I was off that night, Julius came over, and I gave him the pictures in the hallway, because there were people in my house and I didn't want the people to see Julius coming to visit me.

Q. Now I think you told us that he had asked you to have five sets of pictures taken; is that right? Five poses, five sets each?—A. That is right.

Q. How many, in fact, did you have taken at this passport photo shop?—A. I had six sets of pictures taken.

Q. How many did you give to Julius?—A. Five sets.

Q. What did you do with the sixth set?—A. I kept it in the drawer.

Q. Was that set after your arrest given to the FBI?—A. I gave it to the FBI.

Mr. COHN. May we have these marked for identification, please.

(Marked "Government's Exhibits 9-A and 9-B" for identification.)

Q. By the way, I think you gave four of the five to the FBI; is that right?—A. That is right.

Q. Would you look at Government's Exhibit 9-A for identification and 9-B for identification, Mr. Greenglass, and tell us if those are four of the passport pictures you had taken on that Sunday in May of 1950 [showing to witness].—A. That is correct; these are the pictures.

* * * * * * *

Q. Now, I think you told us you handed five of the six sets over to the defendant Rosenberg; is that correct, Mr. Greenglass?—A. That is correct.

Q. What else happened on that occasion? I think you said that was in the hall outside of your apartment.—A. That was in the hall. He said that he would come back and give me the additional money.

Q. Well, had he given you any money up to that point?—A. He had given me a thousand dollars.

Q. When had he given you the thousand dollars?—A. When he first came into the apartment and showed me that Harry Gold had been arrested.

Q. He had given you a thousand dollars in cash; is that right?—A. That is right.

Q. What did you do with that thousand dollars?—A. I gave it to my wife, who paid bills with it and spent it, generally.

Q. Now, go ahead, from the time when you gave him the passport photos he told you he would give you the additional money?—A. That is right.

Q. Any further conversation at that time?—A. No, he just left.

Q. All right, when was the next conversation on this subject?—A. He came back the following week, I believe it was, and it was in the morning, and he came into the apartment, woke me out of bed and put $4,000 in a paper, brown paper wrapping, on the mantelpiece in the bedroom, and he then told me, "Let's go for a walk," because he wants to have me repeat the instructions he had given me.

Q. Instructions he had given you and told you to memorize?—A. That is right. I then went down Columbia Street to Delancey Street, and on the way I met two friends.

Q. What are their names?—A. Hermie and Dianne Einsohn.

Q. Tell us the circumstances of your meeting them.—A. They were across the street from us and Julius said to ignore them but I said I couldn't do that because they are friends of mine and they would wonder why I walked by without saying anything. I crossed the street and Dianne said, "Here is the $40 I owe you," and she paid me by check $40.

Q. Which they had owed you; is that right?—A. That she had owed me; and Julius had crossed the street, walking a little ahead of us, and I then caught up to Julius again. We went down the drive again and he asked me to repeat to him the various instructions he had given me; and I repeated the instructions to his satisfaction.

Mr. A. BLOCH. I object to the last part of the answer, "to his satisfaction."

The COURT. Yes; strike out "to his satisfaction."

A. (Continuing.) I repeated the instructions, and he said that was fine.

Q. Yes?—A. I then—well, that ended the conversation, and he went his way and I went back to my apartment. Later, he came back. I was under surveillance at the time and——

Q. You thought the FBI——

Mr. E. H. BLOCH. I move to strike that out.

The COURT. I will strike that out unless you tell us how you knew.

Q. Did you think you were being followed at that time?—A. I did think I was being followed.

Mr. E. H. BLOCH. I object to it upon the ground that it is not binding on these defendants.

The COURT. Oh, no, no.

Q. All right, let us go to this. You say you had another conversation with Rosenberg; is that right?—A. I had another conversation with Rosenberg.

Q. Tell us what you said and what he said—well, tell us again when did that take place in relation to the last meeting with him?—A. It was—I would say it was—let's say it was like—it is hard to do without saying that I was being—that I noticed some people following me on a Sunday evening; and he came back——

The COURT. Who is "he"?

The WITNESS. Julius came back during that week, which was—oh, I would say, about May—it was June, the 2nd or 3rd or maybe even the 4th; I can't place it exactly, and as he came into the apartment he said, "Are you being followed?" I said, "Yes; I am." He said, "I just came back from up-State New York to see some people, and I was going to Cleveland, Ohio, but I am going—I am not going to go there any more"; and he said to me, "What are you going to do now?" I said, "I am not going to do anything. I am going to sit—I am going to stay right here," and he left.

Q. Did you see him again after that?—A. Only in court here.

Q. Shortly after that, I believe, on June 15, were you arrested by agents of the Federal Bureau of Investigation?—A. I was.

Mr. COHN. May I have this marked for identification, please.

(Marked "Government's Exhibit 10" for identification.)

Q. Now, get back to this $4,000 for a minute, Mr. Greenglass. In what form was that when Rosenberg gave it to you? Was it in paper or something like that?—A. It was in tens and twenties, cash paper.

Q. Was it wrapped in any container of any kind?—A. It was wrapped in brown paper.

Q. Brown paper; is that right?—A. That is right.

Q. Will you look at Government's Exhibit 10 for identification and tell us whether or not you can recognize that as the brown wrapping paper in which the $4,000 was contained [showing]?—A. That is the brown paper, wrapping paper it was contained in.

Mr. COHN. We offer it in evidence, your Honor.

*   *   *   *   *   *   *

The COURT. As I understand it, the testimony was that it was in this brown wrapping paper that Mr. Rosenberg delivered the $4,000.

The WITNESS. That is correct.

Mr. E. H. BLOCH. I will withdraw my objection if I just get one clarification. Will the Court ask the witness whether it is the contention or the testimony of the witness that Rosenberg came with this money wrapped in this bag, that Rosenberg actually carried this bag?

The COURT. Was the money right in that wrapping paper?

The WITNESS. Right in that wrapping paper.

The COURT. Very well.

Mr. COHN. May I show this to the jury, your Honor?

The COURT. Yes.

Q. There are some initials on there, Mr. Greenglass. As far as you know, those were not there—there was no writing on this paper when Rosenberg gave it to you as a container for the money; is that right?—A. No; there wasn't.

(Government's Exhibit 10 shown to jury.)

Q. Now, what did you do with the $4,000?—A. Well, at first I had intentions of flushing it down the——

Mr. E. H. BLOCH. I move to strike that out.

The COURT. Yes, stricken.

A. (continuing) I started to flush it down the toilet bowl.

Mr. E. H. BLOCH. I move to strike that out as not binding upon the defendant Rosenberg, not in his presence.

The COURT. I sustain that. Strike it out.

A. (Continuing) I gave it to my brother-in-law, Louis Abel, to keep for me, and I told him to use it whenever he needs it, take whatever he might need from it, but also that I might need it myself and in such case he would hand it over.

Mr. A. BLOCH. I object to anything he said to his brother-in-law in connection with this money, as pure hearsay.

The COURT. I am going to sustain that objection.

Mr. COHN. So the testimony stands, your Honor, he gave it to his brother-in-law, Louis Abel, and any conversation between him and Abel is stricken?

The COURT. Yes.

Mr. COHN. Very well, your Honor.

Q. Now, following your arrest by agents of the Federal Bureau of Investigation in June you had occasion to talk to them frequently?—A. I did.

Q. Have you had occasion to talk to Mr. Saypol, to me, members of Mr. Saypol's staff?—A. I did.

Q. At any time, in any of those conversations, has any promise of any kind ever been made to you as consideration or anything else for your testifying on this witness stand or telling the facts to any agency of the United States Government?

Mr. E. H. BLOCH. Before you answer, please, I object to it upon the ground that it is improper direct. I also object to the form of the question.

Mr. COHN. I withdraw it.

The COURT. Pardon?

Mr. COHN. I will withdraw the question, your Honor.

The COURT. All right.

Mr. COHN. Mr. Bloch may examine, your Honor.

Cross-examination by Mr. E. H. BLOCH:

Q. Did you just testify that you didn't see your brother-in-law, Julius Rosenberg, from the time that you told him that you were going to stay here and you weren't going to leave the country, until you appeared here in court?—A. I did.

Q. Was that the truth?—A. Well, I had seen him.

Q. Was it the truth?—A. That was not the truth.

Q. You did see him, didn't you, between that period?—A. I did.

The COURT. Where did you see him?

The WITNESS. I saw him in jail.

The COURT. Did you talk to him?

The WITNESS. No.

The COURT. You mean you just saw him; you looked at him and you saw him?

The WITNESS. They brought us into a room together, the authorities at West Street, and told us to stay apart and I didn't say anything to him and he said nothing to me.

The COURT. All right.

Q. Now Mr. Greenglass, I believe you testified that your wife came out to visit you in New Mexico on November 29, 1944, which was at or about the date of your second wedding anniversary; is that correct?—A. I testified that she came out around that date.

Q. Around, about that. When is your wedding anniversary?—A. November 29th is my wedding anniversary.

Q. And when did your wife arrive at Albuquerque, New Mexico?—A. I can't tell exactly when.

Q. When did you first see her at or about that time?—A. It was during a five-day vacation, furlough, and I did not know exactly the date that I got into Albuquerque.

Q. At any rate, there came a time when you met your wife on that five-day furlough—A. That is correct.

Q. Was that the first time that you heard from anybody's lips an invitation to you to engage in spying?—A. That is correct.

Q. And that invitation came from the mouth of your wife; is that not so?—A. That is correct.

Q. There was nobody else present besides you and your wife at that time?—A. That is correct.

Q. Is that correct?—A. That is correct.

Q. I believe you testified you took a walk with her—A. That is correct.

Q. And that is when she told you about spying; is that right?—A. That is right.

Q. The Rosenbergs weren't around, were they?—A. No.

Q. Were the Rosenbergs there?—A. They were not.

Q. And I believe you testified—you correct me whenever you think I am quoting your testimony or paraphrasing your testimony incorrectly—that you became frightened and you told your wife that you would not do it; is that correct?—A. That is right.

Q. And then I believe you testified further the following morning you told your wife that you would do it?—A. That is right.

Q. And you not only told your wife that you would do it, but right there and then when you told her that, you disclosed information concerning some of the personnel at the Los Alamos project; is that correct?—A. It was not right then and there. It was out in the street.

Q. Well, how long after you indicated to your wife that you were willing to engage in this illegal work did you tell her the names of some of the personnel at Los Alamos?—A. It was not long after; during the same day.

Q. It was part of the same conversation, was it not?—A. Almost.

Q. Now, during the time——

Mr. E. H. BLOCH. Withdrawn.

Q. From the time that you told your wife that you were not interested and that you wouldn't do this work, to the following morning when you told her you would, did you consult with anybody?—A. I consulted with memories and voices in my mind.

Q. Physically, did you consult with anybody?—A. No.

Q. Did you see the Rosenbergs during that period?—A. No.

Q. Did you talk to the Rosenbergs by telephone during that period?—A. No.

Q. How old were you at this time?—A. 22.

Q. And when you finally said to your wife the following morning after she invited you to engage in spying, you did this and said this and then disclosed information of your own free will; isn't that correct?—A. That is correct.

Q. You knew at that time, did you not, that you were engaging in the commission of a very serious crime?—A. I did.

Q. You had been briefed and indoctrinated, had you not, in Oak Ridge, Tennessee, when you first came there, and during your stay there, and also at Los Alamos, when you were transferred from Oak Ridge, about espionage; is that not correct?—A. That is correct.

Q. And you were briefed to the extent of having read to you the espionage law; is that correct?—A. Parts of the espionage law.

Q. Do you remember which parts?—A. I haven't the slightest recollection.

Q. Were you told by your instructors at Oak Ridge, Tennessee or at Los Alamos about the penalty for committing espionage?—A. Most likely I was but I don't remember it.

Q. Did it occur to you on November 29, 1944 or November 30, 1944—and I don't want to quibble about the date—at any rate, did it occur to you at the time that you finally said to your wife "I will do this" and then transmitted to her certain information, that there was a possible penalty of death for espionage?—A. Yes.

Q. You knew that?—A. I did.

Q. When you said to your wife "Yes, I will do it"—is that correct?—A. That is correct.

Q. Are you aware that you are smiling?—A. Not very.

Q. Now how long had you been in the Army when you were assigned to Los Alamos?—A. I had been in from April '43 to July—August '44.

Q. That is approximately a year and five months.—A. That is about it.

Q. Is that rough enough?—A. Yes.

Q. Had anybody prior to the time that your wife came down to Los Alamos and invited you to spy, ever made any overture to you to steal information from the military authorities?—A. No one.

Q. And from the time in the latter part of November 1944 until your entire career or during your entire career in the Army, you continued to spy, did you not?—A. I did.

Q. And you received money for that, did you not?—A. I did.

Q. You received $500 from Harry Gold in Albuquerque, New Mexico, for that, did you not?—A. I did.

Q. Did you ever offer to return that money?—A. I did not.

Q. Did you use that money for your own personal use or the use of your wife and the other members of your family?—A. I did.

Q. What did you do with the $500 that you received from Gold?—A. I gave it to my wife and she used it to live on.

Q. Did she deposit it in the bank?—A. She did.

Q. Now Mr. Greenglass, you had been to various camps prior to the time that you were assigned to Los Alamos, isn't that correct?—A. That is correct.

Q. And your wife had visited you at camps before you had been assigned to Los Alamos, is that correct?—A. That is not exactly correct.

Q. Did your wife ever visit you at any of the camps at which you were stationed prior to the time you were assigned to Los Alamos?—A. One camp.

Q. And what camp was that?—A. Pomona, California—Pomona Air Base.

Q. And your wife at that time lived in New York, did she not?—A. She did.

Q. And she took a trip from New York out to California to see you?—A. That is right.

Q. Now when was this?—A. This was the late winter or early spring of '44.

Q. And how long did your wife stay?—A. About two, three months.

Q. And did she have an apartment?—A. Yes.

Q. And did you live with her at that apartment?—A. I did.

Q. And did your wife work?—A. She did.

Q. In or about the camp, in order to sustain herself?—A. She worked away from the camp.

Q. I said in or about the camp.—A. She did.

Q. She didn't work in the camp proper but she worked in some enterprise in the vicinity of the camp?—A. That is right.

Q. And after this incident in the latter part of November 1944 did your wife visit you at any other of the camps—I withdraw that, I am sorry.

Q. You were stationed at Los Alamos, continuously after you had been assigned there up to and including the date of your discharge, is that correct?—A. No; the date of my discharge I was discharged from——

Q. El Paso, Texas?—A. El Paso, Texas.

Q. But otherwise you were always at the Los Alamos project?—A. That is right.

Q. Now, did your wife visit with you and live with you at or about the Los Alamos project or in or about the City of Albuquerque after your five-day furlough was up in the latter part of November 1944?—A. As I said before, in the early spring of 1945 she came to live with me.

Q. And how long did she live with you?—A. She stayed with me until I came home.

Q. And did she work?—A. She worked part of the time; part of the time she was unable to work.

Q. And you were advanced from time to time, weren't you, in your gradings in the Army?—A. I was.

Q. And your salary was increased?—A. That's right.

Q. And that made it easier economically, isn't that correct?—A. To a degree.

Q. Now let me ask you this: You have testified you got a thousand dollars from Rosenberg in 1950.

Q. And you insisted upon that money to pay your debts, isn't that correct?—A. That is correct.

Q. Did you pay your debts?—A. I did not. May I clarify that?
Q. No.
The COURT. Yes.

*   *   *   *   *   *   *

The WITNESS (to the Court). The thousand dollars I spent paying debts of household type——
Mr. E. H. BLOCH. For household type.
The WITNESS (continuing). I mean payments on furniture and things of that nature. The debts I was referring to, when I said about paying debts to Rosenberg, were personal debts incurred because of a business enterprise I was in that I had never even borrowed in the first place.
The COURT. Excuse me, this thousand dollars that you got in 1950, I want you to get this clear in my mind, was that in payment of information which you had at or about that time turned over to Rosenberg or was that in anticipation of your trip out of the country?
The WITNESS. In anticipation of the trip out of the country.
Q. All right. Now it is your testimony now that you used those thousand dollars to pay off certain debts on the household, is that right?—A. That is right.
Q. That was for your own use, was it not?—A. That is correct.
Q. And that was for the use of your family, is that right?—A. That is correct.
Q. When you, as you testified, finally told Rosenberg that you were not going to leave the country and you were going to stay right here——A. Yes.
Q. Did you ever offer to give him back that thousand dollars?—A. I did not.
Q. Now then, I think you testified further that you received an additional $4,000 from Rosenberg sometime in June, is that right—sometime in June 1950?—A. That's right.
Q. Wrapped in that brown bag which is now Government's Exhibit——
The COURT. 10.
Q. 10?—A. That's right.
Q. When you told Rosenberg, as you testified, that you were not going to leave the country, did you offer to return that $4,000?—A. I did not.
Q. And I believe you testified that you turned those $4,000 over to a brother-in-law?—A. That's right.
Q. Do you know now that that $4,000 or a major part of that $4,000 went to pay your lawyer?
Mr. COHN. Your Honor, I merely comment that I was halted in my attempts to develop the history of that $4,000 and where it went, which is quite relevant to the rest of the Government's proof. If Mr. Bloch wants to go ahead and do it, all right.
The COURT. I will let him go ahead if he is——
Mr. E. H. BLOCH. I have a different purpose in doing it.
The COURT. If he is going to open the door, I will let him.
Mr. E. H. BLOCH. I am perfectly willing to open the door.
Mr. COHN. Very well.

By Mr. E. H. BLOCH:

Q. Do you know now that that $4,000 or a major part of it went to pay your lawyer?—A. I do.
Q. Have you been told how much of that $4,000 went to pay your lawyer?—A. Yes.
Q. All of it?—A. All of it.
Q. When did you know that you had a lawyer representing you subsequent to the time of your arrest in June 1950?—A. The FBI said that I could have a lawyer. I called up my——
The COURT. Speak louder, please.
Q. Up, up, up.—A. The FBI agents told me that I could have a lawyer, so I called up my brother-in-law and told him to get in touch with O. John Rogge.
Q. That was your suggestion—A. At my suggestion.
Q. All right. And then after that did Mr. Rogge come down to see you?—A. He sent his partner, Mr. Fabricant and subsequently Mr. Rogge came himself.
Q. Now how long after you had somebody communicate with your brother-in-law to hire Mr. Rogge did Mr. Fabricant come down to see you?—A. The next morning—I should say the same day because it was morning when I got in touch with my brother-in-law.
Q. Do you remember when you were arrested?—A. I do.
Q. You were taken, were you not, from your place of employment and brought down to this building?—A. I was not.

Q. Were you apprehended in your home?—A. I was in my home making a formula when they arrested me.

The COURT. Making what?

Mr. E. H. BLOCH. A formula.

The WITNESS. A formula for the baby.

Q. And what day did the FBI come around to arrest you?—A. It was in the daytime, in the middle of the week, just before I was going to work.

Q. Do you remember what date in June?—A. It was the 15th, I believe, of June.

Q. And after some of the representatives of the FBI came into your home did they tell you that you were under arrest?—A. No.

Q. Did they ask you to accompany them some place?—A. They did.

Q. And where did they ask you to go?—A. I came to this building.

Q. How many representatives of the Federal Bureau of Investigation came around to your house at the time that you were apprehended?—A. Four.

Q. And at what time was this?—A. In the afternoon about 2 o'clock, I suppose.

Q. Now, do you now know the names of the four FBI representatives who arrested you?—A. I do.

Q. And will you name them, please?—A. John Harrington, Leo Frutkin——

Mr. A. BLOCH. I can't hear him.

Q. Counsel at the table cannot hear you.—A. John W. Lewis, and Bill Norton.

Q. You know their first names too, don't you——A. I do.

Q. By now. And what time did you get down here in this Federal Building?—A. I would say it was about 8 o'clock at night.

Q. Eight at night?—A. Yes.

Q. Maybe I misunderstood you. What time did the FBI men come to your home to arrest you?—A. They came at 2 o'clock.

Q. And after they were in your home did they stay with you in your home?—A. Yes.

Q. How long did they stay with you in your home?—A. Till about, I would say, 7:30 or so.

Q. And were you questioned during those five hours at your home?—A. On and off.

Q. Well, when you say "On and off" did some FBI man stay with you during that entire period?—A. No; I ate my lunch, I made the formula——

Q. No, listen to my question, please. Did some FBI man stay with you during this entire five or five and a half hours?—A. No, not all the time.

Q. Did they all leave at one time?—A. They were in other rooms of the apartment.

Q. Now when I talk about your home I am including every room in your apartment. I am going to ask you again, did the FBI men who came down at or about 2 o'clock stay with you in any of the rooms of your apartment for about five or five and a half hours?—A. That is correct.

Q. All four of them?—A. At first there was two and then two more came up.

The COURT. Out of that five and a half hours about how much time would you say was taken up in questioning?

The WITNESS. Maybe an hour.

Q. And what did they do the rest of the time?—A. They were searching my apartment.

Q. Do you know what they took from your apartments?—A. I do.

Q. You watched them search?—A. I did.

Q. Did the FBI man take from your apartment any writing, any document, any paper, any memorandum in the handwriting of Julius Rosenberg?—A. Yes, they did.

Q. What?—A. Some notes that he had from college days.

Q. Anything else?—A. Well, there was——

Mr. COHN. I might say here if the witness knows, your Honor.

Mr. E. H. BLOCH. I am asking him. I tried to lay the foundation.

The WITNESS. There was a trunk full of letters, and I don't know what was in there exactly, and if there was stuff in there, they did take that.

Q. Letters in the handwriting of Julius Rosenberg?—A. Well, I don't know.

Q. Well, I am trying to direct your attention specifically and exclusively——A. There were letters——

Q. Now just listen and try to follow me—exclusively to this, and it is very simple: Did you see the FBI men who came to your apartment on the date they arrested you, take from your apartment any writing, whether in the form of a document or a note or a memorandum in the hand of Julius Rosenberg outside now of the college notes that you say were in his handwriting?—A. No.

Q. And did the FBI men take from your apartment any writing, whether in the form of a document or memoranda or any scrap of paper of any kind in the handwriting of your sister Ethel?—A. I didn't see it.

Q. Well, if you didn't see it you don't know.—A. But it might have been inside that trunk full of letters. We couldn't open the trunk. That is the only reason I didn't see it.

Q. Did the FBI men finally open the trunk?—A. Yes—not there, though; not in my presence.

Q. Did you see them open the trunk?—A. No, I did not.

Q. Did they take the trunk away from your apartment?—A. They did.

Q. Now, so far as I know, outside of the college notes that were made by Julius Rosenberg—when you say "college notes" I am assuming you mean notes made by Rosenberg when he was attending college, is that right?—A. That's correct.

Q. Outside of that writing do you know whether there was in your house on or about June 16, 1950, or June 15, 1950, any writing or written matter in the handwriting of either Julius Rosenberg or Ethel Rosenberg?—A. I didn't know, no.

The COURT. We will take our recess at this point.

(Short recess.)

Q. Now, Mr. Greenglass, when were you inducted into the United States Army?—A. April '43.

Q. Do you remember the date?—A. The 12th.

Q. April 12th. Where?—A. In New York City.

Q. And specifically where in New York City?—A. At the induction center at Grand Central Palace.

Q. Were you drafted?—A. I was.

The COURT. Speak up please, Mr. Greenglass.

Mr. E. H. BLOCH. I am going to stand all the way back here, your Honor; maybe that will help.

(Mr. E. H. Bloch steps farther back to the railing.)

Q. At that induction center in April 1943 did you take an oath?—A. I did.

Q. Did you raise your right hand and together with the other citizens who were inducted into the Armed Services with you swear to certain things?—A. I did.

Q. Do you remember the oath you took?—A. I don't.

Q. At any rate, you know that you violated that oath, don't you?—A. I know it.

Q. And you knew when you said to your wife "Yes, I am going to give you the information," that took place somewhere in the latter part of November 1944, that you were violating the oath that you took here in New York City at the time of your induction—did you know that?—A. I did.

Q. Did it enter your mind?—A. Violation of that oath did not enter my mind.

Q. Did you know that you were besmirching the name and the reputation——

Mr. COHN. Well, I will object to this characterization.

Mr. E. H. BLOCH. I haven't finished.

The COURT. He hasn't finished. I want to hear the question.

Of the uniform of the soldiers of the United States.

Mr. COHN. I will object now, your Honor, to the conclusory terms used in the question. That is clearly improper.

The COURT. I will overrule the objection.

The WITNESS. I did not think so at the time.

Q. The Espionage Act—withdrawn. You had been briefed on the Espionage Act, had you not?—A. I did—I was.

Q. You knew at that time that it was a crime for anybody who had intent or reason to believe that it would advantage a foreign power to communicate any information relating to the national defense of the United States?—A. I knew that.

Q. To somebody else who was unauthorized; particularly a foreign nation or the citizen or subject of a foreign nation? You knew that much about the Espionage Act, didn't you?—A. Yes.

Q. And you knew it was a crime to conspire to violate the Espionage Act in the latter part of November 1944, didn't you?—A. I did.

Q. And you say now that you did not know at that time that when you were giving your wife unauthorized information you were besmirching the uniform——

The COURT. I do not believe——

Q. Of the United States Army?

The COURT. I don't believe the question was whether he knew—whether he believed, wasn't that the question?

Mr. E. H. BLOCH. I will change it to "believe."

The WITNESS. I did not believe that.

Q. You did not believe that.

The COURT. Did you——

Q. Did you believe you were——

Mr. E. H. BLOCH. I am sorry, your Honor.

The COURT. Go ahead.

Q. Did you believe you were doing an honorable or dishonorable thing?—A. I didn't even think of it that way.

Q. All right.

The COURT. How did you think of it?

The WITNESS. I thought of it from what I had—on the basis of the philosophy I believed in. I felt it was the right thing to do at that time.

Q. You felt it was an honorable thing to do; is that what you are trying to tell us?—A. The right thing to do according to my philosophy at that time.

Q. All right, I will accept the word "philosophy." You felt it was the right thing to do?—A. That is right.

Q. And did you continue to think that what you were doing after November 29, 1944, and up to and including the time that you got out of the Army, that you were doing the right thing?—A. I was having my doubts.

Q. When did you begin to have doubts?—A. Almost as soon as I started to do it.

Q. And tell me, if you can—and I want you to fix the time as exactly as you can, what you meant in terms of time when you began to have doubts as to the rightness of what you were doing.—A. I don't understand your question.

The COURT. I do not understand it either.

Q. All right, let me make it easier. Did you begin to have doubts in December 1944?—A. I started to have doubts——

Q. Now will you answer my question?

The COURT. Wait a minute. Let him answer it. It is not an easy question to answer.

Mr. E. H. BLOCH. I want him to give me the exact date.

The COURT. So let him answer it——

Mr. E. H. BLOCH. All right.

The COURT. And please, Mr. Bloch, don't interrupt me while I am talking.

Mr. E. H. BLOCH. All right.

The COURT. You have a bad habit of doing it.

Mr. E. H. BLOCH. I am sorry, I don't mean to do it.

The WITNESS. I started to have doubts almost as soon as I said that I was going to give the information.

Q. And that was in the latter part of November 1944, but you did give information, did you not——A. I did.

Q. From that time on right through 1945?—A. That's right.

The COURT. Now you saw Mr. Rosenberg in January 1950, within a short period of time after you had these doubts that you speak of; did you relate to him on that occasion that you had doubts about the propriety of it?

Mr. COHN. Your Honor said "January 1950."

The COURT. I am sorry—1945. Did you say anything to him about your doubts on the propriety of what you were doing?

The WITNESS. No, I did not say anything to him because——

Mr. E. H. BLOCH. I move to strike out everything after "because."

The COURT. No, let me have it.

Mr. E. H. BLOCH. I respectfully except.

The WITNESS. Because, as I said, when I first started to do it it was one of the motivating factors for doing it.

Q. I am sorry, I cannot hear that at all.—A. I had a kind of a hero worship there and I did not want my hero to fail, and I was doing the wrong thing by him. That is exactly why I did not stop the thing after I had the doubts.

Q. You say you had a hero worship.—A. That is right.

Q. Who was your hero?—A. Julius Rosenberg.

Q. I see. Now tell me, did you have doubts when you passed information to Harry Gold in Albuquerque, New Mexico.—A. I did.

Q. Some time in 1945?—A. I did.

Q. And you say that was in June 1945; is that correct?—A. That's right.

Q. Did you consider at that time that you were passing to him very important information?—A. I did.

Q. And did you consider that that information was secret?—A. I did.

Q. And you knew that that information was unauthorized?—A. I did.

Q. To be communicated to any person outside of the camp itself; isn't that right?—A. That is right.

Q. And did you have any doubts when you took $500 from Mr. Gold for passing that information?—A. I still had doubts.

Q. You still had doubts, but you took the money and you handed it to your wife?—A. I did.

Q. And that money was used for your house and the use of your wife; is that correct?—A. That's right.

Q. And did you have any doubts when you, as you state, got $200 from Julius Rosenberg in September 1945 in New York City?—A. I did.

Q. You took the money?—A. I did.

Q. Did you give that to your wife?—A. I did.

Q. And that was used for your benefit and the benefit of your wife and your family; is that correct?—A. That is correct.

Q. Did you have any doubts when you, as you testified, got a thousand dollars from Mr. Rosenberg in June 1950?—A. I did.

Q. But you took that money and you used it to pay off household debts; isn't that correct?—A. Well——

Q. Did you or didn't you?—A. I did take that money for that reason.

Q. Did you have any doubts when you took that——

The COURT. Are you finished with your answer?

The WITNESS. I would like to answer.

The COURT. Well, we are not going to know when you are finished or not unless you say you want to answer.—A. (Continuing:) About that thousand dollars, I felt that I was giving nothing for this thousand dollars; I had plenty of headaches and I felt the thousand dollars was not coming out of Julius Rosenberg's pocket, it was coming out of the Russians' pocket and it didn't bother me one bit to take it, or the $4,000 either.

Q. You had no qualms at all about taking the $1,000 or the $4,000, did you?—A. Not at all.

Q. Even though that money, as you say, was given to you for purposes of flight?—A. That is right.

Q. You never offered to return that money?—A. I certainly did not.

Q. Even though you had $4,000 of it left and you had determined not to flee?

The COURT. Wait a minute. Wait, wait.

Mr. E. H. BLOCH. I am sorry.

The COURT. Your question is a very clumsy question. It is several questions. It is summation, it is everything all mixed up. Let's break it down.

Did you offer to return the $4,000?

Q. Did you have any qualms about not returning the $4,000 after you had determined not to leave the country?—A. I had no qualms at all.

Q. All right. Now let's go back to the end of your Army career for one moment. You testified on direct examination, did you not, that you received an honorable discharge?—A. I did.

Q. Did you consider that the services that you rendered to the United States during your Army career warranted an honorable discharge?

Mr. COHN. I object to that.

The COURT. I will overrule it. Go ahead.

The WITNESS. I did my work as a soldier and produced what I had to produce and there was no argument about my work, and since the information went to a supposed ally at the time, I had no qualms or doubts that I deserved the honorable discharge.

Q. And you felt at that time you were entitled to an honorable discharge; is that right?—A. That is right.

The COURT. Do you feel that way today?

Q. Do you feel that way now?—A. No; I don't.

Q. When did you change your mind as to whether or not you were entitled to an honorable discharge?—A. I never thought about it until this moment.

Q. Now that you have thought about it, do you believe that you were not entitled to an honorable discharge?—A. In the light of today's events, I was not entitled to an honorable discharge.

Q. Your honorable discharge has never been revoked, has it?—A. It has not.

Q. And so, on the record you still have an honorable discharge from the United States Army?—A. I have.

Q. Feeling as you do now, that you were not entitled to the honorable discharge, do you intend as of this moment, to ask the United States Government to revoke that honorable discharge?

Mr. Cohn. I object to that, your Honor, as incompetent, irrelevant, and immaterial.

The Court. I will sustain it.

Mr. E. H. Bloch. I respectfully except.

By the Court:

Q. Now, do you know, Mr. Greenglass, of your own knowledge, whether it is the policy of the Army not to move with respect to court martial or revoking of honorable discharges, while there are criminal proceedings pending in a court?—A. I have never heard of the Army taking a discharge, an honorable discharge, away from a man unless it was during the time, that it was brought out during the time of the service in the Army. That is my knowledge.

By Mr. E. H. Bloch:

Q. Mr. Greenglass, you now know——A. That is my knowledge.

Q. I can't tell when you finish. I don't want to interrupt.—A. Go ahead.

Q. You know now, don't you, that that honorable discharge was procured from the United States Government by you through fraud, because you concealed your illegal acts; isn't that correct?

Mr. Cohn. Your Honor——

The Court. I think we have gone into the subject enough.

Mr. E. H. Bloch. That was the last question on this.

The Court. I will sustain the objection.

Mr. Bloch. I respectfully except.

Q. Now, tell me, Mr. Greenglass, when did you first meet your wife?—A. I don't remember exactly when I met my wife.

Q. Can you give us the year?—A. I can't give you the year.

Q. Did you meet your wife in New York City?—A. I did.

Q. Do you remember where you met your wife for the first time?—A. We lived in the same neighborhood together for many years and I just don't remember when she started to go with our crowd.

Q. Well, is it a fair statement, then, that you have known your wife since childhood days?—A. I do.

Q. Did there come a time when you proposed marriage to her?—A. I did.

Q. When was that?—A. I——

Q. Approximately?—A. I don't believe I ever proposed marriage. It just was understood we were to be married.

Q. All right, when was there an understanding between you that you were going to get married?—A. I can't give you the year of that, either.

Q. All right. At any rate, when you began to see your wife in a fiancé and fiancée basis, did you love her?—A. I did.

Q. Did you love her when you married her?—A. I did.

Q. And do you love her today?—A. I do.

Q. And you have loved her during that entire time that we have mentioned in the last question; isn't that right?—A. I do.

Q. You love her dearly, don't you?—A. I do.

Q. Do you love her more than you love yourself?

The Court. Oh, I think——

The Witness. I do.

Mr. E. H. Bloch. I am satisfied with the answer.

Q. Do you love your children?—A. I do.

Q. Very much?—A. Yes.

The Court. I dare say that would be a difficult question for any of us to answer.

Mr. E. H. Bloch. I think so.

The Court. Do we love somebody more than we love ourselves?

Mr. Bloch. I could retort, but I am refraining. We are dealing with a peculiar kind of a witness.

Mr. Saypol. I object to that.

Mr. Cohn. I move that be stricken.

The Court. Yes; that will be stricken. I suppose all the Government witnesses are peculiar in the eyes of the defense.

Mr. Cohn. Yes, and vice versa.

Mr. E. H. Bloch. Now, I could say something but I am refraining.

The Court. All right.

Q. Do you bear any affection for your brother, Bernie?—A. I do.

Q. Do you bear any affection for your sister, Ethel?—A. I do.

Q. You realize, do you not, that Ethel is being tried here on a charge of conspiracy to commit espionage?—A. I do.

Q. And you realize the grave implications of that charge?—A. I do.

Q. And you realize the possible death penalty, in the event that Ethel is convicted by this jury, do you not?—A. I do.

Q. And you want to tell——

The COURT. Do you realize also that the matter of penalty is a matter entirely within my jurisdiction, not within the jurisdiction of the jury?

The WITNESS. I understand that, too.

Mr. E. H. BLOCH. That is why I used the word "possible," your Honor.

Q. And you bear affection for her?—A. I do.

Q. This moment?—A. At this moment.

Q. And yesterday?—A. And yesterday.

Q. And the day before yesterday?

The COURT. Well now, how far are you going to go?

The WITNESS. As far back as I ever met her and knew her.

Q. I am sorry, I can't hear you.

Mr. COHN. Mr. Bloch can't hear, your Honor, because he keeps interrupting the witness' answer.

Mr. E. H. BLOCH. Maybe you are right.

The COURT. Well, let's not take this day by day. Let's ask the general question.

Q. At any rate, at present you bear an affection for Ethel?—A. I do.

Q. Do you bear an affection for your brother-in-law Julius?—A. I do.

Q. Now, you testified, I believe, that you had two brothers and one sister.—A. That is right.

Q. Who is the other brother?—A. Samuel.

Q. Is Samuel a full brother of yours?—A. As far as I know.

Q. From the same mother and the same father?—A. That is right.

Q. And Bernie is from the same and mother and same father?—A. That is right.

Q. And Ethel is from the same mother and the same father?—A. That is right.

Q. And were you and Ethel brought up in your parents' home together?—A. Certainly.

Q. In New York City?—A. That is right.

Q. Sheriff Street?—A. That is right.

Q. And did you continue to live in that same house together with your parents until Ethel was married?—A. That is correct.

Q. And how old was Ethel at the time she was married to Julius?—A. It was 1939; it is 10, 11 years ago. I guess she was about 22.

Q. And how old were you at the time?

The COURT. Well, you said she was six years older. You must have been 16

The WITNESS. No, I was about 17, I guess.

Q. You were about 17?—A. Yes.

Q. And after your sister Ethel——

Mr. E. H. BLOCH. Withdrawn.

Q. Now, after your sister Ethel married Julius, did she move out of your mother's house?—A. She did.

Q. And the house where you resided?—A. She did.

Q. And did you continue to see Ethel and your brother-in-law Julius after Ethel moved out of the house?—A. I did.

Q. Did you visit them at their home?—A. I did.

Q. Now, do you remember when your sister Ethel and Julius moved into an apartment at Knickerbocker Village?—A. Do I remember the date?

Q. Yes.—A. I don't remember the exact date.

Q. Do you remember the year?—A. I don't.

Q. If I told you in 1942, would that refresh your recollection?—A. It still wouldn't refresh my recollection.

Q. All right. After you were married to your wife, did you and your wife see Ethel and Julius at your mother's house about once a week?—A. I wouldn't say that. I know——

Q. About?—A. I wouldn't even say that, because I worked——

Q. Did you continue——

The COURT. Let him finish.

The WITNESS (continuing). Because I worked nights at the time, and I saw very few people at the time, but I did see them frequently.

Q. You did see them frequently?—A. Yes.

Q. Did they visit you at your house after you were married?—A. No.

Q. Hardly at all; isn't that right?—A. That's right.

Q. Did you come to their house to visit them?—A. I did.

Q. Frequently?—A. I did.

Q. Now, will you describe to the Court and jury the type of apartment that Ethel and Julius lived in, in Knickerbocker Village?—A. Well——

The COURT. When?

Q. From 1942 on?—A. It was——

Q. Let me clarify it for the Court. Is it not a fact, from your own knowledge, that Ethel and Julius lived in that same apartment in Knickerbocker Village from 1942 right down to and including the date of your arrest in June 1950?—A. That is correct.

Q. Now, will you describe to the Court and jury the kind of an apartment it was, in terms of number of rooms, kind of furniture, and any other matters that you would care to describe about that apartment?—A. Well, as to the furniture, the furniture changed from time to time; she bought extra pieces and stuff of that nature, so I don't recall all the different ways the furniture was set.

The COURT. First tell us about the rooms?

The WITNESS (continuing). The rooms themselves—the apartment was at the end of a long hall, on the 11th floor, in Knickerbocker Village; G–11 is the number.

Q. 10 Monroe Street?—A. 10 Monroe Street. You come through the door; there is a small foyer; you turn to the left. It opens up into a living room, much longer than it is wide, and on the right is a door at the entrance to the living room—I should say, at the foyer to the right there is a door to the bathroom, and in this doorway there is a bathroom, and then through the doorway is a bedroom, and at the end of the living room, on a line with the wall, the wall where the windows are at, is a kitchen.

Q. All right, so that the apartment consisted of a living room, a bedroom, a kitchen, and a bathroom?—A. That's right.

Q. Now, in connection with the kind of furniture that you saw in that apartment, did you notice or did you ever make any remark to Ethel or Julius about the fact that their furniture was second-hand furniture?—A. I never mentioned it to them, but I was in the presence of people who did.

Q. And from your own observation, could you tell this Court and jury whether or not the furniture in their apartment was second-hand furniture, in the main?—A. Some of it was new.

Q. And some of it was second-hand?—A. Some of it was second-hand.

Q. Did you ever know how much rental Julius and Ethel paid for that apartment?—A. I believe I knew, but it slipped my mind. I can't recall it, right now.

Q. Well, if I told you that for many years they paid $45 a month rental, would that refresh your recollection?—A. Yes; that is probably the number, the price.

Q. And, as a matter of fact, while you were in business with Julius, first in the G & R Engineering Company and later in the Pitt Machine Products Co., Inc., there was talk between you and Julius and Bernie about how you were going to get along and survive?—A. That's right.

Q. Isn't that right?—A. That's right.

Q. And in the course of that conversation some talk was made about the various expenses that each of you had to pay, rent, food, and so forth; is that right?—A. That is correct.

Q. Now, does it refresh your recollection as to how much rental Julius and Ethel paid for that apartment at 10 Monroe Street?

The COURT. He said he thought you were approximately right.

Mr. E. H. BLOCH. All right.

Q. Do you feel any remorse now for what you did down at Los Alamos?—A. I do.

Q. Did you feel any remorse, or were you contrite or penitent at or subsequent to the time you did the things you said you did down at Los Alamos?—A. I was.

Q. Is there a change in the quality of your feeling of remorsefulness?

The COURT. Oh, I will sustain that.

Q. Or has there been?

The COURT. I will sustain that objection.

The WITNESS. I don't understand what you mean.

The COURT. Just listen to me. I have sustained the objection.

Mr. E. H. BLOCH. I understand. I am not asking it.

Q. Did you feel more remorseful now than you did down at Los Alamos, with respect to the things you did?

The Court. I will sustain that objection.
Did you have different degrees of remorse?
The Witness. Remorse is remorse.
The Court. All right.

By Mr. E. H. Bloch:

Q. You had remorse at that time?—A. I did.

Q. That is right, and you are remorseful now?—A. That's right.

Q. Now, let's come back to the first time the FBI men came down to see you in June of 1950; was that the first time that you had been interviewed by any of the representatives of the United States Government, including representatives of the FBI, with respect to your activities at Los Alamos?—A. No.

Q. When was the first time after your discharge from the Army that the FBI came around to talk to you about your activities at Los Alamos?—A. In 1950, February.

Q. Was that before or after there was any discussion, as you stated in your direct, about your getting out of the country?—A. I can't place it as to whether it was before or after.

Q. Where did these FBI representatives see or speak to you in February 1950?—A. One man called me up on the phone and he said he would like to see me. He came to my house; he sat down at my table; I offered him a cup of coffee and we spoke—he did not say to me that he suspected me of espionage or anything else—he just spoke to me about whether I had known anybody at Los Alamos, and that was the gist of the whole conversation. He walked out of the house maybe an hour later, and that is all there was to it.

Q. All right now, let's see. Did he introduce himself as a member of the FBI?—A. He did.

Q. Did he ask you any questions, either directly or indirectly, with respect to your knowledge of any illegal activity that occurred at Los Alamos while you were there?—A. I don't recall exactly what the whole conversation was about. It made very little effect on me, because it didn't—I mean, it didn't seem like anything—I mean——

The Court. Were you asked to sign any paper of any kind?

The Witness. No: not at all.

Q. You say you were together with that FBI man for an hour?—A. There was also my wife in the room and my little boy, and I think maybe a conversation might have gotten off on a couple of tangents.

Q. Well, at any rate, whether your wife and children was there, he stayed an hour, did he not?—A. Yes.

Q. Did he tell you what he came for?—A. Well he wanted to know——

Q. Did he tell you what he came for?

Mr. Cohn. He is trying to answer, your Honor.

Mr. E. H. Bloch. It is either a "yes" or "no." I will follow it up.

The Court. Mr. Bloch is a very impatient young man and he wants his answer fast.

Mr. E. H. Bloch. Thanks for the compliment, for calling me young.

The Court. If you will just wait a little while you will get an answer.

The Witness (continuing). He discussed with me—when he came into the house it was very difficult to find out what he wanted. He didn't come out and say that he wanted some information. He just talked around the point. I didn't get what he really wanted to find out.

Q. Was Los Alamos discussed?—A. Oh, yes.

Q. Did he say he was looking for people who might have conducted illegal activity in Los Alamos?—A. No; he just wanted——

Q. He didn't say that?—A. No; he didn't say anything of that kind.

Q. What was the name of that FBI man?—A. I can't remember that FBI man's name.

Q. Did you ever see that FBI man since?—A. I have not seen him since.

Q. Were you frightened at the time that FBI man came down to see you in February 1950?—A. Well, I wasn't exactly calm.

Q. You were frightened, were you not?—A. Not very frightened; no.

Q. I want to ask you again, was Los Alamos mentioned at that first interview that you had with that FBI man in February 1950?

The Court. He did say "yes," it was.

The Witness. Yes.

Q. You didn't tell that FBI man at that time that you had engaged in any illegal activity at Los Alamos, did you?—A. I didn't tell him, but I was pretty well on the verge to tell him.

Mr. E. H. Bloch. I move to strike out the latter part of the answer as not responsive.

The Court. No; it will stand.

Mr. E. H. Bloch. I respectfully except.

Q. Did you mention to him anything about the fact that you had given secrets of the Los Alamos project to any unauthorized person?

The Court. Don't answer that. He has already answered it.

Q. Did you mention the name of your sister Ethel Rosenberg to that FBI man in February 1950?—A. I don't believe I did.

By the Court:

Q. Was there any discussion about Ethel Rosenberg?—A. I don't think she came into the conversation.

Q. I believe you said you weren't asked anything about espionage, were you?—A. No.

Mr. E. H. Bloch. I respectfully except to your Honor's question, and I hesitate to ask the following questions because they are on the same line, and if the Court wants, I will make an offer of proof. I am going to ask him——

The Court. No; you go ahead.

Mr. E. H. Bloch. All right.

By Mr. E. H. Bloch:

Q. Was the name of Julius Rosenberg mentioned by you at that interview with the FBI man in February 1950?—A. I don't believe that there was. I don't remember the conversation that well.

Q. Was your wife's name mentioned?—A. She was sitting right there.

Q. In connection with any activity of an illegal sort, to that FBI man, in February 1950, by you?—A. Of course not. I didn't mention anything like that to the FBI man.

Q. All right. Now then, when was the next time you were interviewed by the FBI?

The Court. And you say that you didn't make any written statement and you weren't asked to sign one?

The Witness. No, nothing; no.

Q. When was the next time you were interviewed by the FBI?—A. When I was arrested, June 15th.

Q. Now, you say that four FBI representatives came around to your house on or about June 15, 1950, at or about 2 o'clock in the afternoon; is that right?—A. That's right.

Q. And they stayed in your house until about 7:00 or 7:30 that evening?—A. That's right.

Q. And during that time they conducted a search of your house?—A. They did.

Q. And during that 5- or 5½-hour period they asked you questions, did they not?—A. They asked me some questions; yes.

Q. Did all of them ask you questions at some point or another?

Mr. Bloch. I will withdraw that.

Q. Did all participate at some point or another in asking you questions during your five- or five-and-a-half hours interview?—A. Well, I don't remember them all standing around me and asking me questions. I just remember that once in a while one of them would ask me a question while they were doing their work.

Q. Now, when they first came in, do you remember who was the first FBI representative to speak?—A. Well, I believe it was John W. Lewis was the first one.

Q. And what did he say, the first words, when he came in?—A. I can't remember, probably "Hello" or "How are you?" Or "Is your name David Greenglass?"

Q. Aside from these preliminaries, what did he tell you about why they came?—A. He said they came—well, this isn't the first words they said. He said they came in connection with leak of information in the security of their—of the United States Government, and they would like to find out what I know about it.

Q. Was the name Harry Gold mentioned during that five- or five-and-a-half-hour interview?—A. No.

Q. Not once. When was the first time that any United States representative brought in the name of Harry Gold?—A. He didn't bring in the name. He just told me of a man that came to see me.

Q. And did he mention where the man came to see you?—A. In Albuquerque.

Q. Did he mention the month he came to see you?—A. Yes; he did.

Q. Now, as a matter of fact, Mr. Greenglass, you had read in the newspapers prior to June 1950 about the arrest of Harry Gold, had you not?—A. I had.

Q. And when you read that information in the newspapers, did you recognize the person who was mentioned as Harry Gold in the newspapers, as the person whom you transferred information at Albuquerque, New Mexico, in June 1945?—A. I did not remember his face at the time.

Q. Now, you said that Harry Gold came to your apartment in Albuquerque, New Mexico, one morning in the early part of June 1945; is that correct?—A. That's right.

Q. Is that right?—A. That's right.

Q. Your wife was there?

The COURT. I think you said it was a Sunday morning.

The WITNESS. He said "some morning in early part of June." O. K., it was a Sunday morning.

Q. A Sunday morning?—A. Yes.

Q. And then you told us about how there was an identification through the Jello box cut; is that correct?—A. That's right.

Q. Then you had a conversation with Harry Gold after that identification had been completed; is that correct?—A. That's right.

Q. Now, where did this conversation take place, in your living room?—A. Yes.

Q. About what time of the morning would you say it was?—A. About ten o'clock, I suppose. I can't place it accurately.

Q. How long did Harry Gold stay in your apartment that morning?—A. Not very long.

Q. About how long?—A. Altogether the whole meeting was about 20, 25 minutes, including the walk, so it couldn't be very long.

Q. At any rate, you were with him for 20 or 25 minutes that June Sunday morning in 1945, both in your apartment and outside in the street; is that right?—A. Yes.

Q. You had a very good chance to look at his face, didn't you?—A. Very good.

Q. You did look at his face?—A. I did.

Q. During that entire 20 or 25 minutes?—A. That is right.

Q. Was there any attempt on his part, so far as you could observe at that time, to conceal his face?—A. No.

Q. It was sunlight, was it not?—A. A bright day.

Q. And Albuquerque usually is a very bright town, is it not?—A. Yes.

The COURT. Don't go into all these details. He said he could have seen it and so forth. You can go on.

Q. Now, that afternoon you also saw Harry Gold, did you not?—A. I did.

Q. That was after you had written down certain information?—A. That's right.

Q. Then he came back and you gave it to him?—A. I did.

Q. That is the time he gave you the $500 in the envelope; is that right?—A. That is right.

Q. Did you have an opportunity to look at his face in the afternoon?—A. I did.

Q. Did you read any account in the newspaper concerning this Harry Gold prior to June 15, 1950, which led you to believe that the person with whom you had dealings was already apprehended and under arrest by the Government authorities?—A. I never would have noted from the newspapers. Julius Rosenberg came and told me that that was the man, and after that I read the newspapers, because I hadn't seen him even when he came up and showed it to me.

Q. Now, you hadn't read the newspapers?—A. Until he had showed it to me.

Q. Until he came around; is that the idea?—A. That is right.

Q. Now, when did you say Julius Rosenberg came around to tell you about the fact that Harry Gold had been arrested—A. It was in the morning.

Q. Do you remember when?—A. It was the day after my wife got out of the hospital with her baby.

Q. And when was that?—A. That is the way I place it. It can be found out. I just don't remember.

Q. Now look——A. I can place it that way.

Q. Now look, your wife gave birth to a baby——A. That's right.

Q. June 16th; six days later she was out of the hospital. The next morning——A. It was May 16th.

Q. Your child was born when?—A. May 16th.

Q. And how long did your wife stay in the hospital after your child was born?—A. Six days.

Q. So that would bring us down to about May 22nd; is that correct?—A. And the next morning is when he brought the paper in.

Q. So you would say that that is about May 23rd?—A. Right.

Q. And you had read nothing——A. I read nothing.

Q. In the newspapers? Let me finish my question—you had read nothing in the newspapers prior to May 23, 1950, concerning the arrest of Harry Gold, or any accounts concerning his activities?—A. I had read nothing about it.

Mr. BLOCH. All right, that is an answer.

Q. Now, then, in this interrogation at your home on June 15, 1950, by these four FBI representatives, did you mention your wife as one of the people who conspired with you to commit illegal acts?—A. I mentioned no one at my home.

Q. You denied, did you not, at that time, to the FBI at your home, that you were guilty of any crime, or that you had engaged in any illegal activity?—A. I denied nothing. They didn't ask me direct questions and I gave answers to the questions they did ask.

Q. Well, let's hear, let's hear some of the questions that they asked and let's find out whether they were directed——A. It was later at night.

Q. No, no. We will come to the night; we will come to the night. Let's stick to the afternoon for a little bit.—A. They wanted to know things, when I had been in the Army, where I had been.

Q. Did you tell them?—A. I did.

Q. All right, what else?—A. What I was doing in the Army.

Q. Did you tell them—A. I did. What I worked at, where I worked.

By the COURT:

Q. Well, did you answer all the questions that were put to you, truthfully?—A. Truthfully, yes.

By Mr. E. H BLOCH:

Q. You didn't lie about a single question during that five and a half hour interval?—A. No. As a matter of fact, I volunteered information.

Q. Not one——

The COURT. He didn't say it was a five and a half hour interview. He didn't say it was five and a half hour interview. Be careful about that. He said he was only questioned for about an hour.

Mr. BLOCH. But the interval was from 2 to 7:30.

The COURT. I don't like the implication.

Mr. BLOCH. I didn't mean to leave any implication, your Honor.

The COURT. All right.

Q. They were there for five, five and a half hours?—A. That is right.

Q. During part of that period they questioned you?—A. Yes; it was the type of questioning that was "Oh, by the way," while they were doing their work.

Q. Now, when they searched your apartment, did you realize that you were a suspect?—A. I would have to be awful dumb not to.

Q. Did you mention during that period any names of people, that you mentioned in your direct testimony here in court? A. I don't recall that I did, except that at one point they found this paper, Julius's notes from high school, and they asked me what it was.

Q. Is it from high school or college?—A. College, I mean. And they saw some math, and they said to me, "What is this? Is this some secret material?" I said, "No, that is notes that my brother-in-law did in—wrote about while he was in college."

Q. In other words, they asked you about some of the notes that Julius Rosenberg had made in college and specifically asked you whether or not that was secret material?—A. They didn't put it that way.

Q. That is, I am using your words.—A. They said, "What is this?" In other words, they wanted an explanation of what the paper was.

Q. Did you give them an explanation?—A. Yes; I told them what it was.

Q. Did you, during that five or five and a half hour period, ask them to give you an opportunity to consult with counsel?—A. I didn't ask them; no.

Q. Did you ask your wife to try to get you counsel?—A. My wife was in the hospital.

Q. Did you call up anybody——A. I did not.

Q. Whether it be a relative or a friend——A. My——

Q. To ask them to procure counsel for you?—A. No, I didn't.

Q. During that five and a half hour period?—A. I did not.

Q. Then, as I understand it, you were taken by these four FBI representatives down to the Federal building here?—A. That is right.

Q. Did you come down here by automobile?—A. That is correct.

Q. Whose automobile?—A. I suppose it belonged to the Federal Bureau of Investigation, I don't know.

Q. It was a private automobile, looked like a private automobile; is that right?—A. Well, it had a radio in it.

Q All right. Finally you got down here. Where did you go when you got down to this building?—A. I went up to the FBI headquarters.

Q. And what floor is that on?—A. It is in the 20's, I guess it is the 29th floor.

Q. 29th floor?—A. Yes.

Q. You have been up on the 29th floor on many occasions, haven't you?—A. Twice.

Q. Now, the first time was this time that we are talking about now. When was the next time?—A. Some weeks ago.

Q. Did these four FBI representatives go up to the 29th floor with you?—A. They did.

Q. Did you have dinner that night?—A. They let me eat my supper before I left.

Q. Would you mind talking a little louder, please?—A. I said, they let me eat my dinner before I left.

Q. Now, after you had your dinner at your home, about what time was it?—A. It was about four, five o'clock in the afternoon.

Q. Did you have anything further to eat that evening?—A. About—I guess, about nine, ten o'clock at night I had some more.

Q. Did any of the FBI representatives in your apartment make any notes or memoranda while you were talking?—A. I can't say for sure whether they did or didn't.

Q. Now, when you got down here to the Federal building in the evening, did you notice whether in the course of the interview any of the FBI representatives made any notes or memoranda?—A. Yes.

Q. Was a stenographer present? When I say a stenographer, I include a stenotypist or anybody who will take down words in a short-cut way. Was any stenographer present during that interview that evening with the FBI representatives?—A. When I told them I would give them a statement, then a stenographer came in.

Q. Now, what time did you tell them that you were going to give them a statement?—A. I don't know exactly what time it was.

Q. Did the room in which you were interviewed have a clock?—A. You have got me there. I don't know either.

Q. Let's see whether we can refresh your recollection. You said you had something to eat around nine or ten o'clock; is that right?—A. I suppose it must have been around that time.

Q. Approximately?—A. I heard some——

Q. If you don't have a clock, we don't expect you to know.—A. No; there is a bell outside that rings certain chimes, I mean, I thought it was about 9 or 10 o'clock.

Q. That is your best understanding. Now, did you tell the FBI that you were going to give them a statement after you had had this repast, at about 9 or 10 o'clock?—A. I don't know when I gave them the statement, when I started to give the statement.

Q. No; when you said to them that you would give a statement?—A. You can't pin point me on when I said I was going to give a statement, because I don't remember those things.

Q. You don't remember that?—A. No.

Q. Were you frightened at the time?—A. No.

Q. Did you ask them to give you an opportunity to get counsel?—A. No.

Q. Did you make any attempt to get counsel?—A. Not until about 2 o'clock in the morning, 1, 2 o'clock in the morning.

By the COURT:

Q. Did they tell you that you could have counsel?—A. Oh, yes.

By Mr. BLOCH:

Q. Did any of those FBI representatives tell you that you were under arrest?—A. No.

Q. In the course of your questioning, did you at any time ask them whether you were under arrest?—A. No, I didn't.

Mr. BLOCH. I don't know whether I asked this, your Honor. My mind is a little fuzzy.

Q. Did you make any attempt to procure counsel that evening?—A. I don't believe so.

The COURT. Did I understand you to say not until 1 or 2 that morning?

The WITNESS. That's right—well, you can consider that "evening."

Q. Well, outside of the time consumed in taking this repast, about 9 or 10 o'clock, would you say that you were continuously questioned by these FBI representatives, from the time you got down here, about 7 or 7:30, to the wee hours of the morning?—A. Some of the conversation was just window dressing; it had nothing—it wasn't pertinent.

Mr. E. H. BLOCH. I move to strike that out, your Honor.

The COURT. No; it will stand.

By The COURT:

Q. What do you mean by that?—A. I mean, the man asked me how I felt, "Do you want a cigarette?" Things like that.

By Mr. E. H. BLOCH:

Q. They were very affable, is that right, they were very polite?—A. That's right.

Q. Is that what you meant by "window dressing"?—A. Also, that he would—I mean, that they would talk about other things, other than what was on hand.

Q. Did those interspersions about other things take very long?—A. Five, ten minutes some time.

Q. Now, outside of this five or ten minutes——A. I said "sometimes." By that I mean that during the conversation it would be broken by these little——

Q. But is it fair to say that outside of these indulgences in the social amenities, you were continuously questioned from the time you came down here at 7 to 7:30 until the wee hours of the morning?—A. Well, if you can consider giving a statement, questioning, yes.

Q. That evening, did you at any time mention your wife's name to the FBI as being one of those who was engaged with you in this illegal work?——A. I did.

Q. And did you mention the name of your sister Ethel at that time?—A. I believe I did mention her name, yes.

Q. Are you sure?—A. I can't remember now what I gave in a statement then. In subsequent statements, I have.

Q. No, no; we will come to subsequent statements. Now please try to concentrate and fasten your mind on this particular—— A. Since I haven't read that statement——

Q. On this particular evening.—A. I haven't read that statement since and I certainly don't know exactly what I put into it.

By the COURT:

Q. Well, did you conscientiously withhold any facts that night?—A. No, I did not conscientiously withhold those facts.

Q. And did you conscientiously tell substantially what you have told in court these past few days?—A. That I did, and in other statements, because I couldn't remember at once.

Q. Well, when you left that evening, was there any understanding that you you would make a subsequent statement?—A. I suppose there was.

Q. I don't want to know whether you suppose.—A. Yes; there was an understanding to that effect. I said to them, "That is to the best of my knowledge at this time."

By Mr. E. H. BLOCH:

Q. Now, let me ask you, so that there will be complete clarification here, and I am confining myself solely to this particular evening of June 15th——

The COURT. Yes; and I wish you would make a little haste with it, because I think you are taking much too long on this subject matter. I haven't interrupted you; I have let you go ahead. I think you ought to make more haste on it, because if you are going to take every one of those incidents—and I think I know what you are leading up to—and you are going to go through it in that detail——

Mr. E. H. BLOCH. I am not going to go through it in that detail insofar as subsequent statements are concerned, but I would like to know particularly about this first statement, the first few days after his apprehension and arrest.

The COURT. Very well.

Q. Are you now stating that you did not withhold conscientiously any information concerning your illegal activities at Los Alamos and elsewhere to the FBI authorities on the evening of June 15, 1950, and the early hours of the morning of June 16, 1950?—A. That is substantially what I mean.

Q. All right. Now, after you finished with FBI representatives, you were taken away from this building, were you not?—A. That's right.

Q. Now, what time were you taken away?—A. Oh, that was—it was 4 o'clock the following afternoon. It was 2 o'clock in the morning that we got through—I mean, so that we could sleep.

Q. Now, where did you sleep that night?—A. I slept in this building, in the nurse's quarters.

Q. Now, by 2 o'clock in the morning had you already given a statement to the FBI?—A. Yes, I had.

Q. Had you signed that statement?—A. Surely.

Q. You did?—A. Yes.

Q. How long in all would you say it took to prepare that statement and have you sign it?—A. I remember remarking that the stenographer was an awful—the typist was an awful slow typist.

Q. And did the stenographer go to another room and type it, or did she type it right there and then did you sign it?—A. I remember her typing something in that room and typing in another room, too, I suppose.

Q. Well, was there one—more than one stenographer or typist?—A. I don't know—no, only one, I believe, just one.

Q. Let us see whether we understand. At any rate, you were asked questions; you gave certain answers; there were a stenographer and a typist there, and whatever you said was put down in writing, is that right, was put down in print?—A. It took quite a while, because I wrote it out in longhand.

Q. Oh, I wanted to get that. You wrote out a statement in your own handwriting?—A. That's right.

Q. Did you sign that statement?—A. I signed that statement.

Q. And did the statement that you wrote out and signed in your own handwriting precede or come subsequent to the time when the stenographer came into the room?—A. I really don't know how she took it. I think maybe she read my notes and typed it off the notes.

Q. You are not sure of that?—A. I am not sure of that.

Q. At any rate, at 2 o'clock in the morning you were all through?—A. Yes.

Q. And you slept in this building?—A. That's right.

Q. Did you stay in this building the following day?—A. Up until I was arraigned and taken over to West Street.

Q. And when were you arraigned?—A. Do you mean the time? It was in the afternoon, I believe.

Q. And that was on June 16th——A. That is right.

Q. 1950 Was that down in the Commissioner's office in this building?—A. That's right.

Q. Did you have counsel at that time?—A. I did.

Q. Who was your counsel?—A. O. John Rogge.

Q. And was Mr. Rogge there personally?—A. He was.

Q. Had you seen any members of Mr. Rogge's firm prior to the time that you saw Mr. Rogge on June 16, 1950?—A. I did.

Q. And who of Mr. Rogge's firm did you see?—A. Mr. Fabricant.

Q. And when did you see Mr. Fabricant?—A. It was in the morning of June 16th.

Q. About what time?—A. I suppose it was during business hours, between 9 and 12 sometime.

Q. Now, did you make a request for counsel at any time after you were taken down here to the Federal Building on June 15, 1950?—A. At about one o'clock or so somebody said to me, "You ought to get a counsel—you ought to get a lawyer."

Q. Was that after you had written out your statement and signed it?—A. Yes.

The COURT. Let me ask you, were you advised before that you had the right to have a lawyer?

The WITNESS. Sure.

The COURT. Were you advised of that fact when you were taken to the building for the first time?

The WITNESS. No.

The COURT. When were you advised of that fact, before you made the statement or during or just when?

The WITNESS. Just about when I started to make the statement.

Q. Now, when you started to make that statement you knew that Gold had put his finger on you, didn't you?—A. I don't get what you mean.

Q. Never heard that expression "put the finger on you"?

The COURT. No; don't answer that.

Mr. E. H. BLOCH. All right, I will withdraw it.

Q. During the course of your interrogation at the Federal Building here that evening on February 15, 1950, did any of the FBI representatives or any representatives of the United States Government tell you that Harry Gold had charged that you were conspiring with him or that you were a conspirator with him in committing espionage?

Mr. COHN. Mr. Bloch said "February." I believe he meant "June."

The COURT. He means June, that is correct.

Mr. E. H. BLOCH. "June" — I am sorry, that is correct.—A. Nobody told me that.

Q. What?—A. Nobody told me that.

Q. You knew——A. They said——

Q. Just a second. You knew when you came down here to the Federal Building on June 15, 1950, that Harry Gold was already under arrest?—A. I did.

Q. Now, in the course of your interrogation on the evening of February 15 of 1950—I am sorry——

The COURT. June.

Q. June 15, 1950, here, did you mention the name Harry Gold first or did somebody representing the Government mention the name Harry Gold first?—A. I don't recall that.

Q. But his name was mentioned?—A. I told—I can't even recall that.

Q. Did anybody tell you at that time that a complaint had either been made or was being made charging you with conspiring with Harry Gold to commit espionage?—A. Nobody.

Q. Now, the following morning you say Mr. Fabricant saw you?—A. He did.

Q. Did you call any relative on the evening of June 15th or in the early morning hours of June 16th from this building, telling him or them that you wanted counsel?—A. I did.

Q. And to whom did you speak?—A. I spoke to my brother-in-law, Louis Abel.

Q. Was that after the stenographer or typist had taken down what you responded to——A. That was after I signed the statement.

Q. The questions propounded to you by the FBI?—A. That was after I signed the statement.

Q. Now, when you were arraigned before the Commissioner on June 16, 1950, in the afternoon, did you know then that you were accused by the United States Government of conspiring with Harry Gold to commit espionage?—A. Repeat that.

The COURT. Read that, Mr. Stenographer.

(Question read.)—A. Nobody ever read it to me until I got in front of the magistrate.

Q. Now, when you got in front of the Commissioner down here in the basement of this building, did somebody read to you the complaint?—A. The District—the U. S. Attorney read the complaint. That is the first I heard of it.

Q. Did Mr. Fabricant of Mr. Rogge's office tell you what the United States Government was charging you with on that day?—A. I——

Q. Pardon me?—A. I don't believe that I can remember that part of the conversation at all.

Q. Was the name Harry Gold mentioned in the conversation you had with your lawyer, Mr. Fabricant, on the morning of June 16, 1950?—A. I can't tell and I don't remember.

Q. You don't remember?—A. No.

Q. And was Harry Gold's name mentioned to you by Mr. Rogge when you saw Mr. Rogge later that day, at the time of the arraignment?—A. You are asking me about a period that was very confused for me, and it is a wonder that so much stayed with me about it.

Q. You don't remember that——A. No, I don't remember that.

Q. What happened in June 1950?—A. No.

Q. But you do remember everything that your wife told you back on November 29th——

The COURT. I will sustain the objection.

Q. 1944?

The COURT. Don't answer that.

Mr. E. H. Block. I respectfully except.

The Court. You know it is argumentative, counselor.

Mr. E. H. Block. I am not at all sure of that, your Honor.

The Court. Well, I am sure of it and I don't want any further argument on it.

Mr. E. H. Bloch. You asked me a question, your Honor, and I tried to respond.

The Court. All right.

Q. At any rate, the Commissioner or Mr. Saypol or somebody before the Commissioner at that arraignment did have read to you or read to you the complaint?—A. That is right.

Q. And that complaint charged you and Mr. Gold, did it not, with conspiring to commit espionage in that you delivered to Harry Gold confidential information from the Los Alamos project and received a sum of money for it in the early part of June 1945?—A. I didn't pay much attention to what the complaint said and you can—I—I don't even believe I heard the words in the complaint.

Q. You didn't hear the words?—A. No.

Q. Were you excited at the time?—A. No, I was just dull.

Q. You were dull?—A. Yes.

Q. Were you dull because you didn't sleep very much the previous evening?—A. It could be one of the contributing factors.

Q. How about now; do you feel sharp?—A. Sharp enough.

Q. Yes.

Mr. E. H. Bloch. I wonder, if the Court please, if this might be a convenient place to stop?

*   *   *   *   *   *   *

We will recess until 2:25.

(Recess to 2:25 p. m.)

AFTERNOON SESSION

David Greenglass resumed the stand.

*   *   *   *   *   *   *

(The following in open court, in the presence and hearing of the jury:)

Cross-examination continued by Mr. E. H. Bloch:

Q. Now, Mr. Greenglass, let's come back for a moment to the night of June 15, 1950, and the early morning of June 16, 1950. How many statements did you sign that evening and the following morning?—A. I signed one statement that evening and the following morning, I mean in this period.

By the Court:

Q. You mean one each time or one?—A. It was one statement. There must have been some copies or something but it is the same statement.

Q. One statement was signed by you?—A. That is right.

By Mr. E. H. Bloch:

Q. And that was the one that you wrote out in your own handwriting; is that correct?—A. That is correct.

Q. Now, in that statement did you refer to the incident in Albuquerque, New Mexico, in the latter part of November 1944, in which you had a conversation with your wife and in which she invited you to commit espionage?—A. I referred to it, yes. I also, in that statement, I gave a general outline of everything I was to say——

Mr. Bloch. Now, if the Court please——

The Witness (continuing). And later statements. I just made that statement as I remembered it then, and any subsequent statements, I had more memory of what I had done and I filled in more.

Q. Now, I understand that, but please, Mr. Greenglass, we are trying to find out specifically just what you said to the FBI.

The Court. Yes; and I don't think you ought to go into such minute detail. Now, I thought we were going to have this long recess for the purpose of trying to shorten this a bit, but instead I notice you are going back again, going back over the thing more minutely than you did before.

Mr. Bloch. All right. I want to make myself clear to the Court. I know that the Court understands the purport of these questions. If I don't ask them in this form, then I may very well have waived my rights to lay the foundation for the introduction of certain documents.

The COURT. Yes, but don't you think you are laboring the point, really?

Mr. BLOCH. I honestly don't, Your Honor. I feel in good conscience that it is my duty to ask these questions. Otherwise, I assure you that I wouldn't ask them. This is not going to take long, Your Honor.

By Mr. E. H. BLOCH:

Q. Did you, in that statement of—we will call it June 15th or June 16th, it doesn't make any difference—down here in the Federal building, mention the incident where Julius Rosenberg is alleged to have come to your house a few days after your first furlough here in New York, in January 1945, and asked for certain information, and you later that evening wrote out that information?—A. I did.

Q. You did mention that?—A. Yes.

Q. And did you mention the fact that the following morning Julius Rosenberg came around and picked up that written information?—A. I don't remember if I did it or not.

Q. You don't remember. And did you mention in that first statement anything about being introduced by Rosenberg to a man on First Avenue, somewhere around January 8th or January 10, 1945?—A. I didn't place the time but I did introduce—say I was introduced to a man.

Q. That answers it. And did you in that first statement detail either roughly or with precision the incident which allegedly occurred in the Rosenberg home in September 1945 on your second furlough where your wife and Ethel, your sister-in-law, and Julius, your brother-in-law—your sister—were grammatically changing your verbiage of the report?—A. I did not make such a statement.

Q. And did you mention in your statement of June 15, 1950, or June 16, 1950, anything with relation to the Jello box incident?—A. I did.

Q. You did. And did you mention in your statement anything about the fact that sometime later—and here I am relying on my memory—whether it is 1946 or 1947 or 1948—at any rate, while you were in business with Julius, that you got an offer to go to college?—A. I don't know if I made it in that statement or not.

Q. You are not sure of that?

The COURT. You say in that statement, did you make it?

The WITNESS. I made it in subsequent statements; yes.

The COURT. Is there anything that you testified to here today that you haven't made in the previous statement?

The WITNESS. There isn't; no.

Q. At any rate, you are not sure as to whether or not you mentioned this invitation to go to college?—A. In the first statement?

Q. I am only talking about the first statement now.—A. That is correct.

Q. Is that your answer, you don't remember?—A. I don't remember.

Q. Did you mention to the FBI that you got $500 from Gold out in Albuquerque in June 1945?—A. I believe I mentioned money; yes.

Q. No—you mentioned money. Did you mention this specific sum of $500?—A. It is—I didn't remember the exact sum of money at the time.

Q. You knew on June 15, 1950, didn't you, precisely how much money you had gotten from Gold?—A. No; I didn't. I had forgotten it and it was just in the subsequent times when I thought it over I remembered how much money it was.

Q. But before June 15, 1950, and during the intervening period from June 1945—that is about five years—did you at any time know precisely how much Gold gave you in the early part of June 1945?—A. At certain times a man's mind is funny. Sometimes I will remember it and sometimes I won't later on. There was no reason to recall it in that period.

Q. You weren't a rich man in June 1945, were you?—A. I have never been a rich man.

Q. No; and $500 was a big sum to you, wasn't it?—A. Pretty big.

Q. And it enabled you and your wife to live and have some luxuries, didn't it?—A. It enabled us to live.

Q. All right. And you want to say now that you did not remember the precise sum of money that Gold gave you in the early part of June 1945?—A. I did not remember the precise sum of money and——

Q. All right, that is an answer?—A.—later times I did remember it and I did put it down.

Q. Now, did you mention to the FBI about $200 being given to you by Rosenberg in September 1945 in his apartment?—A. I did not.

Q. And did you mention to the FBI on June 15th and 16th that you had received a thousand dollars from Rosenberg in February 1950?—A. I did not say that at all.

Q. And did you mention to the FBI——A. Because it was not February 1950 that I received the thousand dollars.

Q. I beg your pardon—I am sorry, you are right. It was June, was it?—A. It was June—it was May 1950 I received it.

Q. Did you mention anything about that thousand dollars?—A. I mentioned it; yes.

Q. That you got a thousand dollars from Rosenberg?—A. That is right.

Q. Did you mention it——A. I did not. I don't believe it was in the statement but in subsequent statements it was there.

Q. Now, please, Mr. Greenglass, let us understand each other.—A. You are saying——

Q. I am talking now solely and directing my inquiries exclusively to what you told the FBI on the evening of June 15, 1950, and what went into that statement.

Mr. COHN. Your Honor——

The WITNESS. Now I——

Mr. COHN. Just a minute, Mr. Greenglass. I see a difficulty right there. He said "what you told the FBI and what went into the statement." I don't think it has been established that everything he told the FBI did go into the statement. It usually doesn't.

Mr. E. H. BLOCH. Mr. Cohn, I agree with you. If there is any unclarity about it in the record I want to clarify the record.

Q. Was there anything in your statement which you signed on the evening of June 15, 1950, here in the Federal Building about the receipt of $1,000 from Rosenberg in the year 1950?—A. I don't believe it was in the statement; no.

Q. All right.

The COURT. But you had told them about it?

The WITNESS. I had told them about it.

Q. And did you have incorporated in your statement or did you yourself incorporate in your statement anything about the receipt of $4,000 from Rosenberg in this brown bag which is marked "Government's Exhibit 10"?—A. I—I had told them about this—what they put in the statement, what they wanted me to put into the statement, what they wanted me to put into the statement in the first thing, they told me was just to make a general statement, that is all.

Q. Now, isn't it your testimony that you wrote out the statement?—A. That is right but it was getting quite late and I mean I am sure that I would have an-another opportunity to write more.

By the COURT:

Q. Do you remember which agent told you just to make a general statement?—A. Well, it was just a general statement.

Q. I say, do you remember which agent told you to make a general statement?—A. I don't remember which one said to make a general statement. It was just a general statement. I said, "Whatever I will remember more I will write as more"—I mean it was getting quite late at the time.

By Mr. E. H. BLOCH:

Q. Well, at any rate, whether it was late or early is it your testimony now that the $4,000 was not incorporated in your written statement of June 15, 1950?—A. I can't say that it was and I can't say that it wasn't. I don't remember if I did put it in or not.

Q. It may have been omitted?—A. It may have been put in, too.

Q. Now, is it your testimony that you mentioned the $4,000 to the FBI representatives that evening?—A. I did and I will tell you how I know that, because when they searched the house they couldn't find it and I told them then later on that I had given it away.

Q. Now, did you tell them that evening to whom you had given that money?—A. I don't remember that either.

Q. Didn't one of the FBI representatives when you mentioned the $4,000 ask you where the $4,000 is?—A. He must have. I just don't remember saying anything to him about where it was.

Q. Well, at that time you had given the $4,000, at least in terms of custody, to your brother-in-law, had you not?—A. I did.

Q. Did you want to conceal from the FBI the name of the person who was holding that money for you?—A. No; I wasn't concealing it.

Q. But you did, didn't you?—A. No; because when I called up my brother-in-law Louis Abel, I told him I was calling my brother-in-law Louis Abel to get in touch with a lawyer. I didn't conceal it from them at all.

Q. Did you tell them either directly or in the conversation that you had with your brother-in-law that evening from this building that your brother-in-law had $4,000 of your money?—A. Well, I tell you, I don't remember actually telling him that or not. I probably did tell it to him. I am not saying that I remember it or not.

Q. But you are not sure?—A. I am not sure that I did or I didn't.

Q. All right, that is an answer. Did you tell the FBI that evening anything about the gold watches and the console table that the Rosenbergs were alleged to have received from Russians?—A. Mr. Bloch——

Q. Well, now, if you don't remember——A. I don't remember.

Q. Well, then, say so. Then say so. I don't expect you to answer——

By the COURT:

Q. Is that all you wanted to say, Mr. Greenglass?—A. No. I wanted to say more but he doesn't allow me to.

Q. Well, I think I am running this courtroom. What is it you want to say?—A. Well, what I wanted to say is all these little details was something I remembered as time went on. It was just a few hours that I was there and I put down what I remembered without trying to conceal a thing.

Q. It wasn't your intention at that time to give every minute detail?—A. Not intention but I couldn't remember every minute detail that had occurred.

By Mr. E. H. BLOCH:

Q. How many questions——A. It is just beyond human ability to do so.

Q. All right, let us see. You got to this building at about 7.00 or 7.30, didn't you?

The COURT. Let us not review that whole story over again. I think we all have a slight bit of intelligence.

Mr. E. H. BLOCH. All right.

The COURT. We all know by now that he got here at 7.00 or 7.30.

Q. Now, did you disclose to the FBI directly——

Mr. E. H. BLOCH. I withdraw that.

Q. That evening, did you ask any of the FBI representatives to go easy on your wife?—A. I did not say any such thing.

Q. Now, Mr. Greenglass, your wife has never been arrested, has she?—A. She has not.

Q. And she has never been indicted, has she?—A. She has not.

Q. And she has not pleaded guilty to any conspiracy to commit espionage, has she?—A. She has not.

Q. And your wife is at the present time home taking care of your children; isn't that right?—A. That's right.

Q. Is that why you hired Mr. Rogge or Mr. Fabricant?—A. I don't see where one question has to do with the other.

Q. I am asking you.

Mr. E. H. BLOCH. I move that the answer be stricken out.

The COURT. The answer is stricken out.

Answer his question.

The WITNESS. I don't understand what you mean.

Mr. E. H. BLOCH. Would you mind rereading the question?

The COURT. Well, he doesn't understand your question. Maybe you can rephrase it for him.

Q. You wanted a lawyer, did you not?—A. That's right.

Q. And you wanted a lawyer, not before you signed a statement, but after you signed a statement; isn't that right—A. I just—it just came up at that time that we were free to get a lawyer and I got a lawyer at that time.

Q. And you were free to get a lawyer, at least you felt free to get a lawyer at about 2 o'clock in the morning, after you had signed your statement; is that correct?—A. That's right.

Q. And then you made attempts to get a lawyer; isn't that right?—A. That's right.

Q. You called up your brother-in-law and you told him, "Get a lawyer," didn't you?—A. That's right.

Q. Or words to that effect. Now, I am asking you whether or not, when you called up your brother-in-law to get a lawyer, you had in your mind that that lawyer should represent your wife?—A. To represent me.

Q. Only you?—A. When I say that, I don't know who else was involved, in the sense that I don't know what the Government was going to do. They were just talking to me.

Q. You involved your wife that night, did you not?—A. Well, I don't know sufficiently enough about the law to realize that I did involve my wife.

Q. Didn't you tell the FBI that night that your wife came out to Albuquerque?—A. That is absolutely true.

Q. And made an invitation to you to commit espionage; you told him that, did you not?—A. That's right.

Q. You also told them, did you not, that she had accompanied you to Rosenberg's house in September 1945; you told them that, did you not?—A. That's right.

Q. You also told them that your wife had received from you the money that Gold had given to you in June 1945; you told them that, did you not?—A. I did.

Q. Isn't that your testimony just within the last couple of minutes?—A. That's right.

Q. All right, and you also told them——

The COURT. Don't repeat it all.

Q. —that your wife——

The COURT. Don't repeat it all, Mr. Bloch. We know.

Mr. E. H. BLOCH. All right, just the last question on this line. I think I will be all inclusive then.

Q. And you also told the FBI, did you not, that night that your wife got one-half of that side of the Jello box, after arrangements had been made for her to go out to Denver and to meet somebody, as a means of identification?—A. What is that?

Mr. E. H. BLOCH. I withdraw the question.

Q. Did you tell the FBI about your wife's participation in the Jello box incident?—A. I did.

Q. All right.—A. But let me point out, as a lawyer—as I wasn't a lawyer, I didn't know it was an overt act or anything else. How was I to know that? I just told them the story as it happened. That was all I was interested in getting out.

Q. You were interested in getting out?—A. I said, all I was interested in was getting out the story. Don't misconstrue my words.

Q. Did you at any time think of your wife while you were down here telling your story to the FBI?—A. Of course I thought of her.

The COURT. Did he think of her in what respect? He must have; he mentioned her name.

Q. Did you think of your wife with respect to the fact that she may be a defendant——A. I did.

Q. In a criminal proceeding?—A. I did.

Q. And was that at least one of the factors which motivated you in getting a lawyer?—A. It didn't motivate me to get a lawyer. At that time I didn't even think about it.

The COURT. Supposing you tell us why did you get a lawyer?

The WITNESS. I got a lawyer to represent me in court, that is all.

By Mr. E. H. BLOCH:

Q. Represent you in court?—A. That's right.

Q. As a matter of fact, when you were finally arraigned the following afternoon and after you had spoken to Mr. Fabricant in the morning and to Mr. Rogge at or about the time of the arraignment, did Mr. Rogge ask for low bail for you?—A. Yes, he did.

The COURT. May I ask what is the relevancy of whether he asked for low bail?

Mr. E. H. BLOCH. I will come to it in just a moment.

Mr. COHN. I don't know what that has to do with this.

Mr. E. H. BLOCH. I am going to come to it right now.

Q. Didn't you tell your lawyers to fight this case for you?—A. I did not.

Q. Now, after you were arraigned, were you then taken to a jail?—A. I was.

Q. And were you taken to the Federal House of Detention?—A. I was.

Q. At West Street and West 11th Street, Manhattan?—A. I was.

Q. What time did you get there?—A. It was in the afternoon.

Q. Were you put in solitary confinement?—A. I was.

Q. How long did you remain in solitary confinement?—A. About three days.

Q. And you were restricted, were you not, in your movements there, to your cell?—A. Yes.

Q. And you weren't permitted during those three days any of the privileges that you subsequently found other prisoners had, with respect to walking on the roof and other recreational facilities that were made available to them?—A. That's right.

Q. Now, during those three days, did you have any visitors?—A. No.

Q. Did you ask anybody for permission to receive visitors?—A. I don't believe I did. The reason I was confined to those three days——

Mr. E. H. BLOCH. Now, I move to strike out any gratuitous remarks of the witness.

The COURT. I will let him give the reason.

Mr. E. H. BLOCH. I respectfully except, your Honor.

The WITNESS (continuing). The reason, I tried to find out, was because there was an erroneous story in the newspapers to the effect that I was going to commit suicide, which was far from the truth; so the keeper, who read the story, felt, well, he wasn't going to take it on himself, so he had me put in solitary and had my laces taken off my shoes and my belt taken away from me so I couldn't commit suicide. That was the whole story. There was no other reason.

Q. Did you tell anybody that you were going to commit suicide?—A. No.

Q. Now, when for the first time did you have a visitor after you were taken to the Federal House of Detention?—A. I don't know exactly. I had a visitor.

The COURT. May I ask what the relevancy of this is?

Mr. E. H. BLOCH. Yes. If you want me to state it in front of the jury, I will, but I think I ought to state it in front of the bench.

The COURT. Come up.

*　　*　　*　　*　　*　　*　　*

By Mr. E. H. BLOCH:

Q. Mr. Greenglass, did you hire Mr. Rogge's firm to defend you against the accusation that was made against you by the United States Government as reflected by the complaint before the Commissioner on June 16, 1950?—A. I hired Mr. Rogge.

Q. Is that your best answer?—A. That is my best answer.

The COURT. But did you hire him to defend you or did you hire him to represent you?

The WITNESS. Well, that is what I was worrying about his question. I said I hired Mr. Rogge.

Q. Now after these three days in solitary were you finally taken out and put into a cell with other prisoners?—A. I was.

Q. And you were treated just the way all other prisoners were treated?—A. That is right; I was.

Q. Is that right?—A. That is right.

Q. When was the first time that you left the Federal House of Detention after you were first brought there?—A. When they arrested Julius Rosenberg, the next day, I was taken out. It was over a week end, so Monday I was taken out.

Q. Do you know when he was arrested?—A. I don't remember the date.

Q. Was it about a month after you were arrested?—A. That's right.

Q. Now, when you were down here in the Federal Building on your arraignment on June 16, 1950, were you aware of the fact that Julius Rosenberg that day was down in this building being interrogated by FBI representatives?—A. I was not aware of the fact.

Q. When did you find out about that?—A. I didn't. I wasn't aware of that fact.

Q. You still don't know it?—A. I mean I know it now.

Q. No. I mean up to the time I directed your attention to that?—A. I don't understand you.

Q. Well, is this the first time that you are hearing that Julius Rosenberg was being interrogated by FBI agents in the Federal Building on June 16, 1950?—A. Well, the FBI agents had never told me that they did.

Q. They didn't tell you? You didn't speak to Julius Rosenberg in this Federal Building on July 16, 1950?—A. July 16th?

Q. I am sorry, June 16th.—A. Speak to Julius Rosenberg? No.

Q. Did you see him in this building on that day?—A. I don't believe I did.

Q. Did anybody ever tell you up to that time that I am now directing the subject to your attention, that Julius Rosenberg was in this building on June 16, 1950, and after a number of hours of interrogation he was permitted to go home?—A. Mr. Bloch, I will say that somebody told me that Julius had been interrogated but he wasn't told as to the date—I wasn't told the date or when that interrogation occurred, but my mother did tell me that on a visit.

Q. And that was within a few days after you were arrested; is that right?—A. Well, I don't know when she told it to me.

Q. You don't remember?—A. I do not remember when she told it to me.

Q. All right. Now while you were in the Federal House of Detention and after you had been released from solitary confinement did you tell any of the prisoners there that you were going to fight this case against you?—A. I had been told by my lawyer to just act like they act, and "Don't make any trouble for yourself."

Mr. E. H. Bloch. I move that the answer be stricken out as not responsive.

Mr. Cohn. I object to it being stricken.

The Court. I will let it stand.

The Witness (continuing). So I——

Q. Did you——A. Whatever they said, I agreed with them.

Q. Did you on your own initiative—— A. I never on my own initiative.

Q. Tell them or any of them that you were going to fight this charge against you?—A. I never told them that on my own initiative.

The Court. Well, you had already confessed, hadn't you?

The Witness Certainly, certainly, but you see the idea was it made it uncomfortable——

The Court. Don't tell me about any ideas.

Q. Now, when was the first time that you left the Federal House of Detention to come down to this building after you were first lodged there?—A. I don't know—a week or so, maybe less. I don't know the exact times.

Q. About a week? Would that be a fair statement?—A. Well, I won't even say that because I don't remember when I came back here.

Q. At any rate, there came a time when you were taken out of the Federal House of Detention and brought down to this Federal Building?—A. That is right.

Q. Were you examined by any of the representatives of the Government at that time?—A. I was.

Q. Was your lawyer present?—A. He had been present at some other times——

Q. I am talking about this time. Try to fasten your mind on the specific date or incident and I think we will get along here.—A. How can I fasten my mind on a specific date or incident? Are you saying when I came down I was interviewed by anybody? Yes, I was.

Q. And was your lawyer present at the time?—A. There were other times my lawyer was present. I don't remember whether it was the third time, fourth time, or the fifteenth time.

Q. If you don't remember, say so. Nobody is going to take your head off. Your best answer is that you don't remember whether your lawyer was present the first time you were brought down here?—A. That is right.

Q. After you were taken from the Federal House of Detention, is that right?—A. That is right.

Q. When did you get down here that day?—A. I don't even remember the day.

Q. And how long were you here?—A. If I don't remember the day I am certainly not going to remember how long I was here.

Q. Do you remember what you talked about to the FBI?—A. When I came down to talk to the FBI I talked about a number of things; whatever their interrogation led to, it loosened the springs of my memory and I was able to remember things that I had forgotten.

Q. Was a stenographer present?—A. Usually I talked to them and they write it down in longhand.

Q. Did you sign a statement that day?—A. I don't remember if I signed a statement that day, but I signed statements, plenty of statements.

Q. From that first time when you were taken from the Federal House of Detention to come down to the Federal Building here how many times in all did you come down from the Federal House of Detention to this building to be interviewed by various representatives of the Government?—A. I don't know.

Q. How long were you in all lodged in the Federal House of Detention?—A. About a month.

Q. And were you then transferred to another prison?—A. I was.

Q. Were you transferred to the Tombs?—A. I was transferred to the Tombs.

Q. And have you been lodged at the Tombs since that time?—A. I have been.

Q. Have you seen your wife while you were lodged either in the Federal House of Detention or in the Tombs at any place outside of those two penal institutions?—A. What was that again?

Q. Did you ever see your wife since your arrest at some place outside the West Street Jail or the Tombs?—A. I have.

Q. Where did you see her?—A. I saw her at the Federal Building here.

Q. Did you also see your children here?—A. I did.

Q. Did you get a special room to see her?—A. The same room that Sobell used when his wife and child came to see him.

Q. And where was that?—A. Right here in the Marshal's chambers.

Q. You mean downstairs?—A. Downstairs.

Q. In the jail?—A. In the jail.

Q. And did you discuss with your wife the testimony that you were going to give on this trial?—A. I have been instructed originally not to discuss testimony with my wife or any other person and I did not.

Q. You did not? Not one word?—A. Not a word.

Q. How many times in all would you say you saw your wife since your arrest?—A. As many times as the Tombs would permit.

Q. Now let us be a little more specific. Give us your best estimate of the number of times?

By the COURT:

Q. By that you mean that she came on the regular visiting days?—A. On the regular visiting days.

Q. There were no special days arranged?—A. No.

Q. For her visit?—A. No.

By Mr. E. H. BLOCH:

Q. You have never been out of the confines of a jail since your arrest, have you?—A. Except to come to the Federal Building.

Q This is still jail; you were always in the custody of the Marshal, weren't you?—A. That is right.

Q. Now, has your wife been receiving any money from the Government of the United States?—A. No.

Q. Have you received any money?—A. No.

Q. Is your wife working?—A. No.

Q. Do you know whether she is on relief?—A. She is not on relief.

Q. How old are your children?—A. One is four and one is nine months.

Q. How old are Ethel's children?—A. One is seven and one is three or four—three.

Q. Can you estimate even roughly how many written statements you have given to the authorities?—A. I would say about six or seven—what do you mean by written? I mean it may not have been written.

Q. Well let us clarify it. How many statements have you given to the FBI or any other of the Government agents or officials where you actually signed your name at the end of certain printed matter?—A. I would say about six or seven.

Q. And how many times in all if you can give a rough estimate besides those six or seven occasions were you questioned where a stenographer was present?—A. I don't remember any stenographer.

Q. Would you say then that all in all you gave only six or seven statements which would include the taking either by notes or otherwise of what you were saying?—A. Mr. Bloch, you will have to clarify that.

Q. All right, I want to clarify it because I want to clarify it for everybody. There were occasions were there not when a stenographer was present when you were asked certain questions?—A. At times there were, yes.

Q. Now, at those times did you sign written statements?—A. I don't think so.

Q. In other words, besides the six or seven written statements there were other occasions when you were asked questions and a stenographer took down the questions and your answers?—A. That is correct.

Q. Is that correct?—A. Yes.

Q. How many times in all would you say there was a stenographer present at which you were asked questions and answers and they were taken down and transcribed?—A. I couldn't give you an estimate.

Q. More than one?—A. More than one what? The time——

Q More than one occasion when that happened?—A. Yes, I suppose so.

Q. More than five?—A. You got me. I really—I can't give you an estimate of times on things like that.

Q. When did you first testify before a grand jury sitting in this court—I withdraw that, if I may, your Honor. You know, do you not, that you have been indicted as a defendant in a criminal proceeding brought by the Government of the United States sitting in a District Court of the United States for the District of New Mexico charging you with the crime of conspiracy to commit espionage?

Mr. COHN. Your Honor, I think Mr. Bloch is going to start going into matters which involve legal technicalities or which have no relevancy to the issues here. I think it is improper cross-examination. If he wants to go to the bench I will be glad to do that.

Mr. E. H. BLOCH. Certainly.

*   *   *   *   *   *   *

Mr. E. H. BLOCH. I withdraw the last question.

Q. In addition to the charge which has been brought against you here and which is the basis for this trial and the issues in this trial, is there another charge pending against you in the United States District Court for the District of New Mexico to which you haven't pleaded and which still is unresolved?— A. Is this within my compass to know this? I mean I don't think I even know it.

The COURT. Now don't ask any questions. He asked you a question.

The WITNESS. But I don't know.

The COURT. If you don't know say you don't know.

The WITNESS. I don't know.

Q. You don't know about that? You pleaded guilty to this charge which forms the subject matter of this trial, did you not?—A. That is right.

Q. You are a defendant?—A. I am.

Q. When did you plead guilty?

By the COURT:

Q. Do you remember the date?—A. No, I can't remember the date.

Q. Why don't you just say that if you don't remember it.

By Mr. E. H. BLOCH:

Q. You pleaded guilty did you not in this building to this charge?—A. I did.

Q. And would it refresh your recollection if I told you that you pleaded guilty in Room 318?—A. Yes.

Q. And would it refresh your recollection if I told you that your lawyer was present at the time you pleaded guilty?—A. That is right.

Q. How many months ago——

Mr. SAYPOL. Just a minute. What is happening here? Is his recollection being refreshed or is he testifying. The form of the question was such that the witness can't give a clear answer. What transpired is a matter of record.

Mr. E. H. BLOCH. All right.

The COURT. Proceed.

Q. Now, does that refresh your recollection as to the date when you pleaded guilty?—A. I pleaded guilty to the charge.

Q. Was that many, many months ago?—A. I pleaded guilty to an indictment earlier and I pleaded guilty to this indictment. I withdrew the plea on the earlier indictment at the same time.

The COURT. This is a superseding indictment.

Mr. E. H. BLOCH. In other words, this is a superseding indictment.

The COURT. Yes.

Q. Now, how many months ago did you first plead guilty to this charge of conspiracy to commit espionage? Do you remember that?—A. Back in last year.

Q. Last year. Have you been sentenced?—A. I have not been sentenced.

Q. Do you believe that by giving testimony in this case that you will be helped in terms of the severity of the sentence to be imposed upon you by the Court?—A. The Court—it is entirely within the discretion of the Court to give the sentence, and whatever I do is just—it depends on the Court and nothing else but the Court.

Q. Now I would be—I would ask you to be good enough to answer my question. Do you believe that the Court will be easier on you——

Mr. COHN. Your Honor, I object.

Q. —because you are testifying here——

Mr. COHN. Your Honor, I object.

Q. —the way you did?

Mr. COHN. I object to the question as to form, the witness' belief.

The COURT. I believe that what he is trying to get at is a motive for his testimony.

Mr. COHN. Yes. I have no objection to a proper question.

The COURT. So I will permit him to ask the question.

The WITNESS. I don't believe that I in testifying will help myself to that great an extent.

Q. When you say "to that great extent," would you like to clarify that for the jury?—A. To any great extent.

Q. Would you say to any extent?—A. To any extent.

Q. All right. Do you believe that by testifying here in this trial that you will help your wife?—A. I don't know what the Government has in mind with my wife and I can't answer for them.

Q. You know, of course, that so far nothing has happened to your wife in terms of any criminal proceedings——

The COURT. That has been answered.

Mr. COHN. It has been answered.

Q. —or brought against her?

The COURT. You don't have to answer that.

Mr. E. H. BLOCH. I respectfully except.

Q. Let me ask you, Mr. Greenglass, did you ever keep a memorandum book or a diary?—A. No, I never did.

Q. So that when you testified here today and yesterday and the day before yesterday, you were relying upon your memory, is that right?—A. Upon my memory, yes.

Q. And solely upon your memory?—A. That's right.

Q. And when you drew the sketches—one of them I believe in June 1950 and the other two a day or two before you testified—and I think they are reflected and marked Government's Exhibits 2, 6 and 7—did you rely solely on your memory in making those sketches?—A. I did.

Q. Now, when was the last day that you worked at the Los Alamos project?—A. 1946, February.

Q. What month?—A. February.

Q. That was about four and a half years ago?—A. That's right.

Q. And you relied solely upon your memory?—A. I did.

Q. During all of your months in jail did anybody go over with you any subject matter which related directly to those sketches that were introduced in evidence here as Government's Exhibits 2, 6 and 7?

Mr. COHN. I don't know what Mr. Bloch means "go over with you." I think if he would clarify that——

The COURT. Try to clarify it for him.

Q. Did you draw any sketches for any of the FBI men or any agents of Mr. Saypol's staff prior to the time you came to testify here?—A. I did.

Q. Was that the one that has been marked Government's Exhibit 2?—A. No.

Q. How many sketches did you draw for them?—A. They were the same sketches, the only thing I used the ruler to accomplish this. The others were freehand.

Q. Freehand?—A. That's all.

Q. Now, were you given any reference books or textbooks while you were in jail since your arrest, relating to any scientific matter?—A. No; I didn't—nobody gave me any.

Q. Did you read any scientific books while you have been in jail?—A. Just science fiction.

Q. That is, of course, not a basic theoretical journal, is it?—A. No.

Q. That is a popular kind of scientific periodical?—A. That's right.

Q. Now, Mr. Greenglass, I believe you testified that you graduated from high school here in New York City?—A. Yes.

Q. And I think you testified that you went to Brooklyn Polytech?—A. Right.

Q. Is that correct?—A. Yes.

Q. How long did you go to Brooklyn Polytech?—A. Six months.

Q. And how many courses did you take during those six months?—A. About eight different courses.

Q. And did you fail——

Mr. COHN. Oh, I object to that, your Honor. What difference does it make?

Mr. E. H. BLOCH. I am coming to a new subject now, your Honor.

The COURT. I assume you are.

Mr. E. H. BLOCH. Yes; and I wish you will bear with me, because I am going to connect this up.

The COURT. All right.

Mr. COHN. Well, I will let Mr. Bloch finish his question. That is as far as I will commit myself at the moment, your Honor.

The COURT. Right.

Q. Did you fail in your subjects?

Mr. COHN. I would now object to that, your Honor. I don't see the relevancy of whether he or anybody else failed in subjects might have and it is certainly not proper cross-examination.

The COURT. Before you answer that question, let me ask you: These sketches that are in evidence, are they the product of your own mind? By that I mean, were you helped by anybody on the outside in drawing those sketches?

The WITNESS. Nobody else, just myself.

The COURT. Did anybody tell you to change any line here or change any line there?

The WITNESS. Nobody told me anything like that.

The COURT. Very well.

Now, you ask your question.

Mr. E. H. BLOCH. Are you permitting it, your Honor?

The COURT. Yes. What subject? Be specific.

Mr. E. H. BLOCH. All right.

The WITNESS. I will tell the story.

Mr. E. H. BLOCH. Look, Mr. Greenglass——

The WITNESS. I was quite young at the time, about 18, and I liked to play around more than I liked to go to school, so I cut classes almost the whole term. Simple.

Q. How many of the eight courses that you took did you fail?—A. I failed them all.

Q. And did you then go to Pratt Institute?—A. That's right.

Q. How long did you attend Pratt Institute?—A. I attended it for one semester and a half, and the half of the other semester I had to work at night, so I had to withdraw from my classes which was allowed by the school, and I went to work at night, and I did not fail those courses. As a matter of fact, I got good marks.

Mr. E. H. BLOCH. Congratulations.

The COURT. Strike that from the record.

Q. Did you ever get a degree in science?—A. I did not get a degree.

Q. Did you ever get a B. S.?—A. I did not.

Q. Did you ever get any engineering degree?—A. I did not.

Q. From any recognized institution?—A. I did not.

Q. Have you pursued any other organized and formal courses, held under the auspices of a recognized educational institution, apart from the Brooklyn Polytech and the Pratt Institute courses that you have mentioned you took?—A. I did not.

Q. Do you know anything about the basic theory of atomic energy?—A. I know something about it; yes. I am no scientific—I am no scientific expert, but I know something about it.

Q. Did you ever take courses in calculus?—A. No.

Q. Differential calculus?—A. I did not.

Mr. E. H. BLOCH. I am just looking for a piece of paper, your Honor.

Q. Or thermodynamics?—A. I did not.

Q. Or nuclear physics?—A. I did not.

Q. Or atomic physics?—A. I did not.

Q. Or quantum mechanics?—A. I did not.

Q. Or advanced calculus?—A. I did not.

*   *   *   *   *   *   *

Mr. E. H. BLOCH. And I might say I never took any of these courses.

Q. Have you read any basic works on any of the subjects that I have just asked you about?—A. No.

Q. Do you know what an isotope is?—A. I do.

Q. What is it?—A. An isotope is an element having the same atomic structure. but having a different atomic weight.

Q. Now, did you learn that in Los Alamos?—A. I picked it up here and there.

Q. When you came to Los Alamos, you were a machinist, were you not?—A. I was.

Q. What was your rating in the Army?—A. T/5.

Q. Had you, prior to the time you came to Los Alamos, done any work as a machinist in the Army of the United States?—A. I had.

Q. Where?—A. At a number of places.

Q. Well, do you mind detailing them, and the length of time that you practiced the trade of machinist in the Army of the United States, prior to the time you got to Los Alamos?—A. I was a machinist at Fort Ord.

Q. For how long?—A. As long as the company was there, I was a machinist there.

Q. For how long was that?—A. Three months. I was a machinist at Southgate, California, in the General Motors Tank Arsenal.

Q. How long?—A. As long as the company was there, I was a machinist there.

Q. How long was that?—A. A period of four weeks.

Q. That is four months so far, right?—A. I was at the P. O. M., Pomona Ordnance Base, three months there; I was at Aberdeen Proving Grounds, three months there.

Q. That is seven; that is ten.—A. Okay, that is ten right there in the Army.

Q. All right.—A. And every other post that I ever worked on, was in, in the Army, I worked as a machinist.

Q. Were you classified in the Army as a machinist?—A. I was classified—I had two classifications.

Q. I am talking about the first one, before your promotions.—A. Before my promotions? I was classified—when you originally come into the Army you have just a basic classification, which means you have taken basic training. After that you have certain skilled classifications. I had two skilled classifications.

Q. What were they?—A. One was automotive machinist and one was machinist and toolmaker.

Q. And when you got to Los Alamos, were you an automotive machinist or a machinist?—A. A machinist.

Q. And you say that you first worked in a certain building called the "C" building—"E" building, I am sorry?—A. "E" building.

Q. Now, how many machinists besides you were in that building?—A. At the time, about four, five, maybe six.

The COURT. Is this a convenient place to break off and recess?

Mr. E. H. BLOCH. I think so.

The COURT. All right, we will take our recess.

May I see counsel, please.

(Discussion at the bench off the record.)

(Short recess.)

Q. I believe, Mr. Greenglass, that before recess we were discussing your job as a machinist in the Army. Now at the E Building how many other machinists were there besides you?—A. There was about four or five or maybe even six.

Q. And did they run up to as high as ten at times?—A. That's right.

Q. And you had an immediate supervisor, did you not?—A. I did.

Q. And his name was what?—A. His name was Demars.

Q. And besides Demars there was Sergeant Fitzpatrick?—A. That is right.

Q. And above Sergeant Fitzpatrick there was Dr——A. Kistiakowski.

Q. Is he the gentleman who testified here?—A. He is not.

Q. Now, when the E shop moved into the Theta Building did the workers in the E shop remain as a departmental unit or were you joined or did you join with other machinists?—A. No; it was the same shop.

Q. Were there any other machine shops in the Theta Building?—A. No other machine shops in the Theta Building.

Q. Now when you were in the E Building the only persons or employees who were working in that building were the five or six or ten mechanics, machinists with their supervisors, is that right?—A. That is right——

Q. Now, how about——A. In working in the building or in the shop? You said "building."

Q. Were there other employees working in the E Building?—A. Yes.

Q. How many others?—A. I can't tell exactly. There was laboratories all over the building.

Q. And how about when you moved into the Theta Building?—A. There were other employees working there, too.

Q. Was the Theta Building a bigger building than the E Building?—A. Well, there was more room for our group in it. It wasn't bigger.

Q. You mean it wasn't bigger physically?—A. It wasn't bigger physically.

Q. Were there more employees working in the Theta Building than in the E Building?—A. I don't know.

Q. Were there more machinists working in the Theta Building than in the E Building?—A. No.

Q. About the same?—A. That is right.

Q. Now with respect to the security regulations at Los Alamos were you searched at all when you came in to report to work in the morning?—A. No.

Q. Pardon me?—A. No.

Q Were you searched at all when you quit at the end of the day?—A. No.

Q. Were there any security police or guards around?—A. There were.

Q. Were they in the E Building?—A. They weren't in the buildings; no.

Q. Were they in the Theta Building—were they outside the building?—A. They were at the entrance to the building—at each entrance.

Q. When you came to work in the morning and when you left at the end of the working day was it necessary for you to pass through this screen of security police?—A. That's right.

Q. Did those security police look over any packages or any other thing that you may have had on your person either coming in or going out?—A. They did that.

Q. And were you frisked—do you know what "frisked" is?—A. I have learned.

Q. Well—you have learned. You learned that in jail. All right, then we both understand the term. Were you frisked when you came in to work in the morning or when you quit in the evening?

Mr. SAYPOL. I think Mr. Bloch ought to be relieved of any unconscious implication that he cast upon himself.

Mr. E. H. BLOCH. I didn't mean it; I didn't mean it.

Q. Frisk means somebody touching your person to find out whether you have something on your person; isn't that right? Isn't that the definition of frisk?—A. That is right.

Q. Now, did you at any time ever take out of the Los Alamos project to your home or to any quarters which you were using for dwelling purposes any blueprint or any sketch during any of the months or years that you worked at the project?—A. No; I didn't.

Q. How long did you continue to work as a machinist until you were promoted to the assistant foreman's position?—A. From about March 1945—I still continued to work as a machinist.

Q. But you supervised other men?—A. Yes. I mean I still——

Q. You were what one would call a working foreman?—A. That is right.

Q. So that when your wife came down to visit you in Albuquerque, New Mexico, in November 1944, and during the time when you received your first furlough in New York in January 1945, you had not as yet been promoted to be an assistant foreman, is that correct?—A. When my wife was out there I was already.

Q. No; maybe I misunderstood you. Let us clarify the record. When you became an assistant foreman, I believe you said sometime in 1945—maybe I didn't——A. I said about March 1945.

Q. March 1945?—A. Yes.

Q. You wife came out to see you for the first time?—A. Oh, in 1944, that is right.

Q. And your first furlough to New York was in January?—A. That is right.

Q. At that time you were still a machinist and you had not been promoted to be an assistant foreman or working foreman, is that right?—A. That is right.

Q. And while you were working as a machinist until the time you were promoted to be an assistant foreman, what color badge did you wear around the project?—A. I would like to explain that a little, Mr. Bloch.

Q. Certainly.—A. When I first came to the Project they had three color badges. There was a white, blue, and red. Now, the blue badge was the one you were supposed to wear if you could know what you were working on but nothing further. The white badge knew practically everything. The red was where the person wasn't supposed to know anything of what was going on.

Q. Correct.—A. Now, these badges were changed sometime during the Project.

Q. During when?—A. During the Project, during the year, that year.

Q. 1944?—A. 1945.

Q. No, no—all right.—A. Now wait, let me explain, and then I will go back.

Q. All right.—A. Now, they changed that. At that time they switched the blue to the red position so if you had a red badge it was what you used to have when you had a blue badge. At the time you are talking of you had a blue badge which was the equivalent later on of the red badge.

Q. So there was a white badge in 1944 and that badge was worn by the top scientists, the real top men in the Project, is that right?—A. Yes.

Q. The more important officials?—A. Yes.

Q. All the most important officials?—A. That is right.

Q. And then came those who were employed regularly at the Project, and this is quite rough, people like you?—A. Yes.

Q. And they wore the blue badge?—A. That is right.

Q. And people who came in sporadically or temporarily to do work on the Project wore the red badge?—A. That is right.

Q. Let us fix our minds on those badges because I want to cover 1944 on. During that year did you procure any information concerning the work at the Los Alamos Project from anybody outside the E Building or the Theta Building?—A. From November 29th after I had seen my wife until January 1 I did get information.

Q. You did? You were still wearing the blue badge?—A. That is right.

Q. Now I want you to name one scientist from whom you received unauthorized information?—A. By that you mean that he knowingly gave me the information?

Mr. E. H. BLOCH. Now may I have Government's Exhibit 1, please?

Q. Now, Government's Exhibit 1 in evidence deals with the regulations governing the conduct of employees at the Los Alamos Project, does it not?—A. That is right.

Q. This you identified. Is there anything in these regulations which prohibit an employee from giving information to another employee outside the official routine run of business?

Mr. COHN. Now, your Honor, I think the exhibit speaks for itself. I tried to inquire into the contents of the exhibit and was stopped on the ground that the exhibit does speak for itself, and I think it does in exact terminology.

The COURT. No. That is a proper question.

(Last question read.)

The WITNESS. I don't know exactly what it says in there because I haven't read it recently, but I suppose it does state something to that effect.

Q. Now, you stated that after your wife came to visit you around November 29, 1944, until the time you got your first furlough in January 1945, you did get information outside what would come to you in the official discharge of your duties as a machinist, is that right?—A. That is right.

Q. And did you procure that information from somebody who was not assigned permanently to the E—I think at that time you were in the Theta Building, weren't you, at the Theta Building?—A. Well, first of all a scientist—it was anybody who was employed up there as a scientist. That could be a GI, a civilian, and I did procure for instance the fact that Baker was Bohr from a man who happened to be a scientist.

Q. What is the name of that GI?—A. William Spindel.

Q. Did you procure any information, to which you believe you were not entitled, from any scientist during this period from November 29, 1944, to January 1, 1945?

Mr. COHN. I object to that.

The COURT. Upon what grounds?

Mr. COHN. I don't think it is relevant.

The COURT. All right. What is the ground? I don't see the relevancy as to whether he got the information to which he didn't think he was entitled?

Mr. E. H. BLOCH. This is on the question of credibility, your Honor. This man is testifying that he is relying solely upon memory and he testified that he procured certain information.

The COURT. You say it is on the question of credibility?

Mr. E. H. BLOCH. That is right.

The COURT. You may go ahead. What is your question now? Put it again.

Mr. E. H. BLOCH. Will you read the question, please.

(Question read.)

The WITNESS. I told about Bohr.

Q. You told us about Bohr already. You said you got that information from this GI, whose name is Spindel. Now I am asking you if you got any information from any scientist working on that project during that period?—A. I was in the room when I heard discussions about implosion effect experiments, implosion effect of lenses, while some scientists were discussing it in the office of the building I was in.

Q. Were you lawfully in that building at the time?—A. I had come in to pick somebody up to go to lunch.

Q. You weren't there unlawfully?—A. No; there was nobody telling me not to go into this room or that room. There are certain restricted areas.

Q. Yes; but you couldn't wander around the development, could you?—A. All over the Tech area, it was perfectly all right for me to go.

Q. Even when you were a machinist?—A. Absolutely. In my case, in my building, in my group, I don't know anybody else.

Q. We are confining ourselves to the time before you became assistant foreman.—A. Yes.

Q. Any other incidents?—A. You said "that month." Well, that is all I remember for that month.

Q. Now, then, after you returned from New York on your first furlough, did you receive any information from any scientist on that project outside of information that would come to you through the discharge of your official duties?—A. Yes; I did.

Q. And from whom?—A. Well, it was in the theoretical physics department.

Q. Where was that located?—A. In the Tech area.

Q. What building?—A. "T" building, probably.

Q. "T" building?—A. Probably, and this man was a mathematician who worked there, gave me a pretty good idea of what the lenses were about; he knew the physics, I mean, involved.

Q. Was this in the course of an official lecture that you attended?—A. No.

*   *   *   *   *   *   *

By Mr. E. H. BLOCH:

Q. Now, I believe on your direct examination you told us, in substance, that you snooped around to get information; isn't that right?

The COURT. Don't shake your head. You had better answer.

The WITNESS. Oh, yes, yes.

Q. And you would make it your business to enter into a conversation or overhear conversations where you could pick up information?—A. That is right.

Q. Is that correct?—A. Yes.

Q. Now, could you give us just two instances of information that you picked up that way?—A. I came into a room; there was a piece of material on the table; I picked it up and I said, "It is an interesting piece of material and it is interestingly machined." The man I spoke to and another man was there said, "Oh, that is neutron source," and explained how it was used, in a conversation. That is one way. That is one instance.

Q. Was that in the tech area?—A. It was in the tech area.

Q. All right, give me another instance—just pardon me, Mr. Greenglass. I don't like to break the trend of thought, but just for clarity, in connection with this first illustration of how you picked up information; were these men, who told you about the fact that this piece of material would have something to do with neutrons, were these men, these employees, top scientists?—A. Now, look, every scientist had a white badge there.

Q. Were they white-badge men? Let me put it that way—A. One was a white-badge man; one wasn't.

Q. All right. Now go to the second instance.—A. Another instance. A man came in to me with a sketch, with a piece of material; said, "Machine it up so that I would have square corners, so I could lay out a lens; come over and pick it up." I would go over to his place; he was a mathematician, a scientist, he had laid it out, and I would say, "What is the idea?" He would tell me the idea.

Q. Tricky like, eh?—A. Nothing tricky about it.

Mr. COHN. Your Honor——

The COURT. Strike that out.

Mr. COHN. I think that should be stricken, your Honor.

Q. Well, you meant to trick, did you not, the person who was talking?—A. Well, I meant to get the information from him.

Q. By trick, didn't you?

Mr. COHN. Your Honor, we have had that many times.

Mr. E. H. BLOCH. All right, I will withdraw it, but I just wanted the Court to know that I wasn't using words loosely.

Mr. SAYPOL. That is why it was objected to, because it wasn't being used loosely.

The COURT. All right.

Q. Now, then, when you were a machinist, were you given a blueprint now and then to work on?—A. That is right.

Q. And were you ever given as a machinist, a job to do without cooperation with other machinists?—A. Surely.

Q. Now, the job that you did was only a part, however, of the matter, or the material that was to be constructed in connection with an over-all blueprint; isn't that so?—A. Sometimes, yes. Sometimes it was something by itself.

Q. And when it was something by itself, wasn't it just the construction of some little metal bar or some other little appliance?—A. A lot of little appliances go into making something bigger.

Q. That is correct; you d:dn't make all the appliances that went into this lens, did you?—A. Of course not—in the lens mold, I made a complete lens mold.

Q. You yourself made a complete lens mold?—A. That is right.

Q. Did you make the complete lens mold that was subsequently assembled at the remote project, at which the detonation went off?—A. I can't tell.

Q. You don't know that?—A. I don't know that.

Q. How long did it take you to make the complete lens mold?—A. Well, the flat type lens mold would take me about twenty-four hours of work.

Q. Were other machinists likewise trying to construct flat lens molds?—A. Certainly.

Q. And did they also work from blueprints?—A. Of course.

Q. Did you ever make a copy of any of the blueprints that were given to you to work on?—A. I made a copy for my own use in the shop.

Q. Did you ever make a copy of any blueprint and take it out of the project?—A. I did not.

Q. Did you ever steal any documents, whether it be blueprints or any other matter, or even including any material, and take it out of the project to your home?—A. I did not.

Q. Or to the place where you stayed?—A. I did not steal anything of that nature. I made a radio; I took the radio out; I showed them the radio on the way out, and it was perfectly permissible to do that. I made a phonograph attachment; I brought that out with me. too.

Q. So that we can be very clear about this now, when you made the sketches for the Government, and particularly the sketches which have been marked as Government's exhibits 2, 6, and 7, you relied solely upon what you remembered you had done and the knowledge that you had accumulated while you were working at Los Alamos prior to 1945; is that right?—A. Prior to 1945?

Mr. SAYPOL. 1946.

Q. 1946.—A. That is right.

By the COURT:

Q. And would you give the same reply with respect to the sketches that you said you turned over to Rosenberg, and that was also a reconstruction of what you carried in your mind of the type molds, that is, as to 2 and 6 anyway, as to the type molds you had made, and then as to the process which is incorporated in 7?—A. That is correct.

By Mr. E. H. BLOCH:

Q. And that was true also of the material that you said you turned over to Gold?—A. That is correct.

Q. Now, tell me, when you worked on the lens mold, or, in fact, when you worked on any piece of apparatus while a machinist, were you given any lectures as to the functions of the particular piece that you were working on and constructing? This was while you were a machinist now.—A. What do you mean by lectures; formal lectures, in a group?

Q. Let's separate all the possibilities. Were you ever given any formal lectures?—A. No.

Q. Were you ever given any informal instructions?—A. Yes.

Q. Concerning their function?—A. Yes.

Q. Now were you ever told their functions in relation to the complete object that was to be constructed?—A. There are different types of lens molds. Some were not used on the bomb itself, and some were just used for experiments.

Q. How many lens molds in all would you say were constructed at the Theta building while you were working there?

Mr. COHN. I object to that as irrelevant, as to how many lens molds were constructed at the Theta building.

The COURT. Need we have that?

Mr. E. H. BLOCH. He said "many." I will be satisfied with the answer "Many."

The WITNESS. Many.

Q. Now Mr. Greenglass, can you sketch for us every lens mold upon which you worked or which was constructed at the Theta shop in Los Alamos?—A. Not everyone but I can draw—sketch a good deal of them.

Q. A good deal of them—showing the developing process and the improvements that had been made; can you do that?—A. The sketches are—well, that was only the improvement in the curve, and I didn't know that. The curve looked the same to me—maybe a little flatter or a little more tapered but I couldn't tell which curve was—I mean it would be very difficult to tell which one was the improvement over the other.

Q. You did not even know the formula for the curvature, did you?—A. That is exactly correct.

Q. What? You had to be a scientist to know the formula, isn't that right?—A. That is right.

Q. Now I would like to direct your attention to the time that you said you came to Rosenberg's house in September 1945. I think you testified—again check me; I am doing this in substance and rather roughly—that you and your wife came there sometime in the evening?—A. September 1945 I came in the afternoon.

Q. Well, when was the time that this Ann Sidorovich was there?—A. That was Janaury 1945.

Q. All right, then let us forget about September and go to January.

In January, or the early part of January, I believe you testified you came to the Rosenberg house in the evening and you met Ann Sidorovich?—A. That is right.

Q. You said you knew her husband?—A. I had known her husband.

Q. Prior to the time that you were introduced to her that evening?—A. That is right.

Q. And that was the first time you met Mrs. Sidorovich, is that correct?—A. That's right.

Q. Were you told where Mrs. Sidorovich lived at the time you were introduced to her?—A. I don't believe that I knew that at the time.

Q. Now you say you did know Mr. Sidorovich?—A. That's right.

Q. How often had you met him prior to the time that you first met his wife?—A. Well, Julius had introduced me to him and I had met him while I was going to school. I met him—I seen him around school and we talked together a number of times.

Q. Don't you——A. I met him on buses.

Q. Would you want to change your answer if I suggested to you that the Sidorovichs did not live in New York City in January 1944?—A. It wouldn't make any difference——

Q. In 1945, I am sorry.—A. It wouldn't make any difference to that because I met her there. I did not know anything about where they lived.

Q. Now we are talking about time. You may have met her there but I am trying to focus your attention on the time. Is it your testimony unequivocally that in January 1945 you met Ann Sidorovich at the Rosenberg's home in Knickerbocker Village?—A. That is correct.

Q. All right. Now I think you testified that Julie Rosenberg told you that he had received from the Russians or from the Russian Government a watch. Did you ever see that watch?—A. He showed it to me.

Q. Describe it?—A. It was a round watch, round dial watch with a sweep second hand.

Q. With what?—A. A sweep second hand—round faced watch with a sweep second hand, and it had—at the time he first showed it to me I believe it had a leather strap.

Q. Did you ever see the watch that you say Ethel got from the Russians?—A. I might have seen it but I didn't—I didn't——

Q. Didn't what?—A. Well, I wasn't told that that was the watch.

Q. Can you describe the watch that you saw on Ethel's hand or any time when she had a watch on her hand in her possession?—A. I can't describe that watch; no.

Q. I think you also said that the Rosenberg's told you or Julie Rosenberg told you that he received a console table from the Russians. Did you ever see that console table?—A. I saw that console table.

Q. Describe it.—A. Well, they had it up against the wall. It is a dark color, mahogany probably. It is wider than that table right there [indicating]—I mean the length.

Q Wider than which table? Do you mean the table against which I am standing [indicating]?—A. Yes. It is longer—it is a little bit wider and it is maybe four feet long, maybe three and a half, four feet long.

Q. Mr. Greenglass——A. And it is——

Mr. COHN. Wait. I would like him to finish the answer.

Mr. E. H. BLOCH. Yes, I want it.

Q. But you are a machinist, you understand that descriptions of lengths don't show up by this table, because that doesn't appear in the record. Then tell us how long it was and how wide it was?—A. I would say it was about—you see, the top of the console table, one side lifted up so it made an L if you had it against the wall, and that is the way I saw it. With the L up against the wall, it was about three and a half feet, maybe three feet long [indicating], except that is the width when the console table is opened up and the part of the table underneath the head or the board on top is about two feet wide.

Q. Now when was the last time you saw that console table in the Rosenberg's house?—A. It was some period between '46 and '49. I can't place it.

Q. Well, do you mind thinking for just a moment about that?—A. I can't definitely say when was the last time I saw it.

Q. Was that the only console table in Rosenberg's living room?—A. Yes.

Q. And was that console table used for eating purposes?—A. That console table was used for photography.

Q. For photography?—A. That's right. Julius told me that he did pictures on that table.

Q. Did you ever—withdrawn. Were you ever at the Rosenberg's house when food was served on that table?—A. I might have been.

Q. I think you also testified that Rosenberg said to you he received a certificate from the Russians?—A. A citation.

The Court. A citation.

Mr. E. H. Bloch. A citation—I am sorry.

Q. A citation. Did you ever ask him to show you that citation?—A. I never asked him to show it to me; no.

Q. Did you ever see it?—A. I did not see it; but he said there was——

Q. No, no—

Mr. Cohn. May we have the balance?

The Court. Go ahead.

The Witness (continuing). There were certain privileges that went along with that.

The Court. Yes, you told us that.

Q. Now did he tell you in any detail how that citation read?—A. I don't believe he did.

Q. Weren't you impressed when Rosenberg told you that he received a citation from the Russian Government?

Mr. Cohn. I object to whether or not the witness was impressed, your Honor.

The Court. I will sustain the objection.

Mr. E. H. Bloch. I respectfully except.

Now, may I have the Jello box exhibits, please?

(Handed to Mr. Bloch by Mr. Cohn.)

Q. Now Mr. Greenglass, let us come to this Jello box incident. I believe your testimony was, substantially, that in September—I am sorry, in January 1945 after Ann Sidorovich left, you and your wife remained along with the Rosenbergs?—A. That's right.

Q. And then this Jello business took place, is that right?—A. That's right.

Q. Now I am going to show you Government's Exhibit 4 [handing witness], and I want you to look at that very, very carefully and after looking at it very, very carefully I want you to tell the Court and jury whether that is substantially similar to the part of the Jello box which corresponds to that that you allege was used that night.

Mr. Cohn. I don't quite understand the question myself, your Honor.

The Court. You are asking him whether this is the same kind of Jello box that was used in 1945.

Mr. E. H. Bloch. That is right. It is not the contention that this was the Jello box that was used?

Mr. Cohn. No.

Mr. E. H. Bloch. That is what I want to know.

Q. I want to know whether it is substantially similar?—A. It is substantially similar.

Q. Is it identical?—A. It is not identical; no.

Q. What is the difference between the box you have in your hand marked "Government's Exhibit 4" and the box that you claim was cut up that night at the Rosenberg house in January 1945?—A. They made a darker colored box at the time. It seemed to me much darker than this, than the way it is now.

Q. Any other differences?—A. I didn't read the Jello box then and I haven't read this one now.

Q. How can you answer or how did you answer intelligently my question before when I asked you whether or not this box was substantially similar to the one that was used at the Rosenberg house?—A. It said Jello on both boxes.

Q. That is the only similarity that you can see, is that right?—A. And the shape of the box is about the same.

Q. Do you notice on that box, Government's Exhibit 4, there is a description of the flavor?—A. I do.

Q. What is the flavor on Government's Exhibit 4?—A. Raspberry.

Q. Is it imitation raspberry?

The COURT. Is that material?

Mr. E. H. BLOCH. It might be.

The COURT. Or is it facetious?

Mr. E. H. BLOCH. No, no. It might be.

The WITNESS. Imitation raspberry.

Mr. E. H. BLOCH. May I say to the Court that I can see what might be coming in the Government's case and I think this is an important piece of testimony that purports to link up my client with Harry Gold.

The COURT. All right, no argument, Mr. Saypol.

Go ahead. Proceed.

Mr. SAYPOL. I would like permission to make one statement.

The COURT. Make it very brief.

Mr. SAYPOL. All right; I certainly can't see how the flavor of the Jello would have anything to do with it.

The COURT. I can't either, but I will let him ask it. It seemed to me like attempted humor but I may be wrong.

The WITNESS. What was the question? (Question read.) This was imitation raspberry flavor.

Q. What was the one you say was used in Rosenberg's house?—A. I really don't know.

By the COURT:

Q. Didn't I understand you to say that they had gone into the kitchen, Mr. Rosenberg, Ethel Rosenberg, and your wife?—A. That is right.

Q. And you were in the living room?—A. Yes.

Q. And they brought out the piece cut up?—A. That is right.

By Mr. E. H. BLOCH:

Q. And they didn't bring out this part of the Jello box which is marked "Government's Exhibit 4," is that right?

*   *   *   *   *   *   *

The WITNESS. But the box itself I never did see. I was given a very good idea what that Jello box looks like and I know what it looks like from previous experience.

Mr. E. H. BLOCH. I would like to introduce Government's Exhibit 4 for identification in evidence.

Mr. COHN. I think it has no relevancy at all, but if Mr Bloch feels that it is going to be helpful in any way I have no objection.

The COURT. It may be received with the understanding that the witness said he didn't see the box or the side of it cut up in the kitchen.

(Marked Government's Exhibit 4.)

By Mr. E. H. BLOCH:

Q. Are you color blind?—A. I am.

Q. Do you know what color this is?—A. I do not. If you will notice Mr. Bloch——

Mr. E. H. BLOCH. I move to strike any gratuitous statements of the witness.

The COURT. Yes.

Q. Now then——

The COURT. Are you going into a new subject?

Mr. E. H. BLOCH. I am continuing on the Jello box, your Honor.

Mr. A. BLOCH. May we step up for a minute?

*   *   *   *   *   *   *

The COURT. We will recess at this point, ladies and gentlemen, until 10: 30 tomorrow morning.

(Adjourned to March 14, 1951, at 10: 30 a. m.)

8. *The Gold Case—Details*

Harry Gold, like David Greenglass, testified at the trial of Julius and Ethel Rosenberg and Morton Sobell. The details of Gold's activities as an espionage courier for Fuchs, Greenglass, and others are brought out in Gold's own words, hereinafter set forth verbatim:

From stenographer's minutes of case 134–245, *United States of America* v. *Julius Rosenberg et al.* Before Hon. Irving R. Kaufman, district judge, United States District Court, Southern District of New York, March 15, 1951.

HARRY GOLD, called as a witness on behalf of the Government, being first duly sworn, testified as follows:

Direct examination by Mr. LANE:

Q. Now Mr. Gold, the accoustics in this particular courtroom are not too good, so if you keep your voice up so that all the jurors can hear, it will help a great deal.

What is your occupation?—A. I am a biochemist.

Q. Now where were you born?—A. I was born in Berne, Switzerland.

Q. When?—A. On December 12th, 1910.

Q. When did you come to the United States?—A. I came to the United States in July of 1914.

Q. And did you come with your family?—A. Yes, I did.

Q. Now, were you naturalized at some time subsequent?—A. Yes. I was naturalized on my father's papers in the year 1922.

Q. Do you have—is your father living?—A. Yes, he is.

Q. Is your mother living?

*   *   *   *   *   *   *

The WITNESS. No, she is not. My mother died in 1947.

Q. Do you have any brothers or sisters?—A. I have one brother.

Q. How old is your father?—A. My father is 75 years of age.

Q. And what is his occupation?—A. My father is a cabinet maker.

Q. And is he occupied at the present time?—A. My father is not working right now. He did work up until about six months ago.

Mr. E. H. BLOCH. Now, if the Court pleases, now I object for a different reason. I object upon the ground that this background material is being elicited for the purposes of building up a sympathy with the witness and has no relation to the issues in this case. I am not saying that is the object of the prosecutor but I think that is the object of the effect.

The COURT. Overruled.

Mr. E. H. BLOCH. I respectfully except.

Q. Now will you briefly sketch your educational background?—A. I attended the public schools in Philadelphia and graduated from the South Philadelphia High School in the summer of 1928.

In 1930 I entered the University of Pennsylvania and left there about March of 1932.

From 1934 to '36 I was a student at Drexel Institute of Technology in the evening. I received the diploma in chemical engineering in June of 1936.

From 1938 to 1940 I attended Xavier University in Cincinnati, Ohio. I graduated from Xavier in June of 1940 and received the degree of bachelor of science, summa cum laude.

At various times from 1936 on, I took a number of specialized technical courses relating to the chemical field, and in addition I had a course in practical psychology.

Q. Now will you tell me something about your background of employment?—A. When I left high school in 19—in the summer of 1928, I was employed by a firm called Giftcrafters—that is one word—a wood-working firm in Philadel-

NOTE.—Counsel for the U. S. Government: Irving Saypol, United States attorney; Miles J. Lane, James Kilsheimer, and Roy Cohn, assistant United States attorneys. Other counsel whose names appear in transcript material represent defendants.

phia. I left the Giftcrafter firm in December of 1928 and began a period of employment with the Pennsylvania Sugar Company. I was employed by the Pennsylvania Sugar Company from January of 1929 over a period of 17 years up until February of 1946. This stretch was broken up on two occasions by leaves of absence so that I could go to college, and on one occasion in December of 1932 I was laid off.

Q. Now you are the Harry Gold, are you not, that is named as a coconspirator in the indictment which is— in the indictment which includes the Rosenbergs and Sobell and Yakovlev?—A. Yes, I am.

Q. That is in the indictment in the instant case.—A. Yes.

Q. Now do you stand convicted of any crime?—A. Yes, I do.

Q. Of what crime?—A. I stand convicted of espionage.

Q. And in what court does that conviction stand of record?—A. That conviction stands in the Federal Court in Philadelphia, Pennsylvania.

Q. Do you recall the date?—A. I pleaded guilty to the charge of espionage on July 20th, 1950.

Q. Now what sort of espionage was it?

Mr. E. H. BLOCH. I object to the question as too general, not binding upon these defendants.

The COURT. I will have to sutain that.

By the COURT:

Q. Was that indictment consummated by a sentence?—A. Yes, it was, your Honor.

Q. There is nothing left open now in that court?—A. There is nothing left open in that court, your Honor.

Q. So far as you know is there anything left open in this court with respect to a plea on your part or conviction?—A. There is nothing so far as I know, your Honor, that is left open in this court against me.

Q. So you will not come before any court any longer for any sentence, so far as you know?—A. So far as I know, your Honor, I shall not.

Q. What was the sentence that was imposed upon you in Philadelphia?— A. I was given a sentence of 30 years in the Federal Penitentiary.

The COURT. Proceed.

By Mr. LANE:

Q. Do you know Anatoli Yakovlev?—A. Yes; I do.

Q. When for the first time did you meet Anatoli Yakovlev?—A. I met Anatoli Yakovlev in March 1944.

Q. Where did you meet him?—A. I met Yakovlev in New York City. It was on the north side of 34th Street between Seventh and Eighth Avenue, and somewhat closer to Eighth Avenue. The exact spot was in front of the bar entrance of a Childs Restaurant.

The COURT. What date did you say that was?

The WITNESS. That was March of 1944, your Honor.

The COURT. All right.

Q. Now, did you have a conversation with Yakovlev at that time?

Mr. A. BLOCH. One minute. I object to it on the ground that it is not within the time circumscribed by the indictment. The indictment is alleged to have begun in June 1944.

The COURT. I suggest that you read the opinion of this Circuit Court in *United States* v. *Dennis, et al.*, which deals with that subject directly.

Your objection is overruled.

Mr. A. BLOCH. Exception.

(Question read.)

The WITNESS. Yes; I did.

Q. And as a result of that conversation what did you do?—A. I continued in my espionage work for the Soviet Union with——

Mr. E. H. BLOCH. I——

A. Anatoli——

Mr. E. H. BLOCH. I am sorry but I have got to get up to object to this. I think it is damaging evidence, and I think it is incompetent, irrelevant, and immaterial, and I believe that the witness is using conclusory words. I submit that this is not binding upon any of these defendants.

*　　　*　　　*　　　*　　　*　　　*　　　*

The COURT. The objection is overruled.

Mr. E. H. BLOCH. I respectfully except.

(Paper marked "Government's Exhibit 11" for identification.)

Mr. E. H. BLOCH. I object further, if the Court please, upon the ground that no proper foundation has been laid for the introduction of the phrase "continuing espionage for the Soviet Union," and I ask that the answer be stricken out.

Mr. LANE. If the Court please, we will connect that very shortly.

The COURT. All right.

Mr. E. H. BLOCH. And I think that is the proper way to do it, your Honor.

The COURT. All right. Why don't you go to it, Mr. Lane.

Q. Now I show you Government's Exhibit 11 for identification and ask you if you recognize the man depicted in the picture.?—A. Anatoli Yakovlev.

Q. Now, for how long a period of time did you work with Yakovlev?

The COURT. Are you introducing that?

Mr. LANE. Not yet, your Honor.

The WITNESS. I worked with Yakovlev for a period of almost three years. From March of 1944 until late December 1946.

Mr. LANE. Now I offer Government's Exhibit 11 for identification in evidence.

Mr. E. H. BLOCH. No objection.

(Government's Exhibit 11 for identification received in evidence.)

The COURT Show it to the jury.

Q. Now, did you have several conversations with Yakovlev after the middle of June 1944?—A. Yes; I did.

Q. And in any of those conversations did Yakovlev identify himself to you?—A. No; he did not. I knew Yakovlev only——

Mr. E. H. BLOCH. I move to strike out everything after "he did not," on the ground the witness has fully answered.

The COURT. All right. He didn't identify himself to you. Very well. Did you know the identity of Yakovlev?

The WITNESS. I knew Yakovlev only; I didn't know him as Yakovlev, your Honor.

Q. What name did you know him by?—A. I knew Yakovlev as John.

Q. Now, for how long a period were you engaged in Soviet activities?

Mr. E. H. BLOCH. I object to the form of the question on the ground that it presupposes a state of facts not proven.

The COURT. On that ground I will overrule it. The witness knows.

By the COURT:

Q. Do you know of your own knowledge whether or not you were engaged in Soviet espionage work?

Mr. E. H. BLOCH. I object to the Court's question.

*   *   *   *   *   *   *

The COURT. The objection is overruled.

Mr. E. H. BLOCH. I respectfully except.

Mr. A. BLOCH. I except.

(Last question read.)

The WITNESS. I was engaged in espionage work for the Soviet Union from the spring of 1935 up until the time of my arrest while I was working at the heart station of the Philadelphia Hospital. This arrest took place in May.

*   *   *   *   *   *   *

Q. Now, during this period of time did you have superiors working with you?—A. Yes; I did.

Q. Who was your Soviet superior in June 1944?

*   *   *   *   *   *   *

By Mr. LANE:

Q. Now, you have identified the picture of Yakovlev; now will you give a description of Yakovlev?—A. Yakovlev was about 28 or 30 years of age at the time that I knew him. He was about 5 feet 9 inches in height; had a medium build, which tended toward the slender. He had dark or dark-brown hair and there was a lock of it that kept falling over his forehead, which he would brush back continually. He had a rather long nose and a fair complexion, dark eyes. He walked with somewhat of a stoop.

Q. Now, you say you had a conversation with him at this meeting which was, I believe, in March of 1944, and that took place in Child's Restaurant on Eighth Avenue?—A. Yes.

Q. Now, as a result of that conversation what did you do?—A. As a result of that conversation I continued my espionage work for the Soviet Union, with Yakovlev as my new Soviet superior.

*   *   *   *   *   *   *

Mr. E. H. BLOCH. All right, I am sorry.

Q. Now, did you meet Klaus Fuchs, Dr. Klaus Fuchs, some time in the middle of June, 1944?—A. Yes, I did.

Q. Where did you meet Fuchs?—A. I met Fuchs in Woodside, Queens.

Q. Did you have a conversation with Fuchs at this time?—A. Yes, I did.

Q. As a result of that conversation what did you do?

*   *   *   *   *   *   *

The WITNESS. As a result of my meeting with Dr. Fuchs in Woodside, in the middle of June, I wrote a report, which I turned over to Yakovlev.

Q. Where did you turn this report over to Yakovlev? And when?—A. This report was turned over to Yakovlev about a week or so after my meeting with Dr. Fuchs. The place was somewhere in New York.

Q. That was in June of 1944; is that correct?—A. That is correct; the middle of June.

Q. Did you have a conversation with Yakovlev at this time?—A. Yes, I did.

Q. Will you tell the jury what that conversation was, as best you can recall?—A. I told Yakovlev—I told Yakovlev that the next time I met Fuchs, Fuchs was going to give me information. This information was to relate to the application of nuclear fission to the production of a military weapon. I gave Yakovlev the exact place where this meeting was scheduled.

Q. Was that the complete conversation?—Did you tell him anything else, that you can recall?—A. That was the complete conversation that I can recall.

Q. Now, did you have a meeting with Fuchs after that, in July of 1944, at 96th Street and Central Park?—A. I had a meeting with Fuchs in June of '44, late in June of '44.

Q. And where was this meeting?—A. The meeting with Fuchs occurred in the area of Borough Hall in Brooklyn.

Q. Did you have a conversation with Fuchs at that time?—A. We had no conversation at all.

Q. Well, what was the rest of that meeting with Fuchs? What did you do as the result of that meeting with Fuchs?—A. As a consequence of the meeting I had with Dr. Fuchs at Borough Hall I turned over to Yakovlev some few minutes after my meeting with Fuchs and still in the area of Borough Hall, I turned over to Yakovlev a package of papers.

Q. Now, did you have another meeting with Fuchs in July of 1944?—A. I had a meeting with Dr. Fuchs in the middle of July 1944.

Q. Now, where did this meeting take place and approximately what time of the day or night?—A. This meeting took place at 96th Street, about 96th Street and Central Park West. The time was the early evening.

Q. And how long approximately did this meeting last?—A. The meeting lasted about an hour and a half.

Q. And did you have a conversation with him during that time?—A. Yes; I had a conversation with Dr. Fuchs.

Q. And as the result of that conversation what did you do?—A. As the result of the conversation I had with Dr. Fuchs I turned over to Yakovlev a report which I had written.

Q. Now, when did you turn over that report to Yakovlev and where?—A. The report to Yakovlev was turned over in New York City and occurred about a week or so after my meeting with Dr. Fuchs.

Q. Now, what did this report contain that you turned over to Yakovlev, do you know?—A. Yes; I do.

Q. What did it contain?

Mr. A. BLOCH. I object to it upon the ground that it is incompetent, irrelevant, and immaterial and hearsay against these defendants and not the best evidence.

The COURT. What about that?

Mr. LANE. Well, your Honor, Lakovlev is one of the codefendants in the case.

The COURT. Yes.

Mr. LANE. The witness here is a coconspirator.

The COURT. Yes.

Mr. LANE. The witness read the report. He knew the contents of the report and he turned it over to Yakovlev, but I will withdraw the question, your Honor, and I will ask the witness if he had a conversation with Yakovlev on this particular occasion.

The WITNESS. Yes; I did.

Q. And what was the conversation?—A. The conversation had to do with the fact that Fuchs had given me further information on the progress of the work going on in New York by a joint American and British project, which project was aimed at producing an atomic bomb. The work was going on somewhere in the area of Church Street, New York.

The COURT. This is in July of 1944?

The WITNESS. Yes. It is the middle of July 1944.

Q. Now, during this period when you were meeting Dr. Fuchs, were you also meeting Yakovlev at regular intervals?—A. During this period I had a regularly scheduled series of meetings with Yakovlev. Yakovlev continually advised and instructed me and we talked over together as to how I should continue my work with Dr. Fuchs.

Q. Now, did you have any particular modus operandi in your connections, in your work with Yakovlev?—A. Yes, I did.

Q. Would you tell the jury, would you relate the details of that?

Mr. E. H. BLOCH. Do I understand that this is pursuant to the directions——

Mr. LANE. Pursuant to a conversation I believe with Yakovlev.—A. I worked in the following manner with Yakovlev: My duties were to obtain information from a number of sources in America and to transfer this information to Yakovlev. The meetings with the sources of information in America were effected in two ways: First, it could be personal introduction; secondly, there was an introduction which was effected only between the American contact and myself and was effected by means of a set of recognition signals. Now, these signals included at least two features: One of them was that there was always an object or a piece of paper involved, on one part or the other, or possibly on both parts; that is the other person in America and myself. In addition there was a code phrase used and this phrase was usually used in the form of a greeting. In all cases when I introduced myself I used a false name, and in all cases I never indicated my true place of residence.

Now, once the introduction had been effected, I proceeded to work: I conducted myself in the following manner: I give the source of information in America—whoever that person was who was going to furnish me the information—I give him a list of the data or material which was desired; secondly, in case there had been a Soviet agent who had preceded me, I would take steps so that the source of information, that the person with whom I was working would first clean up all of the back work. Then thirdly, we would arrange for a series of meetings. These were very precise arrangements.

All of these people from whom I obtained information were not residents of Philadelphia. I had to go some distance usually to meet them. I would arrange for a meeting in the town where they lived or in some other town, and the meeting would be for an exact time at an exact place and there would be an exact schedule for what was to be done during the meeting. In other words, if we were just going to discuss the possibility of obtaining certain types of information, the hazards involved, just how much information should be obtained, and just what source was needed, then a rather long meeting was scheduled. If I was going to actually get information, very usually a brief meeting was scheduled, the idea being to minimize the time of detection when information would be passed from the American to me. In addition to this, I made payments of sums of money to some of the people whom I regularly contacted, and always I wrote reports detailing everything that happened at every meeting with these people, and these reports I turned over to Yakovlev.

By the COURT:

Q. And where would you get the money from that you paid to some of these people for the information?—A. The money was given to me by Yakovlev.

(Paper marked "Government's Exhibit 12" for identification.)

By Mr. LANE:

Q. Now I show you Government's Exhibit 12 for identification and I ask you if you can identify the person in that picture?—A. This is Dr. Klaus Fuchs.

*   *   *   *   *   *   *

HARRY GOLD resumed the stand.

Direct examination continued by Mr. LANE:

Q. Now, Mr. Gold, at the conclusion of this morning's session I believe you were giving an account of a conversation with Yakovlev, in which you were reciting the details of the modus operandi. Now, did you complete your conversation with reference to that particular item?—A. No; I had not.

Q. Will you continue, please?—A. In addition to the details of my operations, of my conduct with the source of information in America, I had a very set pattern which I used in connection with my dealings with Yakovlev. This is how it worked: We had an arrangement not only for regular meetings, but we had an arrangement for alternate meetings, should one of the regular ones not take place, and then, in addition to that, we had an arrangement for an emergency meeting. This emergency meeting was a one-way affair. A system was set up whereby Yakovlev could get in touch with me if he wanted me quickly, but I couldn't get in touch with him because I didn't know where. Yakovlev told me that in this way the chain was cut in two places. The person from whom I got the information in America did not know me by my true name, nor did he know where I lived, nor could he get in touch with me, and I couldn't get in touch with Yakovlev. Yakovlev said this was a good thing.

In addition, Yakovlev and I had a very exact method when we were going to transfer information. It might be that something would have to be copied and then returned. In that case, we had to have a set procedure. On certain occasions we had a definite means by which we transferred information. Such means would include a set-up whereby I would take the information and put it between the folds of a newspaper and Yakovlev and I would exchange the newspapers. The one that I got was just a newspaper. The one that he got had the information between the folds, the information usually being in some sort of an enclosure. In addition to this, of course, we had regular conferences all along at which we discussed my conduct with the people in America who were furnishing me with information, and the final point was that we had a system set up whereby we could act or react very promptly in case there was any sign of surveillance, and this system provided for not only surveillance should I notice any before I came to see Yakovlev but surveillance if we should suspect any while we were actually having a meeting, and also if there should be any suspicious actions around either of us after we had parted, around either of us after we had parted.

Q. Now, does that exhaust your recollection of that conversation?—A. That is about all that I can recall about that conversation, Mr. Lane.

Q. Now, did you have a meeting with Dr. Fuchs in Cambridge, Massachusetts, early in January of 1945?

Mr. E. H. BLOCH. If the Court please, do I assume that this question presupposes that this is part of the conspiracy charged in this indictment?

Mr. LANE. Yes, your Honor.

The WITNESS. Yes; I did.

Q. Did you have a conversation with Dr. Fuchs at that time?—A. Yes; I did.

Q. As a result of that conversation, what did you do?—A. In consequence of the conversation that I had with Dr. Fuchs, I did two things. On the same day that I saw Dr. Fuchs, the same morning that I saw Dr. Fuchs in Cambridge, Massachusetts, I returned from Cambridge to New York City and turned over to Yakovlev a package of papers which Fuchs had given me. About a week later, I wrote a report which I turned over to Yakovlev.

Q. When did you turn this report over to Yakovlev?—A. The report was turned over to Yakovlev somewhere around the second week in January.

Q. Whereabouts did you turn this report over to Yakovlev?—A. It was turned over to Yakovlev somewhere in downtown.

Mr. E. H. BLOCH. Pardon me. Is that January 1945?

Mr. LANE. That is correct.

The WITNESS. January '45.

Q. At that time did you have a conversation with Yakovlev?—A. Yes; I did.

Q. What was the conversation? A. I told Yakovlev that I had received the following information from Fuchs: First, that Fuchs was now stationed at a place called Los Alamos, New Mexico; that this was a large experimental station. It had formerly been a boys' school, a select boys' school. Fuchs told me that a tremendous amount of progress had been made. In addition, he had made mention of a lens, which was being worked on as a part of the atom bomb. Finally, we had set a date, this date to be the first Saturday in June of 1945, and at this date I was to meet Fuchs in Santa Fe, New Mexico.

Q. Did you have a further meeting with Fuchs within the next month or so?—A. Not within the next month or so.

Q. Did you have a meeting with Yakovlev in May of 1945?—A. I had a meeting with Yakovlev in February, early in February of 1945.

Q. Where did this meeting take place?—A. This meeting took place about 23rd Street and Ninth or Tenth Avenue.

Q. Did you have this conversation with him at that time?—A. Yes, I did.

Q. What was the conversation? A. At this meeting Yakovlev told me to try to remember anything else that Fuchs had mentioned during our Cambridge meeting, about the lens. Yakovlev was very agitated and asked me to scour my memory clean so as to elicit any possible scrap of information about this lens. I recall the meeting very well, because I was frightened, it was a bad neighborhood to be in.

Mr. E. H. Bloch. I move to strike that out.

The Witness. I am sorry.

The Court. All right, strike it out.

Mr. Saypol. What is the objection? Is it the fact that he was frightened, if the Court please?

The Court. I am striking out being frightened.

Mr. Saypol. Otherwise I take it the witness may continue?

The Court. That is right.

Q. Does that exhaust your recollection of the conversation at that particular meeting?—A. Yes; it does.

By the Court:

Q. Well, did you give him anything else?—A. What is it?

Q. You say he asked you to search your memory, give him every scrap of information you had. Now, did you give him anything else at that meeting?— A. I didn't give him anything else, that I can recall, at that meeting.

The Court. Very well.

A. (Continuing). I think I put everything down in the report which I gave about the second week in January.

The Court. All right.

By Mr. Lane:

Q. Now in May of 1945, did you have a meeting with Yakovlev?—A. Yes; I did.

Q. Will you try and fix the date as best you can of that particular meeting?— A. The exact date of the meeting that I had with Yakovlev in May of 1945 was on the last Saturday in May, and about—pretty late in the afternoon, around four in the afternoon, I would say.

Q. Where did this meeting take place?—A. The meeting with Yakovlev on the last Saturday in May took place inside of a combination restaurant and bar, called Volks—V-o-l-k—V-o-l-k-s. I think it is, k's, or possibly, V-o-l-c-k-s. In any case, it is at the southwest corner of 42nd Street and Third Avenue, in Manhattan.

Q. Now, will you tell the jury what happened on this occasion?—A. I stood at the bar with Yakovlev and we had a drink. I told Yakovlev that I wanted to take a walk with him, to dispose of the two matters which were the purpose of our meeting. These matters were, one, first, that Yakovlev would be able to ascertain definitely that I was going to see Fuchs on the first Saturday in June of 1945. I was having difficulty at this time getting off from work and he wanted to make sure that I was going. The second point of our meeting was that we had to arrange for two meetings when I returned from Santa Fe, one meeting at which I would transfer information which I was supposed to receive from Fuchs; then there would be a second meeting some time later, at which I would give Yakovlev a detailed report, as well as a verbal account of exactly what would have transpired at this meeting with Fuchs.

Now, Yakovlev told me that he had some matter to discuss with me at length and he said that we should sit in the rear of Volks' cafe, where there was a circular place with some tables in it, fairly secluded. We did so. We sat down there and the waiter brought us a drink. Yakovlev told me that he wanted me to take on an additional mission besides the one to see Dr. Fuchs. He said that he wanted me to go to Albuquerque, New Mexico. I protested——

Mr. E. H. Bloch. I move to strike that out, your Honor.

The Court. All right, strike out "protested."

The Witness. Sorry, your Honor.

The Court. Tell us what you told him.

A. (Continuing) I told Yakovlev that I did not wish to take on this additional task. Yakovlev told me that the matter was very vital and that I had to do it. He said that a woman was supposed to go in place of me but that she was unable to make the trip. He said, therefore, he said that I had to go. I told Yakovlev that it was highly inadvisable to endanger the very important trip to see Dr. Fuchs with this additional task. Yakovlev told me that I didn't understand that this was an extremely important business, that I just had to go to Albuquerque, in addition to going to Santa Fe, and he said, "That is an order"; and that was all. I agreed to go.

Yakovlev then gave me a sheet of paper; it was onionskin paper, and on it was typed the following: First, the name "Greenglass," just "Greenglass." Then a number "High Street"; all that I can recall about the number is that the last figure—it was a low number and the last figure, the second figure was "0" and the last figure was either 5, 7. or 9; and then underneath that was "Albuquerque, New Mexico." The last thing that was on the paper was "Recognition signal. I come from Julius." In addition to this, Yakovlev gave me a piece of cardboard, which appeared to have been cut from a packaged food of some sort. It was cut in an odd shape and Yakovlev told me that the man Greenglass, whom I would meet in Albuquerque, would have the matching piece of cardboard. Yakovlev told me that just in case the man Greenglass should not be present when I called in Albuquerque, that his wife would have the information and that she would turn it over to me. Yakovlev gave me an envelope which he said contained $500, and he told me to give it to Greenglass. Yakovlev told me that I should follow a very devious route on my way to Santa Fe and to Albuquerque. He said that I should go first to Phoenix, Arizona; then to El Paso, and from there to Santa Fe. Yakovlev said that I should do this to minimize any danger of being followed.

The last thing that took place on this last Saturday in May was that Yakovlev and I arranged for two meetings upon my return from the Southwest.

Q. Were you to go to Santa Fe before Albuquerque?—A. Yakovlev told me that I was to go to Santa Fe first.

Q. Was there any route to be followed from Santa Fe to Albuquerque?—A. He didn't specify, your Honor. There is only one way that I know of to get from there. They are 60 miles apart and there is a bus runs between them.

The COURT. All right.

By Mr. LANE:

Q. Do you have the paper on which those instructions were written?—A. I have the paper. When Yakovlev gave me the paper——

Mr. E. H. BLOCH. I move to strike out everything else, your Honor.

The COURT. All right, strike it out.

Q. What happened to that paper, if you know?—A. I was to destroy the paper.

Q. Did you destroy it?—A. I don't know what happened to it. I just remember the emphasis on memorizing what is in it and destroy it.

Q. Now, will you describe that cardboard you received from Yakovlev on that occasion?—A. Yes. It was about, just about an inch in height, possibly two inches—yes, possibly two inches in length and was shaped this way [indicating]. It was rectangular and it was shaped like this [indicating]. I can draw it.

Q. Well, I show you Government's Exhibits 4–A and 4–B in evidence and ask you if one of those represents the cardboard you received from Yakovlev?—A. This piece here where I have my finger.

Mr. LANE. May the record indicate that the witness is pointing at Government's Exhibit 4–B in evidence.

Q. Now, did you go to Santa Fe in June of 1945?—A. Yes; I did.

Q. When did you arrive in Santa Fe?—A. I arrived in Santa Fe on Saturday the 2d of June 1945.

Q. And did you meet Dr. Fuchs?— Yes; I did.

The COURT. What was that date again that you gave?

The WITNESS. The 2nd of June 1945, a Saturday.

Q. Did you have a conversation with Dr. Fuchs?—A. Yes; I did.

Q And how long did that conversation last?—A. My conversation with Dr Fuchs in Santa Fe lasted about 20 minutes to a half hour.

Q. And as the result of that conversation what did you do?—A. As the result of that conversation I did two things: Upon my return to New York I turned over to Yakovlev a bunch of papers which Dr. Fuchs had given me in Santa Fe. About two weeks after that——

Q. No. I neglected to ask you how long this conversation took place in Volks' Cafe with Yakovlev? How long did it last?—A. The conversation with Yakovlev in Volks' Cafe took about an hour; three-quarters of an hour to an hour. It is hard to place it just exactly.

Q. Now, when did you depart from Santa Fe?—A. I left Santa Fe in the very late afternoon.

Q. Where did you go?—A. The 2d of June.

Q. Where did you go?—A. I went by bus from Santa Fe to Albuquerque.

Q. What did you do after you arrived in Albuquerque?—A. I arrived in Albuquerque early in the evening of the 2d of June, and about 8:30 that night went—about 8:00 or 8:30, yes, I went to the designated address on High Street. There I was met by a tall elderly whitehaired and somewhat stooped man. I inquired about the Greenglasses and he told me that they were out for the evening but he thought they would be in early on Sunday morning.

Q. Then what did you do?—A. Then I returned to downtown Albuquerque.

Q. Did you register in any hotel when you were in Albuquerque on this occasion?—A. I stayed that night—I finally managed to obtain a room in a hallway of a rooming house and then on Sunday morning I registered at the Hotel Hilton.

Q. Now, did you register under your own name?—A. Yes; I did.

Q. What name did you use?—A. Harry Gold.

Q. Now, what did you do on Sunday?—That is June 3, 1945?—A. On Sunday about 8:30 I went again to the High Street address. I was admitted, and I recall going up a very steep flight of steps, and I knocked on a door. It was open by a young man of about 23 with dark hair. He was smiling. I said, "Mr. Greenglass?" He answered in the affirmative. I said, "I came from Julius," and I showed him the piece of cardboard in my hand, the piece of cardboard that had been given me by Yakovlev in Volks' Cafe. He asked me to enter. I did. Greenglass went to a woman's handbag and brought out from it a piece of cardboard. We matched the two of them.

Q. Will you describe that cardboard which Greenglass showed you?—A. It appeared to be from the same, part of the same packaged food from which the piece of cardboard that I had had originally been cut.

By the COURT:

Q. When you say Greenglass matched the two, just what did he do?—A. He showed it to me and we put them together, as nearly as I can remember.

Q. How did you put them together? To see whether the ends met, is that what you did?—A. No. Just roughly. I mean you could see at a glance that they were the same thing.

By Mr. LANE:

Q. Now I show you Government's Exhibit 4–A in evidence and I ask you if it resembles the piece of cardboard which Greenglass showed to you?—A. This resembles it, the one that I had my finger on, 4–A, yes.

Mr. LANE. May the record indicate that the witness has identified Government's Exhibit 4–A as the cardboard replica of the piece that was shown to him by David Greenglass.

Q. And the two pieces matched, you say?—A. The two pieces matched.

Q. Now, will you continue the conversation which you had at that time?

(Marked "Government's Exhibit 13" for identification.)

The WITNESS. At this point after we had matched the two pieces of cardboard I introduced myself to Greenglass as Dave from Pittsburgh; that was all. Greenglass introduced me to the young woman who was there and said she was his wife Ruth.—Mrs. Greenglass said that it was coincidence that my first name and the first name of her husband were the same.

Greenglass then told me that my visit to him on this exact day was a bit of a surprise; he had not expected me right on that day, but that nevertheless he would have the material on the atom bomb ready for me that afternoon.

At this point Mrs. Greenglass went into the kitchen to prepare some food. Then I gave Mr. Greenglass the envelope which Yakovlev had given me in Volks' Cafe. This envelope was the one that contained $500. Greenglass took the envelop from me. Greenglass told me that he would have the information ready at about 4 o'clock, 3:00 or 4 o'clock in the afternoon; the exact time I can't recall except that we set it.

At this point Greenglass told me that there were a number of people at Los Alamos that he thought would make very likely recruits; that is, they were also people who might be willing to furnish information on the atom bomb to the Soviet Union, and he started to give me the names of these people, the names of some of these people. I cut him very short indeed. I told him that such procedure was extremely hazardous, foolhardy, that under no circumstances should he ever try to proposition anyone on his own into trying to get information for the Soviet Union. I told him to be very circumspect in his conduct and to never even drop the slightest hint to anyone that he himself was furnishing information on the atom bomb to the Soviet Union.

The last thing that took place that morning was that just as I was preparing to go, Mrs. Greenglass told me that just before she had left New York City to come to Albuquerque she had spoken with Julius. This meeting that we had in the morning on the 3rd of June 1945, this Sunday morning, took about 15 minutes.

Q. Now I show you Government's Exhibit 13 for identification and I ask you if you can identify the people in that picture.—A. Yes. The man with his arm around the woman is David Greenglass. The woman is Mrs. Ruth Greenglass.

Mr. LANE. I offer Government's Exhibit 13 for identification in evidence.

Mr. E. H. BLOCH. No objection.

(Marked Government's Exhibit 13.)

Q. Did you return to the home of the Greenglasses on the same day, Sunday, June 3, 1945?—A. Yes; I did.

Q. And what time did you return?—A. It was about between 3:00 and 4 o'clock.

Q. And who was present when you returned?—A. Mr. Greenglass was there and his wife Ruth was there.

Q. And did you have a conversation at that time?—A. Yes; I did.

Q. Will you tell the jury what that conversation was?—A. Mr. Greenglass gave me an envelope which he said contained the information for which I had come, the information on the atom bomb. I took the envelope. Mr. Greenglass told me that he expected to get a furlough sometime around Christmas, and that he would return to New York at that time. He told me that if I wished to get in touch with him then I could do so by calling his brother-in-law Julius, and he gave me the telephone number of Julius in New York City.

Q. Do you recall now what that number was?—A. I cannot.

Q. Was that all that was said at that time?—A. I told Greenglass that very likely I might be returning to Albuquerque in the early fall of 1945, and if I did so there was a possibility that I might stop in and see him.

Q. And did you receive some information at that time, some papers from Greenglass?

The COURT. At what time?

The WITNESS. I have already related——

Mr. LANE. That Sunday afternoon.

The WITNESS. I have already related that he gave me an envelope which contained, which he said contained information on the atom bomb.

Q. But did you examine it at that time?—A. I did not examine it at that time, no. I just took the envelope and put it in my pocket.

Q. How long a period of time did this meeting consume?

The COURT. How large an envelope was that?

The WITNESS. I would say it was a large letter size, pretty large letter size, as I recall it.

Q. How much time was consumed in this particular meeting?—A. The meeting took only about five minutes, possibly ten at the very most.

Q. And then what did you do?—A. The three of us, Mr. Greenglass, Mrs. Greenglass, and myself left the Greenglass' apartment and we walked along a slanting back street in Albuquerque, and there in front of a small building I left the Greenglasses.

Q. And did you return to New York?—A. Yes; I did.

Q. Immediately?—A. Yes; I did.

Q. En route to New York did you at any time inspect the material which you had received from Greenglass?—A. Yes, I did, on the train from Albuquerque to Chicago and somewhere in Kansas, I believe. I examined the material which Greenglass had given me. I just examined it very quickly. My sole purpose was to get it——

Mr. E. H. BLOCH. I object to what his purpose was.

The COURT. All right.

The WITNESS. I put it into an envelope, into a manila envelope, one of the kind with a brass clasp, and in another manila envelope I put the papers which Dr. Fuchs had given me. I labeled the two envelopes. On the one from Fuchs I wrote "Doctor." On the one from Greenglass I wrote "Other," O-t-h-e-r.

Q. Well, now, can you describe a little better what was in the Greenglass package?—A. Yes. The material given me by Greenglass consisted of three or four handwritten pages plus a couple of sketches. The sketches had letters on them which were referred to in the text of the three or four handwritten pages. The sketches appeared to be for a device of some kind.

Q. When did you arrive back in New York?—A. I arrived in New York on the 5th of June.

By the COURT:

Q. Did you read the descriptive matter that was contained on these other sheets of paper?—A. No; I did not, your Honor. I just—it happened to hit my eye and I leafed through them, is what I recall.

By Mr. LANE:

Q. Now the question was, when did you arrive back in New York?—A. I arrived in New York on the 5th of June 1945, in the evening.

Q. And did you meet Yakovlev?—A. Yes, I did.

Q. On the same day?—A. On the same evening.

Q. Where did you met him?—A. I met Yakovlev along Metropolitan Avenue, in Brooklyn, and somewhere, where Metropolitan Avenue runs into Queens. It was a very lonely place, particularly at the time of night when I met Yakovlev.

The COURT. What time was it?

The WITNESS. It was about 10 o'clock at night.

Q. Now did I understand you to say that this meeting was by prearrangement?—A. This meeting had been arranged at Volks' cafe on the last Saturday in May of 1945.

The COURT. Excuse me. You say there was an alternative meeting arranged also, in the event you didn't meet that evening?

The WITNESS. No, Your Honor, there was an alternative meeting but its primary purpose was to hand over a written report on what had transpired during my trip to Santa Fe. If it should have happened that I hadn't been able to make the meeting with Yakovlev, then the alternative meeting would have had to serve a double purpose.

The COURT. Very well.

Q. How long did this meeting last with Yakovlev?—A. The meeting with Yakovlev lasted about a minute, that was all.

Q. What did you do at that meeting?—A. We met and Yakovlev wanted to know if I had seen the both of them, said, "The doctor and the man." I said that I had. Yakovlev wanted to know had I got information from the both of them and I said that I had. Then I gave Yakovlev the two manila envelopes, the one labeled "Doctor," which had the information I had received from Fuchs in Santa Fe; the one labeled "Other," which had the information I had received from David Greenglass in Albuquerque, on the 3d of June 1945.

Q. When was your next meeting with Yakovlev?—A. My next meeting with Yakovlev was about two weeks later, and——

Q. Where did this meeting take place?—A. This meeting took place at the end of the Flushing elevated line, at Main Street, in Flushing.

Q. What time was it?—A. The time was in the middle of the evening.

Q. Did you have on conversation with Yakovlev?

The COURT. Excuse me. When did you say that was?

The WITNESS. In the middle of the evening.

The COURT. What date did you say that was?

The WITNESS. It was about two weeks after the 5th of June. We are still in June.

Q. Did you have a conversation with Yakovlev at that time?—A. Yes, I did.

Q. Will you tell the jury what that conversation was?—A. I told Yakovlev—Yakovlev told me that the information which I had given him some two weeks previous had been sent immediately to the Soviet Union. He said that the information which I had received from Greenglass was extremely excellent and very valuable. Then Yakovlev listened while I recounted the details of my two meetings, the one with Fuchs in Santa Fe, the one with Greenglass in Albuquerque. I told Yakovlev that Fuchs had related that tremendous progress had been made on the construction of the atom bomb and Fuchs had said that the first trial of an atomic explosion was due to take place in July, in New Mexico. Fuchs and I—I told Yakovlev that Fuchs and I had agreed upon a meeting between the two of us in Santa Fe for September of 1945.

Q. Approximately how long did this conversation last, with Yakovlev?—A. This conversation lasted about two and a half hours. I gave every detail at that time of my two meetings with these people who furnished me with information in the Southwest.

Q. Did you have a meeting with Yakovlev early in July of 1945?—A. Yes, I did.

Q. Where did this meeting take place?—A. The meeting with Yakovlev early in July of 1945 took place inside of a combination seafood and bar, which is at the Broadway stop of the Astoria elevated line.

Q. Did you have a conversation?—A. Yes, I did.

Q. What was this conversation?—A. Yakovlev told me that it was necessary that we have an arrangement whereby some Soviet agent other than himself could get in touch with me. At Yakovlev's instructions, I took from my pocket a piece of paper. This piece of paper was a memorandum sheet, something like an invoice, from the Arthur H. Thomas Company, of Philadelphia, a laboratory supply house, and I tore off the top part of this memorandum sheet, and on the reverse side, at Yakovlev's instructions, I wrote the following words, and I wrote it across as a diagonal of a parallelogram. I wrote, "Directions to Paul Street" and the "Street" is spelled out, S-t-r-e-e-t. Then Yakovlev took this piece of paper and he tore it in an irregular fashion, so that the tear came between the P and the a of the Paul. Yakovlev retained the part of the paper which said "Directions to P" and I was given by Yakovlev the piece of paper which had on it, "aul Street." Yakovlev told me that should I ever receive two tickets in an envelope, in the mail, with no other enclosure inside that envelope, that I should take this to be a signal, that at a definite number of days after the date that was printed on the tickets—the tickets themselves were to be tickets to either a sporting event or a theatrical attraction in New York City; it would have a date on them—now, a certain number of days after that date on the tickets, I was to go to the Astoria—to the Broadway stop of the Astoria elevated line, but before I went there I was to scout the place very carefully for about an hour, to make sure there were no signs of surveillance. Then when I was inside, I was to take a seat at a table in this combination restaurant and bar. My replacement Soviet contact was to stand at the bar where he could observe me, to again watch for any symptom of anything suspicious. Then he would approach me, the new Soviet contact, and he would say, "Can you direct me to Paul Street?" I was seated at the table. I would answer, "Yes, I am going there myself. Come along." First, however, he would show me the piece of paper, that is the first thing he would do, he would show me the piece of paper, and then I would produce the matching half. Then would come the recognition signal or the recognition phrases having to do with the directions to Paul Street.

(Marked Government's Exhibit 14 for identification.)

Q. I show you Government's Exhibit 14 for identification, and I ask you if this is a replica of the sheet of paper which you had on that particular occasion of your conversation with Yakovlev [handing]?

*   *   *   *   *   *

Mr. E. H. Bloch. May I ask Mr. Lane to explain what he means by "replica"?
The Court. Yes.
Mr. Lane. Well, it is a duplicate, I believe, Mr. Bloch, of the same sheet of paper that he had on that occasion, took out of his pocket.
Mr. E. H. Bloch. Does this purport to be the original?
Mr. Lane. No, it does not.
Mr. E. H. Bloch. Oh, I see.

*   *   *   *   *   *

The Witness. This sheet of paper is of the type used by the Arthur H. Thomas Company when someone went there to pick up a package. That was how I had got the original sheet and happened to have it in my pocket.

Q. Is that a duplicate of the sheet you had on that particular occasion?—A. This appears to be a duplicate, Mr. Lane. I just want to see one thing—yes, I was looking at the reverse side to see if there was any printing on there. There isn't.

Mr. E. H. Bloch. Mr. Lane, I wonder whether you can let me look at it, so that I don't get confused.
Mr. Lane. Yes [handing].

Q. Now, will you write on this Government's Exhibit 14 for identification the wording that was on that paper, as best you can recall it, and then tear it in the way that you recall it was torn on that particular evening? [Handed to witness.]—A. The wording will have to be written on the reverse side, because that is the way that it was done. It was on the reverse side, not the side where the printing is, but on the back.

Q. Well, you do it as best you recall it.—A. (Witness writes on Government's Exhibit 14 for identification.)

Q. Have you got the writing on there, Mr. Gold?—A. I have got the writing there.

Q. Now, will you tear it in the way you recall it was torn at that time?—A. There is one difficulty, which was unavoidable here. The actual piece of paper was larger in this direction [indicating], but in order not to touch the imprint here, "Government's Exhibit," I had to crowd the words towards the top a little bit, so that it is actually just a little smaller in this direction [indicating].

Mr. LANE. On the left-hand side. May the record indicate that the witness is pointing to the left-hand side of the exhibit?

Q. Go ahead, rip it, tear it the way it was torn.—A. Well, I will have to tear across this notation.

Q. Go ahead, tear across it.—A. (Witness tears as requested:) Just about like this. Now, this would have to go down a little further. Shall I do it?

Q. You do it.—A. (Witness tears further.)

Q. Now, is that a reasonable facsimile of the way the paper was torn before?—A. Yes, it is.

* * * * * * *

By Mr. LANE:

Q. Mr. Gold, did you have a meeting with Yakovlev in the middle of August 1945?—A. Yes, I did.

Q. Where did this meeting take place?—A. This meeting took place in Brooklyn.

Q. And did you have a conversation at that time?—A. Yes, I did.

Q. Will you tell the jury what the conversation was?

The COURT. Let me get this. When was this?

Mr. LANE. August of 1945, your Honor.

The COURT. All right, proceed.

The WITNESS. The conversation concerned the fact that I was to take, soon to take a trip in September to Santa Fe to meet Dr. Fuchs. I told Yakovlev that since I was going to see Dr. Fuchs I might as well go to Albuquerque and see the Greenglasses. At this time Yakovlev told me that it would be inadvisable to endanger the trip to see Fuchs by complicating it with a visit to the Greenglasses in Albuquerque.

Q. Now, was that the sum and substance of that conversation?—A. That is the sum and stance of the conversation.

Q. Now, did you have a meeting with Dr. Fuchs in September 1945?—A. Yes; I did.

Q. When and where did this meeting take place?—A. The meeting took place in Santa Fe, New Mexico, on the 19th of September 1945. The exact place was on the outskirts of Santa Fe along a road leading out of Santa Fe near a large church.

Q. As the result of this conversation what did you do?—A. As the result of the meeting—of the conversation I had with Dr. Fuchs I returned to New York City and I did two things: I returned on the 22d of September and I missed my meeting with Yakovlev. I got back just too late that evening to see Yakovlev. About 10 days later I met Yakovlev and this time—the original meeting on the 22d was supposed to be a short one. This time I met him in Main Street in Flushing, in Main Street, Flushing, and I turned over to him a package of information which I had received from Fuchs.

Q. Now, at this time did you have a conversation with Yakovlev, in your September meeting with Yakovlev? Was it September or October?—A. This was about 10 days later I would say or two weeks—it was still September, as I recall. Yes, I did.

Q. What was the conversation?—A. The conversation covered the following matters which Fuchs told me: Fuchs had said first that there was no longer open and free and easy cooperation between the British and the Americans at Los Alamos and that many departments which had been formally readily accessible to him were now closed. He said as a result his superiors among the Britains had advised him that he would probably very soon have to return to England.

The second thing was that Fuchs told me that the first explosion at Alamogordo, New Mexico, in July had produced a tremendous feeling of awe among the scientists who had been producing it. He said in addition the attitude of the townspeople had changed quite a bit. Before that they had been regarded as a sort of boondoggling outfit and how they were raised to the level of heroes.

Fuchs told me, the third thing, that he had attended——

Mr. E. H. BLOCH. If the Court please, may we approach the bench because it is in connection with a proposition of law on this Fuchs conversation. I would like to discuss it with the Court and Mr. Saypol.

The COURT. Come up.

* * * * * *

Mr. E. H. BLOCH. Just to clarify the record, is it understood that what the witness is now testifying about is a conversation which he had with one of the defendants?

The COURT. Yakovlev.

Mr. E. H. BLOCH. Yakovlev?

The COURT. Yes.

The WITNESS. I reported to Yakovlev that Fuchs had told me about being present at the first atomic explosion at Alamogordo, New Mexico. Fuchs had said that the flash had been visible some 200 miles away. I told Yakovlev that Fuchs was very worried about one matter: This concerned the fact that the British had gotten to Kiel, Germany, ahead of the Nazis—ahead of the Russians, and Fuchs was very worried, very much concerned over whether the British Intelligence might not discover the Gestapo dossier upon him. I told Yakovlev that Fuchs had said that he, Fuchs that is, had been the leader of the student group, the German student group at the University of Kiel and had fought the Nazis, Nazi storm troopers in the streets of Kiel. Fuchs had said that there was a very complete dossier by the Gestapo upon him and he was greatly troubled by the fact that should the British Intelligence come upon it they would become aware of his very strong Communist background and ties.

Mr. E. H. BLOCH. If your Honor please, although as we as lawyers may understand it quite clearly, I would ask the Court to instruct the jury, to tell the jury at this point that Dr. Klaus Fuchs is not a defendant or a conspirator in this proceeding.

The COURT. Dr. Klaus Fuchs is not a defendant in this case. This testimony is being admitted in connection with the general charge of conspiracy to commit espionage, in which Yakovlev is a defendant.

Mr. SAYPOL. Will the Court supplement that observation with the statement also that this witness is a coconspirator?

The COURT. Oh, yes. The witness Harry Gold is a coconspirator.

The WITNESS. The final item I reported to Yakovlev were the details of an arrangement which Fuchs and I had arrived at, which arrangement concerned the means by which someone would get in touch with Fuchs when he returned to England. The exact details were these: Beginning on the first Saturday of every month after it had been determined that Fuchs had returned to England, at a stop on the British subway, underground in London called Paddington Crescent, possibly Teddington Crescent, 8 p. m., Fuchs was to be on the street at the underground stop, the street level. He was to be carrying five books bound with strings and supported by two fingers of one hand; he was to be carrying two books in another hand. His contact, whoever that would be, was to be carrying a copy of a Bennett Cerf book, Stop Me if You Have Heard This.

Q. Does that conclude your conversation with Yakovlev at that time?—A. That concludes my conversation with Yakovlev.

Q. When was your next meeting with Yakovlev?—A. Yakovlev and I had a regular series of meetings all throughout that fall of 1945.

Q. Do you recall a meeting in November of 1945?—A. Yes; I do.

Q. With Yakovlev?—A. Yes; I do.

Q. Where was this meeting held?—A. This meeting was held in New York City.

Q. And did you have a conversation with Yakovlev?—A. Yes; I did.

Q. Will you tell the jury what that conversation was?—A. I told Yakovlev that Greenglass had said at the time that I saw him in June of that year that Greenglass would be possibly coming home on a furlough about Christmas time. I told Yakovlev the time was drawing near to Christmas and that we ought to make some plan to get in touch with this brother-in-law, Julius, so that we could get further information from Greenglass. Yakovlev told me to mind my own business. He cut me very short.

Q. Now, in January 1946 did you have a conversation with Yakovlev?—A. Yes, I did.

Q. Where did this conversation take place?—A. This conversation took place in New York City.

Q. What was the conversation?—A. Yakovlev at this time told me that I should be very careful, much more careful than ever before. He related to me an incident which had taken place toward the end of 1945. He said that a very important person who had upon him information on the atom bomb had come to New York at the end of 1945 and that he, Yakovlev, had tried to get in touch with that person over a period of time, a period of a few days, but that the man had been trailed by Intelligence men continually, so that Yakovlev had to give up the idea of getting in touch with this source of information.

Yakovlev told me, he was telling me the story as an example of how to conduct myself. He said if it came to a choice it was far better to pass up any chance of getting information than to do anything which would endanger the whole scheme. Yakovlev said that he was very apprehensive.

By the COURT:

Q. Did you know what Yakovlev was doing by way of employment aside from these activities?—A. I did not, your Honor. I had assumed——

Mr. E. H. BLOCH. Wait, if the Court please. I object to what he might have assumed.

Q. Now was there anything else said on that occasion about any future meetings—A. Yes. Yakovlev and I had arranged the following. We made a slight change in our method of meeting. Should a scheduled meeting not take place, then the next meeting which would take place would be at the same time as had originally—at the same place that had originally been planned, but the time was changed and the day of the week was changed, and should this alternate meeting not take place, then the emergency meeting would take place. But I was to use extreme caution about the emergency meeting. I was to go there well in advance, to scout it thoroughly, and to make certain that there was absolutely no surveillance going on.

Q. Now, did you have a meeting scheduled with Yakovlev for February of 1946?—A. Yes, I did.

Q. And where was this meeting scheduled to take place?—A. This meeting in February was scheduled to take place at the Earl Movie Theatre in the Bronx.

Q. Did you keep this appointment with Yakovlev?—A. Yes, I did.

Q. And what happened on that occasion?—A. Yakovlev didn't show up.

Q. When was the next time that you saw Yakovlev?—A. The next time that I saw Yakovlev was on the night of the 26th of December 1946.

Q. And you didn't see him in the interim between February and December of 1946?—A. For a period of about ten months I did not see Yakovlev.

Q. Now, will you tell the jury how you happened to meet Yakovlev in December of 1946?—A. Early in December of 1946 I received two tickets, two tickets to a boxing match in New York City. The tickets had arrived very late for me to keep the appointment. They had been addressed wrongly; they had been sent to 6328 Kindred Street instead of 6823 Kindred Street. So I wasn't able to keep the appointment.

Q. Kindred Street where?—A. In Philadelphia, my home in Philadelphia. At about five o'clock in the late afternoon of the 26th of December the phone rang at the laboratory where I was working in New York City, the laboratory of A. Brothman and Associates. I answered. A voice asked for Harry Gold. I said that I was Harry Gold. The voice said, "Have you been all right?"

Q. Did you recognize the voice?—A. And said, "I am John." I recognized it as that of Yakovlev. I said, I told him that I had been fine. Now, the phrase used by Yakovlev was a code phrase which was employed to indicate whether there had been any suspicious signs about us, about my being watched, and my answer that I was fine was my code answer which indicated that there was no sign of any surveillance going on.

Yakovlev told me that he would meet me at the theater—he didn't specify. He said at the theater at 8 o'clock that night. And I said—I answered that I would be there.

Q. And what happened?—A. By the theater I knew that Yakovlev meant the Earl Theatre where our last meeting had been scheduled.

Q. Now, on that evening did you appear at the Earl Theatre?—A. At exactly eight o'clock I was in the upstairs lounge of the Earl Theater in the Bronx.

Q. What happened there?—A. There I was accosted by a man who showed me a torn piece of paper which Yakovlev and I had prepared on that night in July in Astoria.

Q. Are you referring to Government's Exhibit 14–A and B in evidence [showing]?—A. Yes, I am.

Q. And are you now telling the jury that one part of that exhibit was shown to you?—A. Yes, it was.

Q. And did you have the other part?—A. I had the matching part.

Q. Did you match the two papers at that time?—A. We did not match them. I was just shown this part.

Q. And then what happened?

The COURT. What did you mean by accosted? I don't quite understand what you mean.

The WITNESS. The man who met me was not Yakovlev. He was——
The COURT. Is that all you care to convey by the word accosted?
The WITNESS. No. I would like——
The COURT. Go ahead.
The WITNESS. To convey more. The man was not Yakovlev. He was tall, about 6 feet 2, had blond hair, and a very determined feature. He walked with a catlike stride almost on the balls of his feet, and he told me, after showing me——
Mr. E. H. BLOCH. One moment. I object to what he told him. It wasn't one of the defendants.
Mr. LANE. Well, it isn't necessary, your Honor, to get that. It isn't necessary. It isn't too important.
The COURT. All right. Don't tell us what he told you.
Q. Now as a result of your meeting with this individual whom you just described, what did you do?—A. As a result of the meeting I had with the man I met in the lounge of the Earl Theater I went to the southwest corner of Forty-second Street and Third Avenue, and there later that evening met Yakovlev.
Q. Now, was there anybody with you and Yakovlev at that time?—A. No; there was not.
Q. Did you nave a conversation?—A. Yes; we did.
Q. What was the conversation?—A. Yakovlev asked me if I had anything further from Dr. Fuchs in the interim. I told him I did not. Yakovlev then apologized for not having seen me in almost ten months, but he said it was unavoidable, that he had had to lie low during that period. We took a walk and we sat down at a bar on Second Avenue. There Yakovlev told me that he was very glad that I was now working in New York. He said it would put much less of a strain upon me as regards to meeting my Soviet contact. Yakovlev told me that I should begin to plan now for a mission which he had. He said that he wanted me to go to Paris, France, beginning in March of 1947. Yakovlev told me that I was to go first to London, and from London to take a plane to Paris. He gave me a sheet of paper, on which was typed—a sheet of onion-skin paper—and on it was typed the following information regarding a meeting I was to have in Paris. I was to meet a physicist——
Mr. E. H. BLOCH. If the Court please, this may seem technical but will the witness tell us what happened to that piece of paper? Otherwise I will have to object to it on the grounds it is secondary evidence.
The COURT. What happened to the piece of paper?
The WITNESS. I destroyed that piece of paper just a day or so before Memorial Day of 1947, and just about a few minutes before Agents Shannon and O'Brien of the FBI came to question me about my relationship with Abe Brothman.
Mr. E. H. BLOCH. I must say, your Honor, this presents a rather novel proposition to me in connection with the secondary evidence rule. It is my understanding that where a witness wilfully destroys a document, he can't introduce secondary contents of it. Of course, I am not sure of my position in the light of the context of the charges, but may I make my objection to save the record?
The COURT. All right.
Mr. LANE. Well, your Honor, we withdraw the question and we withdraw the answer.
The COURT. All right. I must say, Mr. Lane, you have been very helpful to the Court.
Mr. LANE. Thank you, your Honor.
Q. Now, was there any discussion at that time about the name Brothman?—A. Yes, there was. Yakovlev began to discuss with me the means by which I should manage to get off from work, in order to make this trip to Paris. I told Yakovlev that once the pressure of work at Abe Brothman and Associates had eased up a bit—and then Yakovlev almost went through the roof of the saloon. He said, "You fool!" He said, "You spoiled eleven years of work." He told me that I didn't realize what I had done, and he told me that I should have remembered that some time in the summer of '45 he had told me that Brothman was under suspicion by the United States Government authorities of having engaged in espionage and that I should have remembered it. Yakovlev threw down on the table where we were sitting, the bar, an amount of money which was about two or three times the actual cost of the drinks which we had had, and he dashed out of the place. I walked along with him for a while, and he kept mumbling that I had created terrible damage and that he didn't know whether it could be repaired or not. Yakovlev then told me that he would not see me in the United States again, and he left me.

Q. Now, during the period between the middle of June of 1944 and December 26 of 1946?—A. Yes.

Q. That was your last meeting with Yakovlev?—A. That is right.

Q. During that period of time did you have any conversations with a man by the name of Dean Slack?—A. I had conversations about—about three or four conversations with a man called Alfred Dean Slack.

Q. As a result of those conversations, what did you do?—A. As a result of those conversations I turned over to Yakovlev information relating to the——

Q. Now, just a minute. I haven't asked you what the contents were.—A. I see, I am sorry.

Q. I merely want to know what you did.—A. I turned over information to Yakovlev.

Mr. E. H. BLOCH. If the Court please, I would be obliged if the Court would just tell the jury that Alfred Dean Slack is not a defendant in this case, nor is he named as a coconspirator.

The COURT. That is right, he is neither a defendant nor a coconspirator.

Q. Now, prior to your association with Yakovlev, were you associated with anyone else?—A. Yes, I was.

Q. Who was that?

Mr. E. H. BLOCH. Would you fix the time, please?

Mr. LANE. Well, I think he has testified that he met Yakovlev in March of 1944. I want to know——

Mr. E. H. BLOCH. Prior.

Mr. LANE. On that day, just before he met, before he became associated with Yakovlev.

Mr. E. H. BLOCH. Well, I object to it upon the ground it is not within the corners of the indictment.

Mr. LANE. I withdraw the question.

*　　　*　　　*　　　*　　　*　　　*　　　*

Q. Now I show you Government's Exhibit 15 in evidence and I ask you to look at the picture attached thereto and tell me if you recognize that individual [showing]?—A. Yes, I do.

Q. Who is it?—A. That is Anatoli Yakovlev.

The COURT. Would you be able to shorten this perhaps if we recessed until tomorrow?

Mr. SAYPOL. No; we can get it right in.

Mr. LANE. That is the last thing I am going to do.

Q. Now, is this man whose picture I showed you on Government's Exhibit 15 in evidence and whom you have identified as Yakovlev, is that the party whom you knew as "John"?—A. Yes, it is.

Mr. LANE. Now, I am going to read to the jury portions of the certified document from the Department of State.

It is entitled "Foreign official status notification," and it has on here the full name, "Anatoli Antonovich Yakovlev."

The second item: "Name of Government and agency or department thereof," and it is the "Consulate General of the U. S. S. R., New York," United States of Soviet Russia.

"Present nationality—U. S. S. R.

"Previous nationality or nationalities—U. S. S. R.

"Place of birth"—I will have to spell that—"B-o-r-i-s-o-g-l-e-b-s-k," that is the "City or town" "State" is V-o-r-n-n-e-z-h, and the "Country" is the U. S. S. R.

"Date of birth—May 31, 1911.

"Port or place, date and manner of last arrival in United States and name of vessel, if any: San Pedro, February 4, 1941, SS. *Ecuador*."

"American visa issued by American Embassy in Moscow."

"Duties terminated: Left New York December 1946 by ship.

"Business address: 7 East 61st Street, New York, N. Y.

"Home address: 6 East 87th Street, New York, N. Y.

"Capacity in which signer is now serving, giving title of position, if any: Clerk.

"Date of assumption of present duties in the United States: February 8, 1941.

"Detailed statement of signer's present and proposed activities, including the place or places of performance and for whom performed or to be performed: Clerical duties.

"Nature and place or places of occupation or employment during the last five years, 1936–1941: Student, engineering—Economic Inst., Moscow."

Then attached hereto is a smaller notation, which is headed: "Department of State

"Washington

"Anatoli Antonovich Yakovlev has notified the Secretary of State of his (or her) status in the United States as an official (or employee) of the Union of Soviet Socialist Republics Government"; and that is dated April 3, 1941.

On the back of that card is the notation:

"The Secretary of State must be notified of any change in the status of the holder of this receipt."

Attached thereto also is another card:

"Department of State," giving the name "Anatoli Antonovich Yakovlev, 6 East 87th Street, New York, N. Y.," and it sets forth the Yakovlev is a citizen of subject of the Soviet Union, "Whose status as an official (or employee) of the Soviet Government has been notified to the Secretary of State by the Embassy all legation of the country by which he is employed."

At the bottom of this particular document is the picture of Yakovlev attached thereto and it also contains Yakovlev's signature.

*   *   *   *   *   *   *

PART II

CHARGES NOT PROVEN IN A COURT OF LAW

1. Charges Concerning Arthur Adams, Clarence Hiskey, John Chapin, and Others

The following excerpt is reprinted, without comment and only for purposes of assisting the reader, from a committee print issued by the Committee on Un-American Activities of the House of Representatives and entitled "Report on Soviet Espionage Activities in Connection With the Atom Bomb" (September 28, 1948; 80th Cong., 2d sess.) :

THE CHAPIN-HISKEY CASE

Sometime in 1938 an individual who was born either in Sweden or in Russia, according to information which the committee has, and who goes under the name of Arthur Alexandrovich Adams, entered the United States from Canada, and in connection therewith furnished a fraudulent Canadian birth certificate.

Adams' admission to the United States for permanent residence was facilitated by the false statements of one Samuel Novick, who is presently the president of Electronics Corp. of America, and has been prominently connected with that company, and with the ownership thereof, since its creation in 1942. During the war Electronics Corp. of America performed secret Government contracts in the amount of some $6,000,000. For a time it was the only contractor engaged in producing certain highly secret items for use in radar installations. Further reference is hereafter made to the activities of Samuel Novick.

Arthur Adams is virtually a charter-member Communist of the Soviet Union. He participated as a revolutionary in the Russian Revolution of 1905, was imprisoned, and is still suffering from the injuries resulting from the beatings he received at that time. As will hereafter appear, all of the persons who were his close contacts in New York during the period 1943–45 testified as to his physical disabilities—they were conveniently ignorant as to the cause of those disabilities.

It has been discovered subsequently that Arthur Adams had several previous sojourns in the United States. In 1920 or 1921 he was in the United States representing himself as a technical engineering adviser to a commission of the Russian Government. Between 1922 and 1927 he was back in Russia. He returned to the United States in 1927 and represented himself on this occasion to be engaged in some Russian governmental business, having to do with the Amo Motor Co., the first automobile work constructed in Russia. Between 1927 and 1932 he was again back in Russia. In 1932 he showed up in the United States representing himself as a member of a purchasing commission sent by the Soviet Government to purchase airplanes from the Curtiss-Wright Co., and as a member of the Aviation Trust of the Russian Government.

In 1936, Adams, accompanied by his wife, reappeared, and for a time visited Adams' wife's sister in New York City. According to the brother-in-law of Adams' wife, they left in 1937, ostensibly to return to Russia. The committee does not know whether or not Adams actually returned to Russia before he appeared in Canada seeking admission to the United States. It was about this time that Samuel Novick facilitated Adams' entry into the United States from Canada by false statements as to Adams' employment by Novick.

According to the brother-in-law of Adams' wife, there had always been something very "hush-hush," as he put it, about Adams' work and the real reasons for his visits to the United States. Adams' work had been similarly characterized by various friends of Adams who were known to his wife's brother-in-law. The brother-in-law testified that he, of course, knew Adams to be a Communist.

The committee has little information as to Adams' activities since his entry from Canada in 1938 to the time it was discovered that Adams was acting as a

163

Soviet espionage agent. It does know, however, that he was endeavoring to conceal the nature of those activities. Sometime in 1941 or 1942, Adams gave $1,875 to one Samuel J. Wegman, now deceased, who was operating a business in Hollywood, Calif., and in New York City, and requested Wegman to use these funds to pay Adams $75 per week and forward the checks to Adams at the Peter Cooper Hotel in New York City. Before his death, when Wegman explained his association with Adams, he stated that he had first met Adams through Julius Heiman. In 1942 Adams represented himself as employed by Wegman as a machine designer, and listed Wegman's business address as Hollywood, Calif.

During the war period Adams was discovered to be actively engaged in espionage activties for the Soviet Government. Those activities included the securing of information with respect to the developments that were being made in the United States in connection with nuclear fission. He was in physical contact with Clarence Hiskey, an atomic scientist assigned to the Manhattan project. This report herein after deals at length with Clarence Hiskey's connection with Adams.

He experienced considerable success in securing data that he desired, for when his room and his effects were secretly searched in 1944 by Government agents, he was found to have in his possession highly secret information regarding the atomic bomb plant at Oak Ridge, Tenn., as well as other vital information regarding the development of atomic energy in other countries. The committee has evidence of at least one direct contact that he made with the Soviet consulate in New York. On October 25, 1944, Adams was observed to leave the home of a lawyer in New York and enter an automobile registered in the name of Pavel Mikhailov. Pavel Mikhailov was vice consul in the Soviet consulate in New York at that time. When Adams entered the car he was carrying a large carrying case. Adams was driven directly to the Soviet consulate.

The importance of Adams in the Soviet hierarchy was testified to by Adams' wife's brother-in-law, who in 1932 had visited Adams and Adams' wife in Moscow. He testified that they had a maid and served the best of food. In short, they were well off. A station such as this in the Soviet Union is limited to persons regarded by the Soviet Government as being of considerable importance.

When Adams' status as a Soviet espionage agent was discovered, he was thereafter under fairly constant surveillance by the security officers of the Manhattan Engineering District, as well as by special agents of the Federal Bureau of Investigation.

Eventually, Adams discovered that he was under surveillance, and in February 1945, made a desperate but unsuccessful attempt to leave the United States.

The committee is presently investigating the complicity of two of the subjects of this report in the attempted escape of Arthur Adams. Contradictory testimony has been given by Government agents and these two witnesses. When the facts can be definitely verified, the committee intends to recommend indictment and prosecution for perjury.

Adams began his attempted escape by leaving the apartment of Victoria Stone, one of these witnesses. He appeared in Chicago with Eric Bernay, the other of these two witnesses.

Adams, from Chicago, proceded to Portland, Oreg., where he attempted to board a Soviet vessel. He was thwarted by the FBI in this attempt; but, because of governmental policy in existence at that time, the full details of which are unknown to this committee because it has not had access to the records, Adams was not arrested. He proceeded immediately back to New York, where he subsequently disappeared and has not been seen or heard of since. According to the committee's best information, he is now in the Soviet Union.

Clarence Francis Hiskey is presently employed as professor of analytical chemistry at the Polytechnic Institute in Brooklyn, N. Y. He was born on July 5, 1912, in Milwaukee, Wis. He graduated from La Crosse Central High School in La Crosse, Wis., in 1929. From 1929 to 1933, he attended the La Crosse State Teachers College. He attended the University of Wisconsin, Madison, Wis., receiving a B. S. degree in 1935, an M. S. degree in 1936, and a Ph. D. degree in 1939.

During the time he attended the University of Wisconsin, he met Marcia Sand, his first wife. An official Military Intelligence report, dated June 5, 1945, makes the following evaluation of Hiskey:

"Hiskey was active in Communist movements while attending graduate school at the university. * * * Allegedly Marcia, subject's wife, was a Communist. It was reported Hiskey had stated 'that the present form of government is no good, the Russian Government is a model and that Russia can do no wrong; if the lend-lease bill is passed this country will have a dictator.' * * * Also remarked that the United States Government should look to Russia for leadership. Hiskey reportedly urged radical-minded young men to take ROTC training to provide for 'possible penetration of the Communist Party in the Armed Forces of the United States.' In various lectures he discussed communism. * * * Investigations conducted in 1942 revealed Hiskey read the Communist publications Daily Worker and In Fact, and he had definite communistic leanings. * * * Hiskey and his wife lived for approximately 2 years with ———— —— whose brother was later president of the Young Communist League (cited as subversive by the Attorney General) at the University of Wisconsin. * * * It was reported that subject and his wife associated with other alleged Communists or Communist sympathizers. Hiskey was said to be an active member of the Communist Party."

The committee makes no findings as to the correctness of such evaluation by Military Intelligence. When Clarence Hiskey and his former wife, Marcia Sand, testified before this committee on September 9, 1948, however, both he and Marcia Sand refused to answer any questions concerning their Communist affiliations and connections while attending the University of Wisconsin, on the basis that they might incriminate themselves. However, Marcia Sand testified, under oath, that she was not presently a member of the Communist Party, but, asked if she had ever been a member of the Communist Party, she refused to answer this question "on the grounds that it may degrade or incriminate me."

While attending the University of Wisconsin, Clarence Hiskey applied for a Reserve commission from the United States Army. On June 18, 1938, he was issued a Reserve commission as a second lieutenant in the Chemical Warfare Service. This commission was given on the basis of his educational background and due to the fact that he had had ROTC training while attending college.

From September 1939 to 1941, Clarence Hiskey was employed as an instructor in chemistry at the University of Tennessee at Knoxville. For 5 months in 1941, he was an associate chemist with the Tennessee Valley Authority aluminum nitrate plant at Sheffield, Ala.

According to the testimony of Hiskey, he then went to Columbia University, New York, where he was engaged as an instructor from September 1941 until approximately September 1942. In the fall of 1942, upon the recommendation of Harold Urey, Hiskey was requested to do research work in connection with atomic energy in a laboratory at Columbia University, known as the SAM Laboratory. The SAM Laboratory worked primarily on the gaseous diffusion process of separating out uranium 235. This process was later known as the K–25 process. The plant which resulted from the research at Columbia University later became known as the K–25 plant at Oak Ridge, Tenn. The major research on this phase of the work was done at Columbia University. According to Army records, Hiskey, while engaged on this project, was in charge of a team of scientists. This project also worked on the development of heavy water.

The code letters "SAM" for the laboratory were taken as the first letters of the words "substitute alloy material" and were purposely chosen to mislead outsiders as to the true purpose of the laboratory.

The gaseous diffusion plant, which resulted from this work at Columbia University, was later constructed at Oak Ridge, Tenn., and details of the process, and its rate of production, are at the date of this report, top secret.

Hiskey, in his testimony, stated that he was a chemist and was engaged in chemical research on the SAM project. In September of 1943 or thereabouts, the laboratory in which Hiskey was engaged was moved to the University of Chicago. In Chicago, Hiskey joined the Metallurgical Laboratory, where he remained until April 28, 1944, when he was ordered to active duty in the United States Army for limited military service.

The importance of the Metallurgical Laboratory toward the development of the atomic bomb is quoted from the Smyth report which states:

"It would be foolish to attempt an assessment of the relative importance of the contributions of the various laboratories to the over-all success of the atomic bomb project. This report makes no such attempt, and there is little correlation between the space devoted to the work of a given group and the ability or importance of that group. * * * Such criteria, applied to the objectives and accomplishments of the various laboratories set up since large-scale work began, favor the Metallurgical Laboratory as the part of the project to be treated most completely.

"In accordance with the general objectives just outlined, the initial objectives of the Metallurgical Laboratory were: First, to find a system using normal uranium in which a chain reaction would occur; second, to show that, if such a chain reaction did occur, it would be possible to separate plutonium chemically from the other material; and finally, to obtain the theoretical and experimental data for effecting an explosive chain reaction with either U–235 or with plutonium. The ultimate objective of the laboratory was to prepare plans for the large-scale production of plutonium and for its use in bombs."

As previously mentioned, Clarence Hiskey was called to active duty in the United States Army on April 28, 1944. It should be noted that during World War II, scientists—and particularly scientists engaged within the Manhattan Engineering District—were deferred pursuant to instructions given selective service boards. The United States Army was cognizant of such instructions in connection with ordering Reserve officers to active duty. The reasons he was called to active duty were given to the committee by a high-ranking official formerly connected with the Manhattan Engineering District. The pertinent portions of his testimony are as follows:

"Colonel X. Hiskey was, I think, a $9,000-a-year man. You can verify that at the University of Chicago, the money being paid by the Government on a project. He was very definitely, in our opinion, a strong suspect, and, then finally we were convinced that he was a subversive agent.

"The CHAIRMAN. You were convinced—what?

"Colonel X. We were convinced that he was a subversive agent. Now, the question was what to do with Hiskey. We had trouble with scientists when we tried to move one. Someone, I think it was Colonel Lansdale, found in Hiskey's record that he had a second lieutenancy in college in the ROTC. Providentially, he had not given up his second lieutenancy, and we called The Adjutant General, and we had him call Hiskey to active duty amidst a great furore that we were doing it deliberately, and so on, and we transferred Hiskey, I think to the Canol project, I think, in Canada, where, in the Quartermaster Corps, he counted underwear until that went out of business. He was then transferred to an outfit in the South Pacific. He was promoted under ordinary steps from lieutenant to captain with no interference from us, and he finally came out of the Army as a captain."

An Army Intelligence report reflects that a letter dated July 4, 1942, recommended that Hiskey not be called to active duty. On April 19, 1943, it was recommended that Hiskey's Reserve commission be revoked. The recommendation on April 19, 1943, was based on a report dated March 10, 1943, in which an investigating agent stated, among other things, that Hiskey's attitude was un-American and his discretion and integrity were questioned. According to the agent, the investigation proved definitely that Hiskey was communistic in his beliefs.

In a letter dated April 13, 1944, the recommendation that Hiskey not be called to active duty was rescinded, and on April 28, 1944, Hiskey was ordered to active duty for limited military service. The reasons for this action have already been explained. He was relieved from active duty on July 18, 1946.

Official records of the Army reflect that Clarence Hiskey was property survey officer at White Horse, Yukon Territory, from May 8, 1944, to August 26, 1944. While at this station, a search of Clarence Hiskey's effects by security officers disclosed that he had in his effects a personal notebook which contained notes that he had made while working on the atomic bomb project at Chicago, Ill., relative to the development of several components of the bomb.

A CIC (Counter Intelligence Corps) agent assigned to the Chicago office of the Manhattan Engineering District was dispatched to White Horse, where he obtained the afore-mentioned notebook and returned it to authorities at the Manhattan Engineering District.

When Hiskey appeared before the subcommittee, he was given every opportunity to clear his record and deny the testimony before the committee that he had given information to Arthur Adams. Instead, he refused to answer these questions on the ground that his answer might incriminate and degrade him. The whole of his testimony is being made public with the making of this report.

The day after that on which Hiskey was called to active duty in the Army, Arthur Adams arrived in Chicago from New York and immediately met with Clarence Hiskey. From what transpired thereafter, as will presently be described, the inference is irresistible that at this meeting Adams told Hiskey that Hiskey would have to develop a contact within the Metallurgical Laboratories to take Hiskey's place; for on the following day Hiskey proceeded to Cleveland. There he went to the hotel of John Hitchcock Chapin, a chemical engineer employed in the Metallurgical Laboratories, who was in Cleveland engaged in a project which, according to Chapin, was even secret within the Manhattan Engineering District project itself.

At that meeting between Chapin and Hiskey, Chapin, upon the urging of Hiskey, agreed to meet with Arthur Adams and furnish him with information as to the progress being made in the development of the atom bomb. So that Chapin would be sure of the individual to whom he was to give the information, Chapin gave Hiskey a key, which Hiskey in turn gave to Arthur Adams for the latter's identification. Chapin admitted to the committee that Hiskey informed him that Adams was a Soviet agent. Some of Chapin's testimony follows:

"Mr. STRIPLING. Before you met Arthur Adams, did you have any conversation with Clarence Hiskey regarding Arthur Adams?

"Mr. CHAPIN. Yes.

"Mr. STRIPLING. Would you give the committee the details of that conversation?

"Mr. CHAPIN. As well as I can remember. Well, I was told that Arthur Adams was a Russian agent, and told by Hiskey, that is——

"Mr. STRIPLING. When did he tell you that?

"Mr. CHAPIN. That was around the spring of—it must have been the spring of 1944; yes.

"Mr. STRIPLING. April 29 or 30—do you know?

"Mr. CHAPIN. It could have been.

"Mr. STRIPLING. It was in April, spring?

"Mr. CHAPIN. Well, it was the spring.

"Mr. STRIPLING. Spring of 1944. All right; what else did he tell you?

"Mr. CHAPIN. I have to think now—he asked me whether I would be willing to meet Arthur Adams at some future time.

"Mr. STRIPLING. What was your reply?

"Mr. CHAPIN. After thinking about it awhile, I said, 'Yes, I would be willing to meet him.'

"Mr. STRIPLING. Why did Hiskey want you to meet Arthur Adams?

*　　　*　　　*　　　*　　　*　　　*　　　*

"Mr. CHAPIN. Well, to the best of my recollection, it would be to discuss whether or not I should hand out any information to Adams on my work.

"Mr. STRIPLING. On your work?

"Mr. CHAPIN. That is vague; I mean the gist of it was that.

*　　　*　　　*　　　*　　　*　　　*　　　*

"Mr. STRIPLING. Was any arrangement made at that time between you and Hiskey for you to meet Adams?

"Mr. CHAPIN. No definite arrangement; that is—well, it was arranged that I would meet Adams sometime probably; no date or anything like that.

"Mr. STRIPLING. Did you give Hiskey a key?

"Mr. CHAPIN. Yes; I did.

"Mr. STRIPLING. Explain the circumstances and the details of the key arrangement to the committee.

"Mr. CHAPIN. The key would be a means of my knowing if Adams ever did get in touch with me—would be a means for my knowing that that was Arthur Adams, or the man that Hiskey had spoken to me about.

"Mr. STRIPLING. You gave Clarence Hiskey a key. Do you recall what kind of a key it was? What it was a key to?

"Mr. CHAPIN. It was an ordinary key. I think it was a key to the basement of our apartment, or something or other. It was an extra key that I had.

"Mr. STRIPLING. But the key was to serve as a so-called instrument of identity; is that right?

"Mr. CHAPIN. Yes.

"Mr. STRIPLING. Did you ever see Arthur Adams?

"Mr. CHAPIN. Yes.

"Mr. STRIPLING. Would you give the committee the details of your meeting with Arthur Adams.

"Mr. CHAPIN. As best as I can. He came around to our apartment—first, he phoned actually; and then——

"Mr. STRIPLING. That was the fall of 1944?

"Mr. CHAPIN. Yes. He phoned and then came around to our apartment sometime after that, and he did not come into our apartment actually, he came downstairs, and I went out and answered the doorbell and went down to meet him, and he gave me the key, and I believe I asked him whether he would come up or not, and he did not and suggested that we meet in a hotel room, or something like that.

"Mr. STRIPLING. Did he suggest that you come to his hotel room?

"Mr. CHAPIN. Well, he must have, because that is where I went; I do not know.

"Mr. STRIPLING. Yes; where did you go with him?

"Mr. CHAPIN. This was another night now; I went to the Stevens Hotel, I think it was.

"Mr. STRIPLING. In other words, you met him on this particular night, and you made arrangements to meet him—was it the following night?

"Mr. CHAPIN. I do not know; I mean, I cannot remember.

"Mr. STRIPLING. But you did go to the Stevens Hotel?

"Mr. CHAPIN. Yes; I did.

"Mr. STRIPLING. It was within a few days, at least?

"Mr. CHAPIN. Yes.

"Mr. STRIPLING. All right; tell the committee what transpired at the hotel.

"Mr. CHAPIN. Well, I went when Adams was there in the room that he told me he would be in, and he suggested that he would like to—this again, I am trying to give the essence of it; I honestly do not remember the details—the essence of it was that he would like to have me give him information on my work.

*   *   *   *   *   *

"Mr. STRIPLING. Yes. How long were you in his room?

"Mr. CHAPIN. Oh, I would guess an hour so so."

(NOTE.—Chapin in his testimony said that he did not furnish any information to Arthur Adams in the Stevens Hotel.)

*   *   *   *   *   *

"Mr. STRIPLING. When did you decide not to cooperate, Mr. Chapin?

"Mr. CHAPIN. Well, I think I said it must have been during the interview; because, after all——

"Mr. STRIPLING. You got cold feet, in other words?

"Mr. CHAPIN. I would say so.

*   *   *   *   *   *

"Mr. STRIPLING. What happened to the key?

"Mr. CHAPIN. The FBI has it, so far as I know."

John Hitchcock Chapin was born August 18, 1913, at Rutland, Vt. He graduated from Cornell University in 1935 with a bachelor's degree in chemistry. He received his Ph. D. in chemistry from the University of Illinois in 1939. While Chapin was attending the University of Illinois he was also employed there as a part-time assistant professor. His first employment started in July 1939 with the du Pont ammonia department in Belle, W. Va. He remained at du Pont until the fall of 1942, when he joined the Manhattan project, being assigned to the SAM laboratories as a chemical engineer, which was located at Columbia University. In the spring of 1943 Chapin was transferred to the metallurgical laboratories at the University of Chicago, where he remained until the spring of 1945.

Chapin first met Clarence Hiskey at the SAM laboratories, at Columbia University. According to Chapin, they did not become good friends until after both had moved to the metallurgical laboratory in Chicago. During the period they knew one another Chapin often discussed with Hiskey the nature of the research in which he, Chapin, was engaged.

When Chapin completed the secret project in Cleveland and returned to the metallurgical laboratories in Chicago, he wrote a letter to Clarence Hiskey, in care of Marcia Sand Hiskey, who at that time was Hiskey's wife. Thereupon Marcia Sand Hiskey wrote her husband, who was at that time stationed at White Horse, Yukon Territory, Canada, that she had received a letter from Chapin which she had forwarded to Adams. Chapin told the committee that this letter to Marcia Sand Hiskey was a prearranged signal whereby Arthur Adams would be informed that Chapin had completed the project in Cleveland and was returning to Chicago.

"Mr. STRIPLING. Now, did you ever write to Mrs. Hiskey?

"Mr. CHAPIN. Yes; I think so. There is a technical point there, but go ahead.

"Mr. STRIPLING. You go ahead and tell me whether or not you wrote to Mrs. Hiskey.

"Mr. CHAPIN. Whether or not I wrote a note to Mrs. Hiskey—a note—I would not swear. I wrote to Mr. Hiskey, sending it to Mrs. Hiskey for forwarding to Hiskey.

"Mr. STRIPLING. What were the contents of the letter?

"Mr. CHAPIN. It was practically zero. I do not know, really. The FBI has it, I think.

"Mr. STRIPLING. Did you tell Hiskey in the letter that you had met Arthur Adams?

"Mr. CHAPIN. The only time that I wrote Mrs. Hiskey must have been—well, before I met Arthur Adams, I think. In fact, I know.

"Mr. STRIPLING. Did you ever write—did you say that you were going to meet Arthur Adams in this letter?

"Mr. CHAPIN. No; I do not believe so.

"Mr. STRIPLING. Was it agreed between all parties concerned that when you wrote this letter to Mr. Hiskey that that would mean that you were ready to see Arthur Adams?

"Mr. CHAPIN. That would mean that I was going back to Chicago; yes.

"Mr. STRIPLING. Yes. That was part of the agreement?

"Mr. CHAPIN. That is correct."

In the fall of 1944 Arthur Adams called at Chapin's apartment house and Chapin came downstairs and talked with Adams. Adams gave Chapin the key which Chapin had some time previously given to Hiskey and suggested that Chapin come to Adams' hotel room at the Hotel Stevens. A day or so later Chapin did go to Adams' hotel room and conversed with Adams for approximately an hour and a half. He denies that in this conversation he gave Adams any information.

Chapin was a very cooperative witness. Hiskey, on the other hand, came to the committee accompanied by counsel and refused to answer all pertinent questions that were put to him, on the ground that his answer might incriminate or degrade him. It is apparent from all the testimony that the committee has, as well as from other reliable information secured from former intelligence officers, that Hiskey was engaged in Soviet espionage in connection with the atom bomb. The testimony of Chapin is clear that he continued his contacts with Arthur Adams after he went into uniform. Hiskey still holds a Reserve commission as captain in the Army of the United States.

Compare, for example, the following from Hiskey's testimony:

"Mr. STRIPLING. Did you ever turn over any information concerning the atomic bomb or the development of the atomic bomb or any scientific research relating to the discovery of the atomic bomb to the individual whose picture I show you?

"Mr. HISKEY. Upon the advice of counsel, I refuse to answer that question on the grounds that it may tend to degrade and incriminate me.

"Mr. STRIPLING. I would like for the record to show, Mr. Chairman, that I have shown the witness a photograph of Arthur Adams.

"Mr. Hiskey, you are aware of the seriousness of the position which you are taking before this committee; are you not?

"Mr. HISKEY. I am aware that this is a serious matter; yes.

*   *   *   *   *   *   *

"Mr. STRIPLING. Did you ever confer with any official representative of the Soviet Union in 1943, '44, or '45?

"Mr. HISKEY. Upon the advice of counsel, I refuse to answer that question on the grounds that it will tend to degrade or incriminate me."

While the military and civilian investigative agencies knew of the contact Chapin made with Arthur Adams and with Marsha Sand Hiskey, no effort was made to remove Chapin from the highly secret chemical research in which he was engaged. Chapin was permitted to continue in charge of his chemical group until May of 1945, when he was released by reason of the reduction in force which resulted from the curtailment of research activities. After a period of employment in the plastics department of the General Electric Co., Pittsfield, Mass., Chapin became employed at the M. W. Kellog Co. The M. W. Kellog Co., specializing in engineering development at this time, was under contract to the United States Air Forces for a phase of secret Air Corps development. Chapin was assigned to work on a phase of this secret development. However, when the Air Forces discovered that Chapin was employed on this project, they immediately demanded that he be relieved from this employment. He is presently employed in a brewery in Newark, N. J.

Chapin, in testifying before the committee, denied under oath that he was now or had ever been a member of the Communist Party. While he admitted that he might have subscribed to or read Communist publications, the committee did not feel that this was a sufficient motive for Chapin to agree to meet a Russian agent for the purpose of turning over to him secret developments on the atomic bomb. Under questioning, he explained as a possible motive the fact that he and Hiskey shared a mutual interest in Russia and that he had often held the desire of teaching or working in Russia at a future date. Chapin related that prior to his agreeing to meet Arthur Adams he strongly felt that the secret of the atom bomb should be shared by all the Allied Nations. When questioned by the committee as to his thought along this line now, Chapin did not think he would be willing at this time to risk the security of the American people by delivering to Russia the secret of the atom bomb.

The meeting in Cleveland between Hiskey and Chapin, the conversation which transpired, the letter from Chapin to Marcia Sand Hiskey, and the subsequent contact between Chapin and Adams in Chicago present a clear case of a conspiracy between Hiskey, Chapin, and Marcia Sand Hiskey, to divulge secret and classified information relating to the atom-bomb project to a Soviet espionage agent. The committee recommends immediate prosecution of the conspirators.

In the opinion of the committee, although John Hitchcock Chapin committed an indictable offense as a coconspirator in matters affecting the security of the United States, it is felt that his participation in the conspiracy is mitigated by the fact that when he appeared before the committee he was cooperative and apparently sincere in his answers to pertinent questions directed to him; whereas in contrast, his coconspirators refused to answer pertinent questions on the ground that to do so might incriminate them. Chapin impressed the Committee as a person of deep sincerity who, in a moment of weakness, had made a vital mistake.

2. Charges Concerning Martin David Kamen

The following excerpt is reprinted, without comment and only for purposes of assisting the reader, from a committee print issued by the Committee on un-American Activities of the House of Representatives and entitled "Report on Soviet Espionage Activities in Connection with the Atom Bomb" (September 28, 1948; Eightieth Cong., 2d sess):

THE KAMEN CASE

Martin David Kamen is presently employed as associate professor of chemistry at Washington University in St. Louis. He was born in Toronto, Canada, in 1913 and was brought to the United States when he was 3 months old. His father was born in Russia. Kamen testified before a subcommittee on September 14. His testimony is being made public concurrently with the making of this report.

From 1936 to 1944 Kamen was staff chemist for the radiation laboratories at the University of California at Berkeley, Calif. From 1941 to 1944 he was assigned to the Manhattan Engineering District's atomic-bomb project. Kamen testified that he believed that his work was of substantial importance to the project. As indicative of his competence, Kamen was the discoverer of carbon 14 and was also the discoverer of an improved method of producing iron 55. He has written a book on his particular specialty, which is tracer research in biology, and has published some 65 articles concerning his various researches.

On July 1, 1944, Kamen proceeded from Berkeley to San Francisco and was met in San Francisco by Gregory Kheifets., the Soviet vice consul in San Francisco, and Gregory Kasperov, the person who was about to succeed Kheifets. Kheifets was leaving for Russia. In fact, he left for Russia 3 days after the incident about to be described took place.

Kamen, Kheifets, and Kasperov proceeded to a restaurant known as Bernstein's Fish Grotto in San Francisco and had dinner together in a booth in that restaurant. The dinner meeting lasted approximately 2 hours and 40 minutes. The dinner meeting was covered by Government intelligence officers. The conversation, or as much of it as could be heard, was taken down. According to the former MED scientist who read the transcript of the conversation, and who testified before the committee, Kamen very definitely gave classified information to Kheifets and Kasperov, including information about the uranium pile in Chicago, and information as to MED activities in other localities in the United States. Certain sections of testimony in this case are being made public concurrently with the making of this report.

While Kamen had numerous friends in Communist and radical circles, there appears to be little if any evidence at this time that his revealing of classified information was willful and deliberate. There is no evidence connecting Kamen otherwise than casually with members of the Communist espionage apparatus that was operating on the Pacific coast.

Ten days after the incident in question, Kamen was requested to resign forthwith from the MED project. He concedes that what he did constituted a gross indiscretion. Whether that was all it was or whether it constituted something more than that, the committee is not prepared at this time to say. Kamen testified that he has been denied a passport to leave the United States on three occasions in the last 2 years.

3. Charges Concerning Steve Nelson, Joseph Weinberg, and Others

The following excerpt is reprinted, without comment and only for purposes of assisting the reader, from a report issued by the Committee on Un-American Activities of the House of Representatives and entitled "Report on Atomic Espionage" (September 29, 1949):

REPORT ON ATOMIC ESPIONAGE (NELSON-WEINBERG AND HISKEY-ADAMS CASES)

SCIENTIST X CASE

This case deals with the activities of that branch of the Communist espionage apparatus which operated on the Pacific coast, particularly within the radiation laboratory of the University of California at Berkeley which was engaged in certain activities in connection with the development of the atomic bomb. This case, in the past, has been identified by the committee as the "Scientist X case." The committee, as a result of an investigation pursued this year, has received testimony identifying the scientist involved in this case as Joseph Woodrow Weinberg.

Previous reports regarding the Scientist X case have identified Steve Mesarosh alias Louis Evans alias Steve Nelson as the Communist espionage agent who was engaged in securing information regarding the development of the atomic bomb from Scientist X.

Steve Mesarosh, or Nelson as he is commonly known, was born in Yugoslavia on January 1, 1903, in a town called Chaglich. He entered the United States on June 12, 1920, accompanied by his mother and two sisters. He gained admission to the United States as a citizen of this country under the name of Joseph Fleischinger, that being the name of his mother's brother-in-law. Nelson's mother and two sisters also gained admission at the time by falsely representing themselves as the wife and children of Joseph Fleischinger. The name of Nelson's mother and the names of her three children were all included on the United States passport issued to said Joseph Fleischinger.

On June 22, 1922, a warrant of arrest in deportation proceedings was issued charging that the subject, his mother, and two sisters had entered the United States without proper passports; that they had entered by false and misleading statements; and that they were persons likely to become public charges at the time of their entry.

A hearing was held under the authority of the warrant of arrest in Philadelphia on October 17, 1922, as a result of which the examining immigration inspector recommended that the aliens be afforded the opportunity to legalize their residence in the United States. It should be noted that during the hearing the United States Government recommended that Steve Nelson, his two sisters, and his mother, be afforded a haven in the United States, even though they illegally entered the country. During the hearing it was brought out that Steve Nelson, his two sisters, and his mother, had taken advantage of opportunities in this country; that Steve Nelson, as well as his sisters, were attending school, and that the entire family had gained employment. In the recommendation of the immigration inspector, it was stated that after examination of the aliens it was decided that the subject individuals were taking advantage of the opportunities offered by this country and undoubtedly would become substantial citizens. On October 30, 1922, the Board of Review entered an order that the warrant of arrest be canceled on payment of head tax if the Department of State would waive passport requirements. On November 14, 1922, the Secretary of State waived the passport and visa requirements in behalf of the subject, his mother, Maria, and his two sisters. Thereafter, on November 27, 1922, the aliens were examined by surgeons of the United States Public Health Service and passed; head tax was collected; and the entry of the subject, his mother, and his two sisters was legalized.

Steve Nelson was admitted to citizenship in the United States District Court, Eastern District of Michigan, Detroit, Mich., on November 26, 1928, and was issued certificate of naturalization No. 2834850.

In evaluating Steve Nelson's entry in the United States and the Government's position in legalizing said entry, the United States afforded a haven for a refugee whose political ideologies in subsequent years dedicated themselves to the violent overthrow of the United States Government by force. It is not definitely known when Steve Nelson joined the Communist Party. However, in an article in the Daily Worker, November 10, 1937, under the byline of Joseph North, dispatched from Valencia, Spain, North stated that while interviewing participants fighting for the International Brigade, he obtained the following information from Steve Nelson:

' The working people of the Soviet Union were passing through a bitter period and Steve joined the Friends of Soviet Russia. On the first anniversary of Lenin's death (1925), he joined the Communist Party at the memorial in Philadelphia.''

This alleged statement by Steve Nelson is noteworthy because, as previously stated, he was granted citizenship on November 26, 1928. If the truth of the article written by Joseph North, which appeared in the Daily Worker, could be established, it is apparent that Steve Nelson was a member of the Communist Party prior to gaining his citizenship and therefore perjured himself when he obtained his naturalization papers.

In 1931 Steve Nelson's importance to the Communist movement was recognized in Moscow and he was called there to attend the Lenin Institute. On August 1, 1931, he filed a passport application with the Department of State in which he requested permission to visit Germany to study building construction. He falsified his passport by stating that he was born in Rankin, Pa., on December 25, 1903. This criminal offense was never prosecuted due to the fact that it was not discovered until the statute of limitations had run. There is further evidence with respect to Mr. Nelson's attendance at the school in Moscow. Mr. William Nowell testified before this committee on November 30, 1939, and he stated that while he was a member of the Communist Party he attended the Lenin Institute in Moscow and that Steve Nelson was in attendance at this school under the name of Louis Evans. Mr. Nowell stated in his testimony that Nelson's prominence in the Communist Party was conspicuous because of his frequent contact with the OGPU (Russian secret police) in Moscow. Additional evidence of Nelson's visit to Russia has been developed by this committee which indicates that in July 1933 Nelson filed with the American consul in Austria a 2-year renewal of his passport, stating that he had resided in Russia from September 1931 to May 1933, and had resided in Germany, Switzerland, and Austria from May 25, 1933. Nelson, when questioned by this committee regarding his attendance at the Lenin school, refused to answer on the ground of self-incrimination.

Official intelligence reports in possession of this committee reflect that Nelson was in China for 3 months in 1933, working for the Comintern in Shanghai, and that a coworker of his was Arthur Ewert, a well-known Comintern agent, who was subsequently sentenced to imprisonment in Brazil for his part in the Communist revolution in 1935. The exact date of the subject's return to the United States from China and the European countries mentioned above is unknown, but in 1934 he contributed an article to the Party Organizer, official organ of the central committee of the Communist Party, United States of America.

During the Spanish Civil War, Nelson received consideral publicity in the Communist press because of the fact that he had risen to the rank of lieutenant colonel in the International Brigade of the Loyalist Army. Nelson returned to the United States in the latter part of 1937 from Spain and became active in the affairs of the Veterans of the Abraham Lincoln Brigade and the American League for Peace and Democracy, both notorious Communist organizations.

Since 1938, Steve Nelson has been a national figure in the Communist Party, as well as a leading functionary in the Moscow-controlled Communist underground.

With reference to Nelson's participation in the Abraham Lincoln Brigade, Nelson applied for a passport on February 13, 1937, and the passport was issued on February 23. This passport was issued to Nelson under the name of Joseph Fleischinger. It is noted on the application form that the name Fleischinger was misspelled in two places by the applicant. This criminal violation likewise escaped the attention of the authorities until the statute of limitations had expired. When questions were propounded to Nelson regarding this false passport, he again followed the current Communist Party line by declining to answer questions and placed himself under the sanctuary of the fifth amendment to the Constitution.

Steve Nelson was so important to the Communist movement and had gained such favor with his superiors that in 1940 he was assigned as organizer for the party in the bay area at the port of San Francisco, Calif. He was also given the underground assignment to gather information regarding the development of the atomic bomb. This assignment was facilitated by Steve Nelson's having met a woman in Spain who had gone to Spain in 1937 to meet her husband, also a volunteer of the International Brigade. Upon arrival in Spain, this woman was informed that her husband has been killed, and she was befriended by Steve Nelson. This woman, upon her return to the United States, moved to Berkeley, Calif., where she became acquainted with and married one of the leading physicists engaged in the development of the atomic bomb.

The Communist Party and the Soviet Government were aware of Steve Nelson's acquaintance with the physicist and attempted to use this as a medium of infiltration of the Radiation Laboratory at the University of California, which was working on the development of the atomic bomb. An investigation of the aforementioned scientist disclosed that neither he nor his wife engaged in any subversive activities and that their loyalty has never been questioned by the Government. Nelson later reported that neither the physicist nor his wife were sympathetic to communism.

Under the guidance of Steve Nelson, infiltration of the Radiation Laboratory actually began in other ways. A cell was developed within the laboratory, consisting of five or six young physicists. The existence of the cell has been established in sworn testimony before this committee. According to a sworn statement by a witness, Giovanni Rossi Lomanitz was the principal Communist Party organizer. The records of this committee also reflect that David Bohm, presently a professor of physics at Princeton University, was also a member of this cell. Upon two occasions, both Giovanni Rossi Lomanitz and David Bohm declined to answer questions regarding their respective memberships in this cell upon the ground that to do so might tend to incriminate them.

In 1942 Steve Nelson gained another promotion within the Communist Party when he was assigned as county organizer at Alameda, Calif. This assignment placed the atomic bomb project under the direct jurisdiction of Steve Nelson for the Communist Party. According to the official files of the Government, when Nelson was under surveillance, he visited the home of Vassili Zubilin, a former secretary of the Soviet Embassy in Washington, D. C., who was then in Oakland, Calif. Zubilin's cover name in the Communist Party was "Cooper." During this meeting, Nelson complained to Zubilin about the inefficiency of two individuals working for the apparatus. These persons have been identified by the committee and their names are being presently withheld from the public. Because of Mr. Nelson's complaint to Zubilin, these individuals were transferred from Alameda County, one to Detroit, Mich., and the other to Los Angeles, Calif.

The details of the meeting between Nelson and Scientist X are set forth as follows:

Late one night in March 1943, a scientist at the University of California, who identified himself as "Joe" went to the home of Steve Nelson, after having made arrangements earlier in the evening with Steve Nelson's wife to meet Nelson at Nelson's home. When Joe arrived at Nelson's home, Nelson was not present but arrived at about 1:30 on the morning of the following day. Upon his arrival at his home, Nelson greeted Joe and the latter told him that he had some information that he thought Nelson could use. Joe then furnished highly confidential information regarding the experiments conducted at the radiation laboratories of the University of California at Berkeley. At the time this occurred, the radiation laboratories at Berkeley were engaged in vital work in the development of the atomic bomb.

Several days after Nelson had been contacted by Joe, Nelson contacted the Soviet consulate in San Francisco and arranged to meet Peter Ivanov, the Soviet vice consul, at some place where they could not be observed. Ivanov suggested that he and Nelson meet at the "usual place."

As a result of the surveillance that was being kept on Nelson, the meeting between Nelson and Ivanov was found to take place in the middle of an open park on the St. Francis Hospital grounds in San Francisco. At this meeting, Nelson transferred an envelope or package to Ivanov. A few days after this meeting between Nelson and Ivanov, on the St. Francis Hospital grounds, the third secretary of the Russian Embassy in Washington, a man by the name of Zubilin, came to the Soviet consulate in San Francisco. Shortly after his arrival, Zubilin met Nelson in Nelson's home and at this meeting paid Nelson 10 bills of unknown denominations.

When Nelson testified before the committee in September 1948, he refused to answer all pertinent questions on the ground that his answers would tend to incriminate him. During this interrogation, he was asked whether he was acquainted with Vassili Zubilin of the Soviet Embassy and he refused to answer on the ground that to do so might incriminate him.

EXTRACTS FROM INTELLIGENCE REPORTS—SCIENTIST X CASE

During the course of the committee's investigation of the Scientist X case, certain information contained in reports made by intelligence agents was obtained by the committee. An extract from one of these reports reads as follows:

"A very reliable and highly confidential informant advised that certain instructions had been given by Steve Nelson, who was at the time a member of the national committee of the Communist Party of the United States, to the scientist identified herein as Joseph W. Weinberg, a research physicist connected with the atomic bomb at the University of California, at Berkeley, Calif. The instructions were that Weinberg should furnish Nelson with information concerning the atomic bomb project so that Nelson could, in turn, deliver it to the proper officials of the Soviet Government. Nelson advised Weinberg to furnish him any information which he might obtain from trustworthy Communists working on the atomic project; he, Nelson, being of the belief that collectively the Communist scientists working on the project could assemble all the information regarding the manufacture of the atomic bomb. Nelson told Weinberg that all Communists engaged on the atomic bomb project should destroy their Communist Party membership books, refrain from using liquor, and use every precaution regarding their espionage activities."

At the time of this meeting, according to an extract from an intelligence report, Weinberg furnished Nelson with information regarding the experiments which had been conducted in connection with the development of the atomic bomb at the radiation laboratories of the University of California. The information furnished Nelson by Weinberg was taken down in the form of notes by Nelson.

An extract from a report filed with the committee states that Weinberg, while employed on the atomic project, had as his closest associate Giovanni Rossi Lomanitz, David Joseph Bohm, Max Bernard Friedman, and Irving David Fox, all of whom have refused to answer questions propounded by the committee regarding Communist Party activities and associations on the ground of self-incrimination.

Regarding the identity of the scientist as the person who furnished information concerning the atomic bomb to Steve Nelson in March of 1943, the following is an extract from testimony given to the committee during the month of August 1949 by James Sterling Murray, presently assistant to the president of the Lindsay Light & Chemical Co., West Chicago, Ill., and formerly officer in charge of security and intelligence in the San Francisco, Calif., area for the Manhattan Engineering District, which was the division of the United States Army charged with the development and production of the atomic bomb:

"Mr. MURRAY. A highly confidential informant informed our office that an unidentified scientist at the Radiation Laboratories had disclosed certain secret information about the Manhattan engineering project to a member of the Communist Party in San Francisco, and this confidential informant went on to say that such information was transmitted to the Russian consulate in San Francisco and later was on its way to Washington, D. C., and later out of the country in a diplomatic pouch. This was the only allegation we had to begin with, but through information which the confidential informant was able to supply us on the background of the particular scientist, we finally narrowed it down and definitely fixed the scientist as Weinberg."

In addition to the identification mentioned above, it should be pointed out that a number of persons who were engaged in the investigation of the Scientist X case have been interrogated by the committee, and/or its staff, and the identification made by Witness Murray has been concurred in by these other persons.

On Tuesday, April 26, 1949, Steve Nelson was again a witness before the Committee on Un-American Activities. On this occasion, Joseph W. Weinberg was brought face to face with Steve Nelson, and when Steve Nelson was asked the question as to whether he was acquainted with Weinberg, he refused to answer on the ground that to answer might tend to incriminate him.

Upon two occasions, Joseph W. Weinberg, in appearances before the committee, specifically denied having furnished any information regarding the atomic bomb to Steve Nelson. This is in direct contradiction to the testimony of James Sterling Murray and other witnesses who have appeared before the committee.

The committee, during its investigation, devoted a great deal of time toward establishing the true facts regarding a meeting which was held in the home of Joseph Weinberg in Berkeley, Calif., in August 1943. According to information furnished by witnesses before the committee, this meeting was attended by Bernadette Doyle, who was secretary to Steve Nelson during the period he was the Communist Party organizer for Alameda County, Calif.; Steve Nelson; Giovanni Rossi Lomanitz; Irving David Fox; David Bohm; and Ken Max Manfred, formerly known as Max Bernard Friedman. As will be shown, Joseph Weinberg was present in his apartment in Berkeley, Calif., at the time this meeting was held. All of the persons mentioned as attendants at this meeting were employed by the Radiation Laboratory of the University of California, at Berkeley, with the exception of Bernadette Doyle and Steve Nelson. All of these persons were reported, during the course of the committee's investigation, as being members of the Communist Party.

All of the persons mentioned as having attended the meeting in the home of Joseph Weinberg in August 1943 were subpenaed as witnesses before the committee, with the exception of Bernadette Doyle. Witnesses Lomanitz, Nelson, Fox, Bohm, and Manfred declined to answer questions regarding this meeting upon the ground that to do so might tend to incriminate them. Joseph W. Weinberg was questioned regarding this meeting upon two occasions by the committee, and denied that such a meeting had ever been held in his home, and further denied that he knew or had ever been acquainted with Steve Nelson and Bernadette Doyle.

Altogether Joseph Weinberg has appeared before the committee upon three occasions, and pertinent extracts of the testimony given by him upon these appearances before the committee are being set forth as follows:

"QUESTION. I show you a photograph of an individual and ask if you have ever seen this person.

"Mr. WEINBERG. No; I have not seen him before.

"QUESTION. Mr. Chairman, I have shown the witness a picture of an individual known as Steve Nelson.

"Have you ever known this person under any name other than Steve Nelson?

"Mr. WEINBERG. So far as I can recollect, no.

"QUESTION. Mr. Weinberg, are you certain you have never seen this individual whose picture I have shown you [showing protograph to Mr. Weinberg]?

"Mr. WEINBERG. Within reason, I am.

"QUESTION. I have shown the witness a picture and he said he is reasonably certain he has never seen Steve Nelson. I am sure the witness is aware of the penalties of perjury. I am sure his counsel has advised him of the penalties of perjury before a committee of Congress.

"You are now viewing two pictures of Steve Nelson. That is not Mr. Nelson's real name. He is known under various names, but I ask you if you have ever seen that individual, if you ever saw him in the years 1942, 1943, 1944, or 1945.

"Mr. WEINBERG. So far as I know now I have never seen him. I don't think it is necessary to call your attention to the fact that you ask me about events that happened 5 years ago and that I have a very large circle of very casual acquaintances. Within those reasonable limits I would say I have not seen him.

"Question. You have never seen that individual?

"Mr. WEINBERG. So far as I am aware.

"Mr. TAVENNER. Now, I believe you met Steve Nelson in Washington on April 26, 1949. Did you meet him prior to that time or had you met him prior to that time?

"Mr. WEINBERG. No.

"Mr. TAVENNER. Now, let me ask you if on or about the 17th day of August 1943 Steve Nelson came to your home at the address which you have just given, in Berkeley, Calif.?

"Mr. WEINBERG. I remember no such occasion.

"Mr. TAVENNER. Do I understand that you merely do not recollect or that you deny that he came there to vist you and that you saw him there?

"Mr. WEINBERG. The situation was more or less this, and perhaps I should take a moment to explain. At this time I had a very wide circle of acquaintances there. There were many people who dropped into my house. There were many students who brought friends and introduced them and who promptly walked out of my life thereafter. I exempt such possibilities when I say I do not recollect the occasion. That is, I would not be prepared to state emphatically and with absolute certainty that no such person ever dropped into my house. I would certainly be prepared to state emphatically that I had nothing significant to do with him at the time.

"Mr. TAVENNER. Do you know Bernadette Doyle?

"Mr. WEINBERG. No; I do not.

"Mr. TAVENNER. Is it not a fact that on the 17th day of August 1943 Steve Nelson and Bernadette Doyle visited you at your house?

"Mr. WEINBERG. I certainly don't remember any such visit. Or at least I don't remember the occasion.

"Mr. TAVENNER. I did not get the last answer.

"Mr. WEINBERG. I have no specific memory of such a visit or that it mattered or that a person by the name of Doyle was ever introduced to me.

"Mr. NIXON. Mr. Chairman, the witness has qualified several answers of late that he does not remember a visit that "mattered," and I think the record should be clear that that is not responsive to the question. Whether the visit mattered or not is not the point. It is whether or not these people visited him.

"Mr. WEINBERG. Well, then, sir, to the best of my ability, I would answer the question with a qualification that I do not remember such a visit, in the interests of saying strictly what I am qualified to say.

"Mr. WALTER. In other words, the visit made no impression?

"Mr. HARRISON. Nothing happened that would have made any impression on your mind?

"Mr. WEINBERG. That's correct. That was the intent of my remark that it did not matter.

"Mr. TAVENNER. Was there another occasion in 1943 on which Bernadette Doyle met you at the front door of your house and at which time you had a short conversation with her?

"Mr. WEINBERG. I remember no such occasion.

"Mr. TAVENNER. You still state that you have never met Bernadette Doyle?

"Mr. WEINBERG. Not to my knowledge.

"Mr. TAVENNER. Do you know now where Bernadette Doyle lived?

"Mr. WEINBERG. No; I do not.

"Mr. TAVENNER. Have you ever been to her house?

"Mr. WEINBERG. No.

"Mr. TAVENNER. Did you ever attend any of the meetings of the Young Communist League?

"Mr. WEINBERG. No.

"Mr. NIXON. Would you remember if Mr. Nelson had come to your house and spent an hour or so alone with you?

"Mr. WEINBERG. I think if Nelson had come to my house and introduced himself as some sort of high-ranking Communist and spoke to me for any length of time I would remember that occasion.

"Mr. CASE. Was that where you met Steve Nelson?

"Mr. WEINBERG. I certainly don't remember meeting Steve Nelson at that meeting or any other meeting."

Upon one occasion when Joseph Weinberg appeared before the committee, he was brought face to face with Steve Nelson, and the following is an extract from the testimony relating to a confrontation between Nelson and Weinberg upon this occasion:

"Mr. RUSSELL. Mr. Nelson, will you stand, please? Mr. Weinberg, will you face Mr. Nelson, the gentleman in back of you?

"Mr. Nelson, I ask you whether you are acquainted with this individual, Mr. Joseph Weinberg?

"Mr. NELSON. I refuse to answer that question on the ground that it may tend to incriminate me.

"Mr. RUSSELL. Mr. Weinberg, I ask you whether or not you are acquainted with Mr. Nelson, the gentleman facing you?

"Mr. WEINBERG. My only recollection of Mr. Nelson is a picture I have seen in the papers.

"Mr. RUSSELL. Are you acquainted with him?

"Mr. WEINBERG. I am not acquainted with him.

"Mr. RUSSELL. Have you ever met Mr. Nelson?

Mr. WEINBERG. I do not recall ever meeting Mr. Nelson. I do not believe I have ever met him."

With reference to the associations of Joseph W. Weinberg, Steve Nelson, and Bernadette Doyle, the following extracts from testimony presented to the committee by several witnesses on this point are being set forth herein. This testimony is in direct conflict with that of Joseph W. Weinberg:

"Mr. VELDE. You are satisfied that Dr. Joseph W. Weinberg was employed by the University of California in January 1943, and that he was engaged in work on the Manhattan project at that time?

"Mr. MURRAY. Yes. We were satisfied he was actively engaged in work on the project.

*   *   *   *   *   *   *

"Mr. MURRAY. On August 12, 1943, we were conducting physical surveillance of Joseph W. Weinberg, and at approximately 5 o'clock in the afternoon a highly confidential informant advised us that there was to be some type of a meeting at Weinberg's home that evening, at which Steve Nelson and Bernadette Doyle would be present. I immediately instituted surveillance of the entire area by the agents assigned to our office, to watch the visitors in the Weinberg home, and I myself stationed myself next door to the Weinberg home. I believe it was located on Blake Street in Berkeley, Calif.

"At approximately 9 o'clock I observed a man known to me to be Steve Nelson, and a woman known to me to be Bernadette Doyle, approach the Weinberg home and enter therein. After their entry into the Weinberg home I, in the company of Agents Harold Zindle and George Rathman, went to the roof of the apartment house which was immediately next door to the Weinberg home, and from an observation post on the roof I was able to look into the second-story apartment of Weinberg.

"I noted Weinberg, Steve Nelson, and Bernadette Doyle, in company with at least five other members, some of whom were employed by the Radiation Laboratory, seated around a table in the dining room of the Weinberg apartment.

"At approximately 9:20 p. m. Weinberg came to the window and attempted to adjust the window, it being a very hot and sticky night. He had some difficulty in raising the window, or lowering it, or something, and Steve Nelson came over to help him, at which time I was able to get a good look and identify him.

"Mr. VELDE. Just a minute, if you please. Do you have a picture of Weinberg? I think at this point possibly you had better have him identify it.

"Mr. APPELL. We have a newspaper picture.

"Mr. RUSSELL. While Nelson and Weinberg were at the window, did you observe whether or not any conversation took place between the two individuals?

"Mr. MURRAY. I did observe some conversation, but I think it only had to do with the window adjustment at that point. I observed them sitting around the table, at which time the conversation appeared to be very serious.

"Mr. RUSSELL. Do you recall the other persons around the table in Weinberg's apartment at this meeting you are describing?

"Mr. MURRAY. I don't recall all. I know Giovanni Rossi Lomanitz, David Bohm, Irving David Fox, Max Friedman. I know Max Friedman was there, but for a very short time. He was the first one to leave.

"Mr. RUSSELL. What other agents of the Manhattan Engineering District accompanied you on the occasion of this surveillance?

"Mr. MURRAY. Special Agents Harold Zindle and George Rathman.

"Mr. RUSSELL. Will you spell Rathman, please?

"Mr. MURRAY. R-a-t-h-m-a-n.

"Mr. RUSSELL. These two agents were also assigned to the Manhattan Engineering District; were they not?

"Mr. MURRAY. Yes; they were. I was their immediate superior.

"Mr. RUSSELL. Did you maintain a surveillance of the Weinberg apartment?

"Mr. MURRAY. Yes; we did. I believe the meeting broke up at about 10:15 p. m., at which time we saw a general shaking of hands and a general showing of disposition to leave, at which time I ran down to the street floor again and observed Nelson and Doyle leaving together. They turned east on Blake Street, and I turned east on Blake Street also, and was immediately in front of them. We proceeded up the street approximately 100 feet in that fashion, at which time I thought, for the purposes of the record, that I should make some face-to-face contact with Mr. Nelson, and so I swung on my heel and started west on Blake Street, and in so doing I touched the shoulder of Nelson. We both immediately pardoned each other, and I continued west on Blake Street, and my surveillance of the entire proceeding was at an end at that point.

"Mr. APPELL. Mr. Murray, I show you a picture and ask you if you can identify the person on the left as you look at the picture as being that of Steve Nelson?

"Mr. MURRAY. Yes; that is Steve Nelson as slightly older than when I knew him.

"Mr. APPELL. And that is the individual you bumped into on Blake Street in Berkeley, Calif.?

"Mr. MURRAY. Yes.

"Mr. VELDE. Let that be marked "Murray Exhibit 1" and received in evidence.

"Mr. RUSSELL. When you bumped into him, that was after he had left the residence of Joseph Weinberg?

"Mr. MURRAY. Yes.

"Mr. APPELL. I show you a picture that appeared in the Washington Post as of September 22, 1948, and ask if that is the individual you observed in the Blake Street residence with Steve Nelson?

"Mr. MURRAY. Yes. I identify the picture as the picture of Dr. Joseph Weinberg, and as the individual who was in his own apartment sitting around the table with Mr. Nelson.

"Mr. APPELL. And the individual you saw standing at a window of the apartment together with Steve Nelson, attempting to fix the window?

"Mr. MURRAY. Yes; that is right."

This witness furnished additional information regarding the association of Steve Nelson and Joseph Weinberg which is not being set forth in this report.

The following is a signed statement obtained by committee investigators from George J. Rathman, whose name appears in the testimony of James Sterling Murray as one of the agents attached to the Manhattan Engineering District who accompanied him on the surveillance regarding the meeting held by Nelson and certain other persons in the apartment of Joseph Weinberg in Berkeley, Calif.:

"I have been interviewed concerning a surveillance I conducted along with Harold Zindle and James Murray, who were attached with me to the Manhattan Engineering District as special agents, Counterintelligence Corps, in Berkeley, Calif.

"On or about August 17, 1943, at approximately 8:45 p. m., Harold Zindle, Murray, and myself arrived at an apartment house adjacent to the residence of the subject of this surveillance, Joseph W. Weinberg. During this surveillance I had occasion to observe the subject, Joseph Weinberg; a man identified to me as Steve Nelson; a woman identified to me as Bernadette Doyle; together with four or five additional persons whom I could not identify due to my point of observation, engaged in conversation. At approximately 9:45 p. m., Joseph Weinberg and the man known to me as Steve Nelson appeared at the window of the second-story apartment of Joseph Weinberg, closing the window and lowering the shade. At approximately 10 p. m. on the night of this surveillance Harold Zindle, Murray, and the undersigned left the roof of the apartment house and proceeded to the street where Murray and the undersigned saw Steve Nelson and Bernadette Doyle walking west on Blake Street from the direction of the subject's residence.

"I am certain if I could observe Steve Nelson personally today that I would be able to identify him as the person who was present in the second-story apartment of Joseph Weinberg on the night of surveillance.

"I have read the above statement, and to the best of my knowledge and belief this statement is true in every respect.

"G. J. RATHMAN."

The following is an extract from the testimony furnished the committee by William S. Wagener, who was also attached to the Manhattan Engineering District:

"Mr. WHEELER. Do you know an individual by the name of Bernadette Doyle?

"Mr. WAGENER. Yes, sir.

"Mr. WHEELER. Did she contact any scientists employed by the radiation laboratory?

"Mr. WAGENER. Joseph Weinberg.

"Mr. WHEELER. On how many occasions?

"Mr. WAGENER. Just once.

"Mr. WHEELER. Will you describe the meeting?

"Mr. WAGENER. One evening we were on physical surveillance, and we saw this woman whom we identified as Bernadette Doyle go up to the door of Joseph

Weinberg and talk to him for a few minutes. She departed and got in her car. She had her car parked a block or so away. She got in her car and drove away.

"Shortly after, Weinberg and his wife came out, got in their car, and drove around very suspiciously, stopping here and there, and apparently like they were going to contact someone, but they apparently did not meet the individual, whoever it was."

The following is an extract from the testimony furnished by Col. John L. Lansdale, Jr., who was also attached to the Manhattan Engineering District during the time agents of that district were conducting an investigation of Joseph Weinberg. Other than the statement set forth herein, Colonel Lansdale was unable to provide the committee with further information because of the Executive order prohibiting him from furnishing information to a congressional committee:

"Mr. RUSSELL. Are you familiar with the name Steve Nelson?

"Colonel LANSDALE. Yes, sir.

"Mr. RUSSELL. Do you recall the names of any scientists who were contacted by Steve Nelson?

"Colonel LANSDALE. At least one; yes.

"Mr. RUSSELL. Who was that?

"Colonel LANSDALE. Joseph Weinberg."

It is to be noted that the testimony of James Sterling Murray regarding the association of Bernadette Doyle and Joseph Weinberg is corroborated by the signed statement of witness George J. Rathman and the testimony of William S. Wagener.

During the committee's examination of witnesses Mr. and Mrs. Paul Crouch, it was developed that both of these individuals had attended Communist Party meetings with Joseph Woodrow Weinberg. Mrs. Crouch recalled one particular meeting of the Young Communist League which Weinberg attended and which was held in a private home. Witness Paul Crouch, who was the predecessor of Steve Nelson as Communist Party organizer in Alameda County, Calif., identified Weinberg as a person who had attended the meeting of the Young Communist League, mentioned by his wife, and at least two or three other meetings of the Young Communist League which were held in a private home in Oakland, Calif. Neither of these witnesses, however, knew Weinberg's name. The testimony of Paul and Sylvia Crouch regarding their identification of Joseph Weinberg is being printed and made a part of this report.[1]

The committee has additional evidence regarding the meeting held in the home of Joseph W. Weinberg during the month of August 1943, but for obvious reasons all of the committee's evidence is not being incorporated in this report.

RECOMMENDATION FOR PERJURY PROSECUTION

Based upon the testimony set forth above and the testimony of other witnesses appearing before the committee, it is the committee's opinion that Joseph Weinberg made untruthful statements upon the three occasions he appeared before the committee.

Testimony before the committee and signed statements furnished to committee investigators indicate that Joseph Weinberg did not testify truthfully when he said:

1. That he did not know Steve Nelson.

2. That he did not know Bernadette Doyle.

3. That he had never attended any meeting of the Young Communist League, and that he had never been a member of the Communist Party.

It is recommended that the Attorney General convene a special grand jury in the District of Columbia for the purpose of hearing certain witnesses whose names will be furnished by this committee, and who have knowledge of the untruthful statements made by Joseph Weinberg.

WITNESSES INTERROGATED INCLUDED THOSE EMPLOYED ON ATOMIC BOMB PROJECT

The committee has interrogated the following persons who were attached to the radiation laboratory at the University of California at Berkeley and who, while employed there, were performing work on the atomic bomb:

---

[1] See hearings regarding Communist Infiltration of Radiation Laboratory and Atomic-Bomb Project at the University of California, Berkeley, Calif., vol. II (Identification of Scientist X).

Giovanni Rossi Lomanitz, formerly professor of physics at Fisk University, Nashville, Tenn.

David Bohm, who is presently assistant professor of physics at Princeton University, Princeton, N. J.

Ken Max Manfred, formerly known as Max Bernard Friedman, who is presently attending the University of California at Berkeley as a result of a scholarship granted him by the University of Puerto Rico amounting to $2,000 per year. Manfred is attending the University of California at Berkeley by virtue of a leave of absence granted him by the University of Puerto Rico, where he is employed as an assistant professor of physics, in order to obtain a degree of doctor of philosophy.

Irving David Fox, who is presently employed as a teaching assistant by the University of California at Berkeley.

Joseph Weinberg, who is presently employed as an assistant professor of physics at the University of Minnesota. Weinberg was born on January 19, 1917, at New York City, N. Y. He attended public schools in the Bronx, N. Y., and graduated from the DeWitt Clinton High School in New York City in 1932. During the years 1932 to 1936, he attended City College in New York City. In 1937, he attended the University of Michigan during the summer session, and during the academic year of 1938–39, he attended the University of Wisconsin. During the years 1939–43, he attended the University of California, from which he received a Ph. D. degree.

Frank Friedman Oppenheimer, whose resignation from the University of Minnesota as an assistant professor of physics was recently accepted by the university.

Robert R. Davis, who was formerly employed by the Manhattan Engineering District at the University of California Radiation Laboratories.

Upon two occasions, witnesses David Bohm and Giovanni Rossi Lomanitz declined to answer questions regarding Communist Party membership and activities on the ground that to do so might tend to incriminate them. Witnesses Manfred and Fox declined to answer questions regarding Communist Party membership and associations on the occasion of their appearance before the committee on the ground that to do so might tend to incriminate them. Witness Robert R. Davis, on the occasion of his appearance before the committee, testified that he had been recruited into the Communist Party by Giovanni Rossi Lomanitz. Frank Friedman Oppenheimer, upon the occasion of his appearance before the committee, admitted former membership in the Communist Party but declined to answer any questions pertaining to the Communist associations of other individuals.

Of the persons mentioned in the above-quoted excerpt, Irving David Fox, Giovanni Rossi Lomanitz, and Steve Nelson were indicted on December 4, 1950, by a Federal grany jury in Washington, D. C., for contempt of Congress. The charges are based upon the refusal of these three individuals to answer questions posed by the House Committee on Un-American Activities. In each case the defendant has pleaded not guilty.

At the present writing, a Federal grand jury has before it the question of an indictment of Joseph Weinberg for perjury.

Regarding Robert R. Davis, the following is excerpted from a newspaper article which appeared on the front page of the Washington Post on June 11, 1949:

(By Arthur Edson, Associated Press reporter)

A wartime worker in a west coast atomic laboratory testified yesterday that he became a Communist after he was taken to a party meeting by an atomic scientist.

The worker, Robert Davis, told the House Un-American Activities Committee that the scientist was Giovanni Rossi Lomanitz.

At one period during the war they both were employed at the important University of California radiation laboratory in Berkeley.

Davis said that Lomanitz asked him to go to a meeting, that he went with him to two or three, and that at one of the meetings Davis joined the party. Mrs. Davis joined, too, he said.

The price of a Communist card in capitalistic terms: 50 cents * * *.

Davis, who now lives in New York, followed Lomanitz to the stand.

Then Louis J. Russell, the committee's senior investigator, asked Lomanitz to come forward once more.

As the two men stared at each other, Davis said:

"He looks familiar."

He said he believed this was the same man he knew at the laboratory, but he refused to make the identification positive, saying that it had been 6 years since he had seen him.

Davis said that near the end of March 1943 he was transferred to the Los Alamos, N. Mex., atomic plant.

But before he left he burned the two Communist cards, his wife's and his own.

Never, Davis said, had he looked upon himself as a real Communist. He had joined out of curiosity, to see what went on, and he had found the meetings "very educational."

If he had known what it would lead to, he said, he wouldn't have joined.

He later was asked to resign at Los Alamos, he said, because the Atomic Energy Commission considered "my character and associations questionable." * * *

4. Testimony of David Hawkins

The following article appeared in the New York Times of January 28, 1951:

PROFESSOR, ONCE RED, HELD LOS ALAMOS JOB

WASHINGTON, January 27 (UP)—David Hawkins, a professor of philosophy, became an administrative aide at the Los Alamos (N. Mex.) atom bomb plant in May 1943, soon after he drifted out of the Communist Party, the House Un-American Activities Committee disclosed today.

Professor Hawkins, now at the University of Colorado, said he joined the party at the University of California in 1938 and left it in the spring of 1943. He said he withdrew from the party when he realized he did not want it involved in his life.

Mr. Hawkins' testimony before the House committee was taken in closed session December 20 and 21. He was called in the committee's investigation of reported Communist infiltration at the University of California's radiation laboratory and atomic bomb project. The professor and his wife, Frances Pockman Hawkins, who also testified, refused to name any persons they had known to be Communists, unless their names already had been cited by the committee.

As administrative aide at Los Alamos, Mr. Hawkins helped draw up personnel regulations, made out draft deferment forms and acted as a liaison officer representing the atom bomb plant in its relations with the Army engineers and the town of Los Alamos.

The committee also disclosed testimony of Dr. Kenneth May, former student and teacher at the University of California. Dr. May admitted having been a Communist from 1936 to 1942.

Mr. Hawkins and Dr. May said they knew Dr. Joseph W. Weinberg, who has been identified by the committee as the "scientist X" who gave atomic material to Communist couriers. Both denied any knowledge that Dr. Weinberg was a Communist or that there was a Communist cell in the radiation laboratory.

A New York Times article of March 14, 1951, describes courtroom testimony of David Greenglass and also contains the following extract:

Yesterday's testimony recalled the recent case of Dr. David Hawkins, philosophy professor, who told the House Un-American Activities Committee last December that be became an administrative aide at Los Alamos soon after quitting the Communist Party in 1943.

Dr. Hawkins held the title of "Historian, Los Alamos project for developing the atomic bomb." As historian, his work gave him access to basic facts about the plant. Dr. Hawkins, according to the committee, admitted that he joined the Communist Party in 1938 at the University of California and left it 5 years later.

183

5. Allegations Concerning Wartime Exports

The following self-explanatory document is from the files of the Joint Committee on Atomic Energy:

JANUARY 9, 1950.

REPORT ON EXPORT OF ATOMIC MATERIALS TO THE SOVIET UNION IN 1943 AND 1944

(Prepared by the staff of the Joint Committee on Atomic Energy)

I. INTRODUCTION

On December 2, 1949, in a radio interview with Fulton Lewis, Jr., George Racey Jordan, a former United States Army Air Force major, alleged that shipments of uranium, heavy water, and engineering documents referring to Oak Ridge, Tenn., were shipped through Great Falls, Mont., to the Soviet Union during 1943 and 1944. In this and subsequent statements on the radio, in the press, before the House Committee on Un-American Activities, and to Joint Committee staff members, the exact charges varied but the following are the principal allegations:

1. 500 pounds of uranium were shipped to Russia through Great Falls, Mont., in the spring of 1943;

2. 1,150 pounds of uranium were shipped from Canada to Russia through Great Falls, Mont., in either late fall or winter of 1943;

3. 1,200 pounds of uranium were shipped to Russia through Great Falls, Mont., shortly after June 1, 1944, under the personal supervision of Col. Anatole Kotikov, Russian military representative stationed at Great Falls in connection with lend-lease shipments;

4. Carboys of heavy water were shipped to Russia through Great Falls, Mont., between January 1943 and June 1944;

5. At least one folder containing an engineering blueprint of some device related to atomic energy and Oak Ridge was shipped to Russia in a black suitcase under diplomatic immunity as air freight from Great Falls, Mont., in the spring of 1944, with an attached note on White House stationery which contained the following: "Had a hell of a time getting these away from Groves" and signed "H. H." (allegedly referring to Gen. Leslie R. Groves of the Manhattan Engineer District and Harry L. Hopkins, Presidential assistant and lend-lease expediter.)

6. On one occasion, Mr. Harry L. Hopkins personally directed Major Jordan by telephone to expedite a special shipment of uranium to Russia and to keep the matter confidential.

7. On December 6, 1949, Fulton Lewis, Jr., stated in a broadcast that the individual responsible for the shipments of atomic materials which Mr. Jordan had described was former Vice President Henry Wallace.

8. On December 14, 1949, Royall Edward Norton, a retired Navy chief petty officer and student at Chapel Hill, N. C., was interviewed on Mr. Lewis' radio broadcast and indicated that on one occasion a document relating to the atomic structure, possibly of element "92," went to Russia in a Navy lend-lease aircraft.

II. COMMITTEE FILES

The files of the Joint Committee on Atomic Energy concerning this subject commence with a letter request to the Atomic Energy Commission, on August 9, 1948, for information pertaining to uranium shipments to Russia between 1943 and 1945. A reply on August 19, 1948, included a summary prepared by Dr. Philip L. Merritt of the AEC New York office who, during World War II, had been in charge of screening requests received by the War Production Board for raw materials relating to the activities of the Manhattan Engineer District. This summary revealed that 200 pounds of uranium oxide, 220 pounds of uranium nitrate, and a quantity of impure uranium metal of less than 25 pounds were shipped with authorization from both Manhattan Engineer District and WPB under export license. A supplemental letter from the AEC on September 15, 1948, stated that 500 pounds of uranium oxide and 500 pounds of uranium nitrate were shipped to Russia by the Eldorado Mining & Refining Corp., Ltd., from Canada in 1943.

III. BACKGROUND

Lend-lease.—Under the lend-lease program, a total of $46,000,000,000 worth of matériel of varied description was supplied to wartime allies of the United States. Of this amount, $11,000,000,000 worth, or 25 percent of all lend-lease aid, went to the Soviet Union. The shipments included basic raw materials, explosives, shells and guns, medical supplies, chemicals, combat vehicles, airplanes, and complete alcohol, synthetic rubber, and petroleum cracking plants, together with the requisite engineering drawings, operating and maintenance manuals, spare parts lists, and other pertinent documents. This data is contained in State Department Document No. 2759, European Series 22, titled "Soviet Supply Protocols" and in the reports to Congress of the Lend-Lease Administration.

The Lend-Lease Administration assumed that military aircraft such as the B–25, of which a considerable number were furnished to Russia, would be comparatively useless if not accompanied by blueprints, drawings, specifications, maintenance and operational manuals, and spare parts. Since the B–25 at this time was one of the principal medium bombers of the United States, data concerning it was classified to avoid its falling into the hands of the enemy. The documents for this particular plane would make a stack 3 feet high. A large volume of blueprints, documents, and papers was required for such lend-lease items as the oil refinery and the synthetic rubber plants shipped to Russia.

While this provides an explanation of the heavy flow of documents, drawings, blueprints, and other papers sent to Russia, it also suggests the difficulty of definitely establishing that no single piece of paper relating to the atomic energy program was surreptitiously taken or sent out of the country. Atomic energy documents were, of course, under the control of the Manhattan Engineer District.

Diplomatic mail.—Diplomatic mail enjoys immunity from search. During the war there was a great volume of such mail between the United States and the Soviet Union. In the light of the information available as a result of the Canadian spy trials, Russia is known to have displayed an interest in the atomic energy project prior to its public announcement. It is by no means inconceivable that, were the Russians able to obtain classified documents from the Manhattan project, such documents might have been shipped out of the country in diplomatic pouches. However, the committee staff was unable to locate definite evidence that such a shipment actually did occur in connection with the allegations which are the subject of this report.

It is possible, however, that documents relating to particle accelerators, such as cyclotrons, could have been obtained from one of the institutions owning and operating them, and carried out of the country. Such information was widely distributed prior to the war due to the fact that cyclotrons were then in operation in this and other countries. At least one was known to have been built in Russia. The term "5-foot walls of water and lead," referred to by Mr. Jordan as appearing on a document shipped to Russia could describe a standard shielding technique used in conjunction with cyclotrons.

Uranium.—Uranium has been known and used to a minor extent in industry since the Curies produced radium at the turn of the century. As a byproduct of radium production, the supply of uranium exceeded demand until 1942. Industrially it was used to color ceramics, in photography, and as an analytical agent. Experiments with uranium as an alloying metal in steel were carried on during this period but were not commercially successful, primarily because other materials such as vanadium were cheaper and more satisfactory. Prior to World War II, the world use of uranium amounted to several hundred tons per year.

Before January 26, 1943, there were no unusual United States Government controls on delivery, price, or use of uranium; but on that date, its ceramic and photographic uses were forbidden by WPB Order No. M–285. Other legal restrictions were not applied as the Manhattan District did not wish to call attention to this material. Instead, an attempt was made to gain control of the supply of uranium indirectly.

Heavy water.—Prior to World War II, heavy water was a scientific curiosity produced at several research laboratories throughout the world. In the United States, the Stuart Oxygen Co. of San Francisco produced about 10,000 grams (about 3 gallons) a year which were sold commercially to universities for use in laboratories and cyclotrons. Heavy water had no industrial or military use prior to 1943, and no legal restrictions were placed on its movement within or out of the country prior to passage of the McMahon Act.

The Manhattan District did not place any restrictions on the movement of heavy water, as this material was not then considered of primary importance in the atomic project. Under the provisions of the McMahon Act, after the war, the Atomic Energy Commission requested the Department of Commerce to prohibit its exportation.

IV. HISTORY OF SHIPMENTS

Four authorizations for shipment to Russia of materials which could be related to an atomic energy program were granted during the war. Two of these were for uranium oxide and uranium nitrate totaling 1,420 pounds. A third was for 25 pounds of uranium metal. The fourth was for 1,000 grams of heavy water. The total amount of uranium oxide and nitrate authorized was shipped by commercial suppliers following standard procedures. The uranium metal actually shipped was 2.2 pounds (1 kilo) procured through commercial channels after considerable technical difficulty. All three shipments of uranium materials were paid for under the provisions of the Lend-Lease Act. A shipment of 1,000 grams of heavy water was made through normal commercial channels on an export license. Another shipment of 100 grams of heavy water was made without export license, there being no license requirement for either shipment. Both shipments were paid for, not with lend-lease funds but with funds supplied by Russia herself, as lend-lease authorities had taken the position that heavy water was not in the list of materials authorized to be financed by the United States. The authorized shipments were cleared by Lend-Lease, the WPB, State Department, and the Manhattan Engineer District to the extent required by existing regulations. The records indicate it was felt that refusal to ship might have been more informative to the Russians than any help they could derive from the small quantities of material requested. A subsequent Russian request for nine long tons of uranium oxide and nine long tons of uranium nitrate was rejected because of "unavailability" of the material.

Shipment No. 1.—On January 26, 1943, WPB Conservation Order M–285 directed that all trade in uranium or uranium compounds, alloys, or mixtures for certain industrial purposes be suspended. On January 28, 1943, the Soviet Purchasing Commission directed a request to the Office of Lend-Lease Administration for 25 pounds of uranium metal, 220 pounds of urano-uranic oxide, and 220 pounds of nitro-uranyl. The request also notified the authorities of an intention to file application for 8 tons of each of the two uranium salts. The Soviet Purchasing Commission was informed on February 15, 1943, by letter from WPB, that these shipments could be approved but that a precise statement as to its intended operational use would be required. On March 6, 1943, a letter application for an export permit for 200 pounds of oxide and 220 pounds of nitrate was submitted by the Soviet Purchasing Commission to the Lend-Lease Administration for "making armaments." Another letter application on March 7, 1943, from the Soviet Purchasing Commission requested an export license for 500 pounds of oxide and 500 pounds of nitrate for "manufacture of ferro-uranium compounds which will be used in the production of armaments."

In the meantime, on March 4, 1943, the Lend-Lease Administration received a letter application from "Chemator, Inc." (Chematar, Inc.) for clearance to ship uranium compounds to the Soviets. A letter reply from Lend-Lease on March 6, 1943, informed Chematar, Inc., that clearance would be granted provided the necessary data on exact quantities and source were filed.

The Soviet Purchasing Commission letter of March 6 bears one notation dated March 8: "Cleared with Mr. Lund of WPB and General Groves of the U. S. Army Engineers." Export license No. C–1537205 was assigned on March 9, 1943, for this shipment.

Chematar, Inc., acting on the order placed with it by the Soviet Purchasing Commission, ordered 200 pounds of uranium oxide and 220 pounds of uranium nitrate from the S. W. Shattuck Chemical Co., Denver, Colo., on March 15, 1943. The material was shipped by rail to Gore Field, Great Falls, Mont., from Denver on March 23, 1943, on Shattuck order No. 9239. The shipment was contained in four boxes and identified on an Air Force outgoing tally sheet dated April 5, 1943, as "Chemicals" with a gross weight of 691 pounds. A signed receipt was obtained from a Russian official for the material.

Shipment No. 2.—On April 10, 1943, General Groves was informed by the Lend-Lease Administration of the request for 500 pounds each of uranium oxide and nitrate submitted by the Soviet Purchasing Commisson on March 7, 1943. He agreed to reach a decision by the following Monday morning, according to the daily report of the Soviet Supply Division of the Lend-Lease Administration. On April 14, 1943, a letter was addressed to the Soviet Purchasing Commission from the Lend-Lease Administration rejecting the request for a total of 1,000 pounds of uranium compounds. Telephone calls were received by the Lend-Lease Administration from the Soviet Purchasing Commission on April 15, 1943, to protest the denial of clearance and to state that "they had been offered large quantities" although the denial was on the ground that the material was not available. This view was expressed again in a letter on April 20, 1943, from the Soviet Purchasing Commission to General Wesson, Senior Assistant Administrator for Lend-Lease. A formal reapplication was submitted on April 23, 1943.

In a classified file memorandum dated May 1, 1943, the Lend-Lease Administration officer, through whom most of these exchanges had been carried on, wrote that as a result of telephone conferences between General Wesson and General Groves, the previous decision not to approve the Soviet request was reversed by General Groves and General Wesson, and that it had been decided to allow the Soviets to proceed under the export license to obtain the particular stocks of 500 pounds of urano-uranic oxide and 500 pounds of uranium nitrate for which they had previously applied. A note appended to the reapplication, dated April 22, 1943, indicated that General Wesson had decided that this export license should be approved and issued as soon as possible.

On April 23, 1943, General Wesson addressed a letter to the Soviet Purchasing Commission approving the request for 500 pounds of uranium oxide and 500 pounds of uranium nitrate. License No. C–1643180 was assigned on the same date.

The records of the Canadian Radium & Uranium Corp., New York, disclose a letter from Chematar, Inc., dated April 27, 1943, confirming an interest in obtaining 1,000 pounds of uranium oxide and nitrate. On May 1, 1943, another letter from Chematar, Inc., placed a firm order for 500 pounds of each material for shipment to Great Falls, Mont., for export under license No. C–1643180.

The Canadian Radium & Uranium Corp. placed a firm order with Eldorado Gold Mines, Ltd., Canada, on May 6, 1943, for 500 pounds of oxide and 500 pounds of nitrate for shipment to Col. A. N. Kotikov for the Soviet Purchasing Commission at Gore Field, Great Falls, Mont. The purpose was stated as: "The uranium oxide will be used for making ferro-uranium compounds, which in turn, will be used in the production of armaments. The uranium nitrate will be used for medical purposes. * * *" By invoice No. U–664 of the Eldorado Gold Mines, Ltd., on May 21, 1943, the oxide was shipped in 5 cases containing 100 pounds each of black powder, and the nitrate was shipped in 10 cases containing 50 pounds each of yellow powder. The total value of the order was $2,455. The packing cases were made of wood. The oxide boxes each had a gross weight of 117 pounds and the nitrate cases each had a gross weight of 100 pounds. The gross shipping weight of the order was 1,585 pounds. The shipment went by air freight from Great Falls, Mont., to Fairbanks, Alaska, on June 10, 1943, and was accepted there by the Russian representative for transshipment to the Soviet Union.

Shipment No. 3.—The Soviet Purchasing Commission included in its request to the Lend-Lease Administration on January 29, 1943, an item of 25 pounds of uranium metal. On February 4, 1943, a similar request was submitted by the same organization in a requisition to the International Division, Services of Supply, War Department. This requisition was denied by letter to the Soviet Purchasing Commission from the Division Director on March 9, 1943, because of "unavailability." The request to the Lend-Lease Administration was granted in their letter of February 15, 1943, conditioned upon more adequate explanation of the ultimate use of the material. The Soviet Purchasing Commission telephoned the Lend-Lease Administration on March 9, 1943, according to that organization's daily reports, and urgently requested approval of a uranium-metal shipment. On March 10, the Soviet Purchasing Commission called again and stated that the Manufacturers Chemical Co., New York, could supply the metal out of stock and that approval was necessary to avoid a black-market transaction. A letter from the Lend-Lease Administration to the Soviet Purchasing Commission on March 12, 1943, stated in part: "There is no objection on the part of this office to the placing of an order for 25 pounds of uranium metal. * * *" A carbon copy of this letter was sent to General Groves on March 15, 1943.

The letter from the Soviet Purchasing Commission to General Wesson of the Lend-Lease Administration on April 20, 1943, again requested authority to buy "25 pounds of uranium metal for which we also have a proposition." (In a letter to the Manhattan Engineer District from the Lend-Lease Administration on April 23, 1943, it was stated that "General Groves has advised General Wesson that the particular request for 500 pounds of urano-uranic oxide and 500 pounds of uranium nitrate can be approved. * * * In addition, it has been agreed that an application for 25 pounds of uranium metal will be entertained if submitted.") General Wesson replied to the Soviet Purchasing Commission on April 23, 1949, following a telephone conference with General Groves, that an application for the uranium metal would be entertained "provided you can locate a source of supply." An application was received dated April 27, 1943, for metal valued at $1,875. Export license No. C–1653288 was assigned on April 29, 1943.

The Soviet Purchasing Commission placed its order for the uranium metal with Manufacturers Chemical Co., which in turn placed it with A. D. MacKay, a chemical supply house in New York, on April 13, 1943. This firm placed the order with Hugh S. Cooper Metallurgical Laboratory in Cleveland, Ohio. The exact date and document for this order has not been available.

On March 31, 1944, the Soviet Purchasing Commission directed a letter to the Secretary of War in further effort to obtain 8 tons each of uranium oxide and nitrate and 25 pounds of uranium metal. Reference was made to the correspondence on this matter during 1943. At a conference of representatives of the International Division and the Production Division of Services of Supply, War Department, on April 7, 1944, General Groves indicated that he desired to handle this new request in its entirety. The memorandum recording this conference contains the International Division recommendation that consideration be given to a solution designed "to avoid arousing Soviet suspicions."

The Secretary of War informed the Soviet Purchasing Commission by letter on April 17, 1944, that "our supply of this material is not sufficient for us to comply with your request * * *." There is no record of a cancellation of export license No. C–1653288, assigned on April 29, 1943, to the Soviet request for 25 pounds of uranium metal.

A sample of uranium metal was sent on September 9, 1944, by Hugh S. Cooper Metallurgical Laboratory, Cleveland, Ohio, to the Bureau of Standards for assay. The assay was prepared on September 11, 1944. The results were considerably at variance with assays of uranium metal used by the Manhattan engineer district according to an interview statement on December 8, 1949, by Dr. Philip L. Merritt, former Raw Materials Director for MED and present AEC Raw Materials Chief in New York. A series of three shipments of varying assay, totaling slightly over 6 pounds of uranium metal, was made to A. D. MacKay by Cooper Metallurgical Laboratory in September and December 1944 and January 1945. Out of this material A. D. MacKay selected 1 kilo (2.2 pounds) of uranium metal for application against the original Soviet order. This shipment was made on February 16, 1945, under export license No. C–1653288, assigned on April 29, 1943. The material was packed in a wooden container, had a gross weight of 3 pounds 6 ounces, at a cost of $314.60. It was shipped to Washington, D. C., for transshipment by diplomatic courier pouch to Moscow.

Shipment No. 4.—The daily report for the Division of Soviet Supply of the Office of the Lend-Lease Administration contains an entry on September 22, 1943, that a Soviet "requisition for heavy water for laboratory purposes was discussed and referred to Mr. Moore for investigation." A second entry on the same day states that Mr. Pauley (Henry M. Pauley, Foreign Requirements Unit, Chemicals Division, War Production Board) was immediately notified of the request.

The Lend-Lease Administration requested the Soviet Purchasing Commission on September 28, 1943, to supply the end use for heavy water and stated that the Lend-Lease Administration was unable to find that it was used for war purposes in the United States. On September 30, 1943, representatives of the Soviet Purchasing Commission came to the Lend-Lease Administration to discuss, among other things, the request for heavy water.

On October 11, 1943, the Lend-Lease Administration wrote the Soviet Purchasing Commission to the effect that Lend-Lease policy covered only materials essential to the war effort and that the heavy-water request would have to be supported by further definition of the intended end use. The letter stated that unless a war purpose could be established, and request for heavy water could not be construed as coming within the materials covered by Lend-Lease provisions and could not be financed under its arrangements. Again, on October 14, 1943, the chronological files of the Lend-Lease Administration bear a notation that representatives of WPB, the Soviet Purchasing Commission, and Lend-Lease met to discuss certain items, among them heavy water.

On November 16, 1943, the Division of Soviet Supply, Office of Lend-Lease Administration, assigned program license No. 339 and release certificate No. C–366 to a Soviet export license application for a cash purchase of heavy water. The.program license report covering the period October 1–December 31, 1943, showed.program license No. 339 for 1,000 grams (2.2 pounds) of heavy water with a shipping weight of 41 pounds and a value of $3,250. This report bears a covering memo with the word "Noted" and the initial "W", followed by a typed name, "General Wesson".

The records of Chematar, Inc., show that on August 23, 1943, a month before the first notation in the Lend-Lease chronological file, a telephone call was received from the Soviet Purchasing Commission in Washington, D. C., pertaining to a potential order for 1,000 grams of heavy water. The firm order was placed on October 21, 1943, after Lend-Lease had told the Soviet Purchasing Commissions that Lend-Lease credits could not cover the heavy-water requisition. The heavy water was obtained without delay from the only commercial manufacturer in the United States, the Stuart Oxygen Co. of San Francisco. It amounted to about 1.2 quarts and was packed in 40 Pyrex ampoules, each containing 25 grams net weight, and then in cotton in four wooden boxes each holding ten ampoules. The four wooden boxes were contained in a larger box with a gross weight of 41.12 pounds. The box was strapped and sealed at the time of delivery to the Soviet Purchasing Commission in Washington, D. C., on November 4, 1943.

The 1,000 grams of heavy water were shipped to Russia through Great Falls, Mont., on November 29, 1943, with a notation on the United States Air Force airway bill "Shipped 12–2–43." A release certificate, dated November 15, 1943, and signed by the head of the Lend-Lease Division for Soviet Supply and by the head of the Chemicals Division for the Soviet Purchasing Commission, is in United States Army records.

Chematar, Inc., solicited another order from the Soviet Purchasing Commission in January 1944 without success. A second order was received on January 16, 1945, without solicitation, for 100 grams. This amount of heavy water was obtained from Stuart Oxygen Co. on February 7, 1945, and it was forwarded to the Soviet Purchasing Commission on February 14, 1945. This package consisted of a single wooden box containing four sealed Pyrex ampoules packed in cotton. The box was wrapped in heavy paper and sealed with wax, had a gross weight of 6 pounds, and a value of $350. Payment for the shipments of heavy water were by check from the Soviet Purchasing Commission. No Lend-Lease or Air Force records have been located for this shipment.

There is no record of any order or directive by MED, Lend-Lease, WPB, or other interested agency forbidding the shipment of heavy water during the war. Neither shipment involved the use of carboys such as Mr. Jordan has stated he saw at Great Falls, Mont. Shipments to Russia through Great Falls of sulfuric acid in carboys have been noted during the period 1943–44 on Air Force shipping tickets.

Ferro-uranium.—On February 9, 1943, the Redistribution Division. WPB, regional office in Cleveland, Ohio, directed a letter to the Manhattan Engineer District, with regard to approximately 65 pounds of ferro-uranium in the possession of the Latrobe Electric Steel Co., Latrobe, Pa. (It was used by this company and others to produce high-quality alloy steel. It was expensive to use and, about this time, uranium was replaced by ferro-chrome and ferro-vanadium because of its cost and scarcity.) The purpose of the letter was stated as the return of this material into war production. The Manhattan District contacted the Lend-Lease Administration by letter on April 17, 1943, and passed along a copy of the letter relating to the ferro-uranium. It was suggested by the Manhattan District that the Soviet Purchasing Commission's request for uranium for experimental work on alloys for armaments might be satisfied by this ferro-uranium.

On April 23, 1943, the Lend-Lease Administration replied to the War Department that Generals Groves and Wesson had indicated approval would be forthcoming for the Soviet request for 500 pounds each of uranium oxide and nitrate and that Lend-Lease assumed General Groves would keep the correspondent,

Colonel Crenshaw of the Manhattan District, informed of any details regarding his decision.

In the meantime, the Lend-Lease Administration had advised the Soviet Purchasing Commission, on April 14, 1943, that its application for export permission on 500 pounds each of uranium oxide and nitrate was impossible to clear but that "we may be of assistance in obtain.ng uranium compounds other than those specified, which nevertheless are suitable for the production of armaments." The lend-lease officer who signed this letter was William A. Moore. He stated in an interview on December 9, 1949, with a joint committee staff member that this was an indirect reference to the ferro-uranium at Latrobe. No direct Soviet reaction to the offer was received. The material was not obtained for the Soviet Purchasing Commission and no further reference to it is contained in records or files which have been examined.

Documents.—In an interview with a joint committee staff member on December 5, 1949, General Groves, former chief of the MED, stated that he had no knowledge of any shipment to Russia during the war of any "maps, documents, blue prints, engineering drawings, or scientific data" originating with the Manhattan Engineer District. Investigation into the records of the various organizations concerned have not revealed any evidence of shipment of documents bearing on the atomic energy program, such as Mr. Jordan mentioned in his testimony before the House Un-American Activities Committee on December 5, 1949.

V. OTHER ALLEGATIONS

Harry L. Hopkins.—The investigation into the alleged Hopkins note which Mr. Jordan reported seeing in a black suitcase on its way to Russia has been developed by interviewing persons who said they had been in a position to know how Harry Hopkins worked, what stationery he used, and how he signed his initials.

On December 9, 1949, Maj. Gen. James H. Burns, who had been executive officer to Mr. Hopkins on the Soviet Protocol Committee and chairman of the Soviet Lend-Lease Shipping and Supply Committee, was interviewed in the Office of the Secretary of Defense. General Burns recollected that the only specific item which Mr. Hopkins "insisted on following" in the Soviet lend-lease program was the number of planes delivered to the Russians.

Dr. Vannevar Bush, wartime head of the National Defense Research Council, was interviewed in his office at the Carnegie Institute on December 7, 1949. He stated that Mr. Hopkins' knowledge of the atomic energy program stemmed from his interest and assistance in working out the international relationship problem with Great Britain and Canada. Dr. Bush knew of no detailed interest in the program on the part of Mr. Hopkins.

Mr. Sidney Hyman, personal secretary to Mr. Hopkins from September 1945 until the latter's death, was interviewed on December 7, 1949. Mr. Hyman's job was to sort and summarize the material in almost 100 filing cabinets containing "the Hopkins papers." He said Mr. Hopkins never permitted any documents or papers to be destroyed. This applied even to penciled notes. Mr. Hyman says he never saw the initials "H. H." and stated Mr. Hopkins invariably initialed his papers "H. L. H."

Mr. Isador Lubin, assistant to President Roosevelt and Chief of the Statistical Division of the Munitions Assignment Board, was interviewed in New York on December 9, 1949. Mr. Lubin had an office adjoining Mr. Hopkins' during the war and frequently shared secretarial service with him. Mr. Lubin said he could not recall Hopkins mentioning General Groves or the atomic energy project. He stated that Hopkins was security-minded and never discussed classified matters over the telephone. Personal memos, Mr. Rubin said, were signed "Harry" and all others initialed "H. L. H." Mr. Rubin added that he could not recall ever seeing any White House stationery with an individual's name printed on it during the war. Hopkins had some special private stationery which Mr. Lubin helped him secure and which was printed "Harry L. Hopkins, Washington, D. C." Mr. Lubin said there was no indication that the stationery came from the White House.

Mr. Robert Sherwood, intimate of Mr. Hopkins and author of the book Roosevelt and Hopkins, was interviewed in New York on December 9, 1949. Mr. Sherwood expressed disbelief that Hopkins ever went outside a chain of command to do a job. He was of the opinion that Hopkins would have called General Marshall or General George if he wanted to expedite an air shipment. Mr. Sherwood has never seen White House stationery with Hopkins' name printed on it. He said he never saw anything signed "H. H." and never heard Hopkins mention General Groves. In this connection, General Groves testified before the Committee on Un-American Activities of the House of Representatives, December 7, 1949, as follows:

"* * * At no time, to the best of my recollection and belief—and I am sure I would have remembered it—did I ever meet Harry Hopkins, talk to him on the phone, receive any letters from him or write any to him, or have any dealings with anyone who pretended to be talking for him. There may be letters on file that are contrary to that, but if there are, they were of a routine type. I can find no one from the people who were the closest to me during that period in the office who remembers any such contact. I do know, of course, that Mr. Hopkins knew about this project. I know that. But as far as any dealings with me, or, as far as I know, with any members of my staff, they didn't occur.

"Furthermore, I think it is important to realize that our organization, both in Washington and in New York, was so closely knit, and things were under such a tight centralized control, that I can't imagine any request or effort by Mr. Hopkins along that line ever occurring without my knowledge, unless someone in the office was lacking in the integrity he should have had, and we never discovered that in our organization. * * *"

In an interview by joint committee staff members with Mr. Jordan in New York on December 9, 1949, he stated he was uncertain as to the exact phraseology, initials, and stationery which he had described on the Fulton Lewis broadcast of December 2, 1949, as "a letter on White House stationery" with "the name Harry Hopkins printed in the upper left-hand corner."

Henry A. Wallace.—In his radio broadcast on December 5, 1949, Fulton Lewis, Jr., stated that "the individual who overruled General Groves' blockade and ordered the atomic materials sent to Russia" was Henry A. Wallace, former Vice President of the United States.

In a telephone interview from his home in South Salem, N. Y., on December 9, 1949, Mr. Wallace stated that he "never told Groves to do anything," and that he had talked to General Groves only once on atomic matters and that this conversation was about a progress report and had nothing to do with shipments to Russia.

Mr. Wallace was a member of the three-man top policy group delegated by the President in the fall of 1941 to determine the course of the atomic-energy project. In this capacity, Mr. Wallace reviewed the report of Dr. Vannevar Bush on June 13, 1942, recommending expansion of the project. Out of this report, after its approval by the President, grew the Manhattan district headed by General Groves. While unable to attend the first policy conference on September 23, 1942, Mr. Wallace did receive the first progress report from General Groves. Mr. Wallace and General Groves do not recollect any further contact between Mr. Wallace and the atomic-energy project.

In the interview with a joint committee staff member on December 5, General Groves stated: "All our dealings on lend-lease were with General Wesson." In his testimony before the House Un-American Activities Committee on December 7, 1949, General Groves testified that "at no time did Mr. Wallace bring any pressure to bear on me directly, and at no time was I aware that any indirect pressure was brought to bear by him."

Royall Edward Norton.—In connection with Mr. Norton's reference on the radio broadcast of December 14 to a blueprint chart of the atomic structure of element "92," it may be recalled that the first "Periodic Chart of the Elements" was promulgated in 1870 by the Russian Chemist Mendeleeff. This and subsequent revisions and refinements are standard equipment in most high-school and college chemistry classrooms. They are available from publishers of scientific literature and at most technical libraries.

These charts describe the electron shell groups, the isotopes, the weight, size, configuration, crystal lattice, boiling point, density, and other characteristics of each element. It is conventional practice to reproduce these charts or parts thereof in blue or ozalid prints to hang on laboratory walls.

VI. SUMMARY

Review of the data examined and interviews conducted in connection with this inquiry indicates that two shipments totaling 1,420 pounds of uranium salts and one shipment of 2.2 pounds of uranium metal were made to Russia under lend-lease with the knowledge and consent of all agencies and parties concerned in 1943 and 1944. In addition, two shipments of heavy water, totaling 1,100 grams, were made to Russia during the war within existing legal provisions for such exports.

The joint committee staff could find no indication that documents, maps, blueprints, or classified papers other than those authorized in connection with the wartime agreements with Russia were shipped during the war through Great Falls, Mont. At the same time, the volume of material in transit and the existence of channels covered by diplomatic immunity through which unauthorized shipments might have reached Russia preclude any positive statement that nothing ever went to Russia without approval via Great Falls.

PERSONS CONTACTED DURING INQUIRY

Arneson, R. Gordon, Special assistant to Under Secretary of State
Bolling, Maj. Gen. A. R., Deputy Chief, Army Intelligence Division
Burman, Lawrence C., Assistant Chief, Raw Materials Division, Manhattan Engineer District
Burns, Gen. James H., executive officer to Harry Hopkins
Bush, Dr. Vannevar, Top Policy Committee
Carroll, Gen. Joseph, Air Forces Intelligence
Consodine, William, Manhattan Engineer District security officer
Dow, Frank, Commissoner of Customs
Farrell, Gen. Thomas, Deputy to General Groves, Manhattan Engineer District
Foster, C. W., president, American Pacific Industrial Co., New York
Groves, Gen. Leslie R., Manhattan Engineer District
Hazard, John, Secretary to Soviet Protocol Committee, State Department
Hillenkoetter, Admiral R. H., Director, Central Intelligence Agency
Hyman, Sidney, Harry Hopkins' personal secretary after retirement
Johnson, Lt. Col. Allen C., Assistant to General Groves, Manhattan Engineer District
Jordan, George Racey, Major, United States Army Air Forces
Lansdale, John, Manhattan Engineer District security officer
Lubin, Isadore, Assistant to President Roosevelt
Merritt, Dr. Philip L., Chief, Explorations Branch, NYAEC
Moore, William A., Senior Administrator, Soviet Supply Lend-Lease
Nichols, Maj. Gen. K. D., Assistant to General Groves, Manhattan Engineer District
Nichols, Louis B., Assistant Director, FBI
Perlman, Philip, Solicitor General, Department of Justice
Potter, J. Seward, S. W. Shattuck Chemicals Co., Denver, Colo.
Pregel, Boris, president, Canadian Radium & Uranium Co.
Rosenberg, Herman, Chematar, Inč., New York
Sherwood, Robert C., biographer of Harry L. Hopkins
Silver, Philip, lieutenant, United States Army Liaison at Fairbanks, Alaska
Symington, W. Stuart, Secretary of the Air Forces
Truesdell, George E., records custodian, Russian Branch, Lend-Lease
Volpe, Joseph, Jr., general counsel, AEC
Wallace, Henry A., Vice President of the United States
Witsell, Maj. Gen. E. F., Adjutant General, United States Army

GOVERNMENT AGENCIES FROM WHICH RECORDS DATA WAS SECURED OR SOUGHT

Atomic Energy Commission
Bureau of Customs
Department of the Air Force
Department of the Army
Department of the Navy
Department of State

Foreign Economic Administration
Lend-Lease Administration
National Archives
Soviet Protocols Committee
War Production Board

PRIVATE COMPANIES CONTACTED DURING INQUIRY

Canadian Radium & Uranium Corp., New York, N. Y.
Chematar, Inc., New York, N. Y.
Hugh S. Cooper Metallurgical Laboratory, Cleveland, Ohio.
A. D. MacKay Co., New York, N. Y.
Manufacturers Chemical Co., New York, N. Y.
S. W. Shattuck Chemicals Co., New York, N. Y.
Stuart Oxygen Co., San Francisco, Calif.

PART III

NONESPIONAGE CASES

1. The Case of the Plutonium Souvenir (Dr. Sanford Simons)

Born in New York in 1922, Sanford Simons was graduated in 1944 from the Missouri School of Mines and Metallurgy. After doing research work at Battelle Memorial Institute, he was stationed at Los Alamos as an enlisted man in the United States Army from August 1944 to March 1946. He remained at Los Alamos as a civilian scientist until July 1946. Following a period of private employment as a consulting engineer, he joined the research staff of Denver University. When arrested, he was engaged in upper atmosphere research as part of a Denver University contract with the Air Force.

On August 22, 1950, the Federal Bureau of Investigation arrested Simons and charged him with illegal possession of fissionable materials—a crime for which the McMahon Act fixes a maximum penalty of 5 years in prison and a $5,000 fine. Specifically, it was charged that while at Los Alamos, the 28-year-old scientist had stolen a glass vial containing plutonium, and had buried it under his home. It was not alleged that Simons had any connection with an espionage network, nor was it charged that he was associated with Communist or subversive organizations.

Simons declared that he had always liked to collect "mineral samples," and said that he took the plutonium as a "souvenir." He also stated that he had taken samples of uranium from Los Alamos for the same reason, and that the Federal Bureau of Investigation had removed the uranium from his home 2 years previous to his arrest. He said he had hidden the radioactive plutonium under his house because he feared that one of his children might come in contact with it.

On October 20 Simons entered a plea of guilty. At his trial, held on November 22, his attorney declared that curiosity alone had prompted Simons' thefts. The trial judge ruled, however, that Simons' action "did constitute a crime and he did know better."

Simons was sentenced to 18 months in Federal prison. He did not appeal the sentence.

2. *The Missing Brief Case*

On August 8, 1950, Mr. Frank Greenlees, Assistant Director of the Minister of Supply's atomic research plant at Risley, England, traveled by train from London to Blackpool. He carried with him a suitcase containing secret atomic energy documents. Mr. Greenlees left the case in the corner of his compartment during a 3-minute absence while he walked along the coach corridor. When he returned, the case was gone.

Three days later, the suitcase—with all documents intact—was located at Folkestone, on the English Channel. William Ralph Wakeham was arrested the same day and charged with theft of the suitcase. The police were satisfied that the theft was not linked to espionage, and that Mr. Wakeham was an ordinary sneak thief, unaware of the contents of the stolen case. The day following his arrest, Wakeham slashed his wrists and swallowed a razor in an unsuccessful suicide attempt.

He subsequently plead guilty to the charge of stealing the suitcase and was sentenced to 4 years in prison.

3. The Souvenir Photos

The following cases all relate to the theft of classified documents and photographs by enlisted men serving with the United States Army at the Los Alamos Laboratory, during the war. All involved charges under title 18, United States Code, sections 100 and 234, Theft and Unlawful Removal of Government Records. None of these cases was prosecuted under the McMahon Act, and at no time was it alleged that the participants in these crimes were connected with any espionage network, or motivated by any desire to pass on restricted data to a foreign nation.

1. *Arnold F. Kivi.*—While serving as an enlisted man at Los Alamos, Kivi removed highly classified photographs and retained them in his personal possession. Following investigation by the FBI he was arrested and pleaded guilty. On October 16, 1947, he was sentenced to 18 months imprisonment in the Federal Correctional Institution of Danbury, Conn.

2. *Alexander Van der Luft.*—From 1944 until March 1946, Luft was stationed at the Los Alamos project. Investigation revealed that during this period he had removed highly classified documents. On August 21, 1947, he pleaded guilty to this charge and was sentenced to 4 years probation.

3. *Ernest Dineen Wallis.*—Upon his departure from Los Alamos in March 1946, Mr. Wallis removed classified negatives of official photographs. He was indicted at Albuquerque on July 24, 1947, and in April of the following year, was given a suspended sentence of a year and a day by the Federal court in Chicago.

4. *Fred Gerard Michaels.*—While serving at Los Alamos from December 1943 until June 1946, Michaels made photographs for his personal use from classified negatives in the possession of Wallis. He was arrested on February 13, 1948. On April 1, 1948, the United States attorney at Albuquerque dismissed the charges filed against Michaels and the case was closed.

5. *Ernest Lawrence Paporello.*—During his stay at Los Alamos, Paporello removed classified photographs. He was arrested and arraigned on January 15, 1948. He pleaded not guilty to the information filed against him and on March 11, 1948, was sentenced to 6 months' imprisonment and fined $250 on each of two counts, with the prison sentences to run concurrently.

6. *George Wellington Thompson.*—Thompson served as an enlisted man at Los Alamos from August 1943 to September 1945, and was subsequently employed as a civilian at Los Alamos and Sandia until March 1947. The investigation of Wallis and Kivi revealed that Thompson had in his possession classified negatives and photographs, which had been removed from Los Alamos. He was arraigned on January 13, 1948, and pleaded not guilty. On March 11, 1948, Thompson was convicted and fined $125 on each of two counts.

4. The Falsified Security Questionnaire

Ernest Jewell Koch was intermittently employed by subcontractors of the Atkinson & Jones Construction Co. at Hanford, Wash., between February 1948 and April 1950.

He was never an employee of the Atomic Energy Commission, and he never received a clearance and never had access to restricted data.

Like all other persons employed by the Atomic Energy Commission or by contractors participating in the project, Mr. Koch was required to fill out a personnel security questionnaire when he made application for employment.

In filling out this questionnaire, Mr. Koch concealed the fact that he had been a member of the Communist Party.

Title 18, section 1001, United States Code, Fraud and False Statements Chapter, provides:

Whoever, in any matter within the jurisdiction of any department or agency of the United States knowingly and willfully falsifies, conceals, or covers up by any trick, scheme, or device a material fact, or makes any false, fictitious, or fraudulent statements or representations, or makes or uses any false writing or document knowing the same to contain any false, fictitious, or fraudulent statement or entry, shall be fined not more than $10,000 or imprisoned not more than 5 years, or both.

On January 9, 1951, Mr. Koch was convicted under the statute and fined $250 and placed on probation 3 years.

196

APPENDIX

The following is quoted from Parliamentary Debates, House of Commons Official Report, Session 1950 (472 H. C. Deb. 5s, page 744, March 13, 1950):

COMMUNIST PARTY, GREAT BRITAIN

45. SIR W. SMITHERS asked the Prime Minister if, in view of the result of the Fuchs trial, he will introduce legislation to outlaw the Communist Party in Britain.

The PRIME MINISTER (Mr. ATTLEE). No, sir.

SIR W. SMITHERS. In view of the alarm and despondency created in most parts of the world by this trial, and in view of their records, will the right hon. Gentleman, as a first step, ask for the resignation of the Chancellor of the Exchequer, the Secretary of State for War, and the Minister of Defence before they can do any more damage to this country; and will he remember the Ides of March?

Mr. BLACKBURN. Is my right hon. Friend not aware that Dr. Fuchs was first given a security appointment by a Conservative Minister of Supply, Sir Andrew Duncan, at a time when the "Daily Worker" had, in fact, been suppressed.

The PRIME MINISTER. No, Sir, that was not the Minister.

The following is quoted from Parliamentary Debates, House of Commons Official Report, Session 1950 (472 H. C. Deb. 5s, page 1228, March 16, 1950):

ALIENS

NATURALISATION

10. Mr. NORMAN BOWER asked the Secretary of State for the Home Department if, in future, he will publish the names of sponsors of candidates for naturalisation.

6. Mr. MARTIN LINDSAY asked the Secretary of State for the Home Department if he will state the names of those who sponsored Dr. Fuchs for British nationalisation.

Mr. EDE. There is no statutory requirement that an application for naturalisation should be supported by sponsors and I do not think that any useful purpose would be served by the publication of the names of the persons who, in fact, support the applicant. The assumptions underlying these questions seem to be that great weight is attached to the standing of the sponsors and that the Secretary of State has to rely to a large extent on the information supplied by them. In fact, the most careful inquiries are made through the police and by other means and where necessary other Government Departments are consulted before he reaches his decision. No application would be granted merely on the recommendation of sponsors however distinguished.

SURGEON LIEUT.-COMMANDER BENNETT. On a point of Order. May I, with all the humility due from a new Member, ask you, Mr. Speaker, if Question No. 6 is in Order, as I have no reason to understand that the Government ever intended to "nationalise" Dr. Fuchs?

Mr. SPEAKER. Question No. 6 was answered with Question No. 10.

NOTE.—The materials in the appendix are reprinted for purposes of furnishing additional information and background. No responsibility is assumed for the accuracy or authenticity of such materials; and expressions of opinion or statements of conclusions, where they occur, are not to be imputed to the compilers of the appendix.

The following is quoted from Parliamentary Debates, House of Commons Official Report, Session 1950 (472 H. C. Deb. 5s, page 1545, March 20, 1950) :

DR. FUCHS

46. Mr. HENRY STRAUSS asked the Prime Minister whether His Majesty's Government received any warning regarding Dr. Fuchs from His Majesty's Government in Canada when the Canadian Royal Commission was sitting in 1946.

The PRIME MINISTER. No, Sir.

Sir W. SMITHERS. May I ask the Prime Minister whether he has read the Report of the Royal Commission on the spy trial in Canada, and whether he is aware that in that report the Russian Ambassador in Canada was implicated, that this same Ambassador is now Ambassador here in Britain and that M. Zarubin has now gone to Russia; and is that the reason?

The following is reprinted from Parliamentary Debates, House of Commons Official Report, Session 1950 (475 H. C. Deb. 5s pages 1383–1384 and 1985, May 18, 1950) :

DR. KLAUS FUCHS

42. Mr. G. THOMAS asked the Secretary of State for the Home Department to what extent permission has been granted to foreign representatives to question Dr. Fuchs; and whether he will make a statement.

69. Mr. DONNELLY asked the Secretary of State for the Home Department what are his regulations governing the questioning of prisoners in British jails by police of other countries; and to what countries are these facilities granted.

72. Mr. WYATT asked the Secretary of State for the Home Department what conditions will govern the interrogation of Dr. Klaus Fuchs, at present in one of His Majesty's jails, by officers of the United States Federal Bureau of Investigation; and what precedents there are for interrogation of His Majesty's prisoners in England by officials of a foreign Power.

74. Mr. FERNYHOUGH asked the Secretary of State for the Home Department if he will make a statement to the House in connection with the permission which has been granted to officials of the United States Federal Bureau of Investigation to question Dr. Klaus Fuchs, the atomic scientist at present serving a 14 years' sentence of imprisonment in His Majesty's prison.

Mr. EDE. I would refer my hon. Friends to the reply given on 11th May to my hon. Friend the Member for Northfield (Mr. Blackburn), to which I have nothing to add.

Mr. THOMAS. Is my right hon. Friend aware that this distasteful departure from normal procedure is watched with some anxiety by the public, and can he say whether the prisoner concerned will be legally represented when foreigners are questioning him in our prison?

Mr. EDE. This prisoner's activities were also exceedingly distasteful, and it is necessary that the State should take such steps as it can to protect itself against such activities. This man need not answer any questions unless he chooses to do so. He will not be interviewed unless he expresses a willingness to be interviewed.

Mr. FERNYHOUGH. Does not my right hon. Friend think that this is rather a reflection upon our own M.I.5 and Scotland Yard, and would they not have been capable of interviewing this man and getting from him, if he is prepared to volunteer it, the information which the Americans are seeking?

Mr. EDE. No, Sir. There is no reflection on the police services of this country in this matter, but some of the offences which this man has committed were committed in the United States of America, and I think it is desirable, in the interests of both countries, that as far as possible his activities should be investigated.

Mr. DONNELLY. What are the normal regulations governing cases of this kind? Could my right hon. Friend also say to what countries these facilities are normally extended?

Mr. EDE. This is the first time that such facilities have been granted because this is the first time that such an offence has been committed.

Mr. Sydney Silverman. Can my right hon. Friend say why it was not thought sufficient for the information which the United States police want to get from this man to be put to him by our own police forces? Can he also say whether any arrangements have been made to provide himself with a copy of any statement which may now be made?

Mr. Ede. A British officer will be present if this interview takes place, and it will be conducted according to British practice. It is important—I should have thought that my hon. Friend the Member for Nelson and Colne (Mr. S. Silverman) would have recognised it—that some questions should be followed up by supplementary questions and it is, therefore, desirable that a person who can supply the information for an appropriate supplementary question should be there to put the question.

Mr. Hector Hughes. Will my right hon. Friend take care that at any such interrogation there will be present responsible British scientists, so that full advantage may be taken of any further information elicited?

Mr. Ede. No, Sir, I do not think that that would be desirable. I think that if an officer representing the British police or the British security service is present that will be enough.

The following is quoted from Parliamentary debates, House of Commons officially reported, Session 1950 (475 H. C. Deb. 5s, pp. 567, 568, May 11, 1950).

Dr. Klaus Fuchs

45 and 46. Miss Irene Ward asked the Prime Minister (1) whether he can give an assurance that steps have been taken to ensure that the spy operations of the four names, other than Dr. Fuchs, sent by the Canadian authorities, have ceased ;

(2) why action was not taken to remove Dr. Fuchs from his position in atomic research following the notification of his name as a spy by the Canadian authorities.

48. Mr. Hollis asked the Prime Minister what representations were received from the Canadian Government after the Canadian spy trials concerning the unreliability of Dr. Fuchs; and what action was taken on those representations.

The Prime Minister (Mr. Attlee). I have nothing to add to the statement made in another place on 5th April by my noble Friend the Lord Chancellor, of which I am sending the hon. Members copies.

Miss Ward. Is the right hon. Gentleman aware that the Canadian Minister of Exteranl Affairs has recently repeated the information which, he asserts, he gave to His Majesty's Government, and could the Prime Minister state the date on which he received this information and what action was taken on its receipt?

The Prime Minister. If the hon. Lady will study the reply which was given by my noble Friend in another place, she will see that she is under some misapprehension as to the exact facts.

Earl Winterton. Can we have an assurance from the Prime Minister that, assuming that some of us raised the question of this very mysterious action in Debate in this House, we should not be told that the Prime Minister or Lord President were not prepared to answer because a full statement had been made in another place? This is a rather important constitutional point.

Mr. Emrys Hughes. In view of all this horror and indignation about spying, could the Prime Minister assure us that the £3 million which is spent on our Secret Service is not spent on bribing people of other countries to spy?

Mr. Nigel Davies. In the interests of national security will the Prime Minister assure the House that in future no one will be employed on atomic research who is known at any time to have been a Communist or fellow traveller?

The Prime Minister. I am quite sure if the hon. Member studies the full account which was given by my noble Friend in another place he will see that this was a wholly exceptional position, in which no one had any reason to suspect that this person was a Communist except from a vague allegation by Nazis of a great many years before.

Sir Waldron Smithers. May I ask the Prime Minister if, in dealing with these difficult questions, he will seek the advice and guidance of the Secretary of State for War, with his tremendous and unusual qualifications?

The following article from the New York Times magazine of March 4, 1951, is reprinted by permission of the New York Times Co. and the author, Rebecca West:

THE TERRIFYING IMPORT OF THE FUCHS CASE

ONE YEAR AFTER HIS SENTENCING WE SEE HE UNITED EXPLOSIVE KNOWLEDGE AND AN IMMATURE MIND

(By Rebecca West)

LONDON.—It is a lie that there is no new thing under the sun. The past had no product to match Dr. Klaus Emil Fuchs, who was sentenced to 14 years' imprisonment at the Old Bailey on March 1, 1950, for having handed information about the A-bomb to an agent of the U. S. S. R. He represents a danger to humanity such as it has never had to face before, and "humanity" is the right word. For, though he was punished for having given by theft to the U. S. S. R. the power to inflict damage on the populations and territory of the West, he is a threat to the U. S. S. R. as much as to any other part of the world. Nobody is in a position, anywhere on this globe, not to feel frightened by the menace disclosed by the existence of Dr. Fuchs. This is not just because he is a Communist spy; it is because he was a particular kind of man. An odd kind of man.

Some measure of his oddity is given by the opening of the statement he made to the security officers on his detention. He began by giving them the date of his birth and assuring them that he had had "a very happy childhood." Now, British policemen seem much milder than American policemen and are certainly more stolid. But it is unlikely that they looked at Dr. Fuchs in a manner suggesting that it would take a weight off their mind if they could learn that he had not been unhappy when he was a small boy. It is unlikely, too, that most people, charged with a crime involving long-standing and heartless fraud and certain to cause hideous consequences, would fail to recognize that society might have other anxieties which it would like to settle first. This is a strange bird.

It is a pity that the authorities have not given the public a fuller warning of what Dr. Fuchs did, and of what he is. W. L. Laurence has said in The Hell Bomb that Fuchs' admission of what he told the Russians about the A-bomb and the H-bomb has not been published for security reasons. The argument is that it is not certain whether Fuchs told the Russians all that he says he did, and that it is possible that he may actually have put into his admission certain facts which he had not told the Russians, so that if it were published they would then learn some fresh information. But Mr. Laurence very sensibly pronounces this great nonsense, on the ground that Fuchs obviously told the Russians all he knew and is unlikely to have had some barrel scrapings by him for use in his admission.

In any case, it is not necessary to know the details of the information given by Dr. Fuchs to the U. S. S. R. to realize his deadly significance. But there are three things which the authorities ought to make plain. First: the general nature of the information he gave to the Russians, whether it related to the H-bomb as well as the A-bomb, how many scientists had worked to get that information, the size and cost of the equipment they had used in the process, and how many years of work and how much expense he saved the Russians by giving them this stolen information. Second: his remarkable eminence as a scientist. And third: the poverty of his general intelligence and the immaturity of his character. It is this third fact, the nature of the man, which makes him more terrible than flood or lightning or any of the familiar cataclysms.

To understand Dr. Fuchs we must note that this is no case of the godless scientist cradled in materialism. Klaus Emil Fuchs came from a pious home His father, Emil Fuchs, was a preacher well known in Germany since the beginning of the century; first as a Lutheran pastor and then as a Quaker. He was a true mystic, illumined by the love of God, and his courage in earthly affairs was superb. He was the first pastor to join the Social Democratic Party, and between the wars he was well known as a speaker for a group known as the Religious Socialist. He defied the Hohenzollern rule and defied Hitler. He was also a loving husband who made a delightful home for his sons and daughters. But in the opinion of some of those who liked him best he was not very intelligent, and his writings show that he was intensely egotistical and self-satisfied. His virtues are so great that it would be foolish to mention his failings, were it not that they have a bearing on his son's career.

Klaus Emil, born in 1911, was his youngest son. He studied at Leipzig University, and then, when his father was made professor of religious science at a teachers' training college in Kiel, moved to the university there. At both places he was deeply involved in the useless and silly and violent political activities by which German undergraduates did so much to destroy the coherence of their own country and the peace of the world. College is a grand place for political discussions and a terrible place for political action. When college students go in for deeds, not words, cold-blooded adults get hold of them and without mercy use them as catspaws.

In Germany at that time the Communists were indulging in a campaign against the Social Democratic Party, although they should for obvious reasons have joined with the Social Democrats and the various schools of Liberals in an unbroken Popular Front against the Nazis. Their secret reason for this was a tragic and ridiculous miscalculation: they wanted Hitler to come to power, in the mistaken belief that the Nazi regime would collapse immediately and leave Germany ripe for capture by communism. But they put up a noisy and hypocritical pretense that they were attacking the Social Democratic Party not because it was doing too much against the Nazis to suit them but too little.

Klaus Emil was completely taken in by this fraud, and very active under its influence. The political follies committed in the dying Weimar Republic are as unpalatable as yesterday's melted ice cream, but Klaus Emil's career must be followed because it led him and us to our present situation. He ran about with the high-speed inconsistency characteristic of German political life. He joined the students' section of the Social Democratic Party, but left it because the party supported a policy of naval rearmament, and he had been brought up to be a pacifist. But very soon afterward he joined a society with a mixed membership of Social Democrats and Liberals, which was in fact a semimilitary organization with a taste for street fighting.

Then he moved to Kiel and went back to the Social Democratic Party, but presently left it again and offered himself as a speaker to the Communist Party without joining the party, and at the same time became a member of an organization, much frowned upon by the Social Democrats, in which rebel members of their party joined with Communists in the dangerous game of fraternizing with those students belonging to the Nazi Party whom they thought "sincere" and possible converts.

This was a nasty organization in which everybody was trying to double-cross everybody else. Then, when the Communists had so greatly weakened the Social Democratic Party that it could do nothing to fight the Nazis, Klaus Emil left it in disgust at the impotence and joined the Communist Party. The record reads like a recipe for mincemeat, but produced nothing wholesome.

When Hitler came into power in 1933 Klaus Emil was engaged in a complicated and futile campus intrigue, in which he showed a great deal of courage, particularly considering that he was of feeble physique, but little sense and even less fastidiousness. When he was in the train on his way to Berlin to attend a secret conference of anti-Nazi students he read of the burning of the Reichstag, saw that the hunt of the Communists had begun, took the hammer and sickle badge out of his coat, and went into hiding.

He was presently drawn into the operations of a mechanism which was one of the most brilliant achievements the Communist Party has to its name. After they had helped the Nazis get into power, they worked to get control of the organizations set up to care for the refugees from Nazi tyranny in all the countries to which they fled. They then saw to it that the Communist refugees received preferential treatment, that the non-Communist refugees were exposed to Communist propaganda and learned to look on Communists as their benefactors, and that the Communists and non-Communist refugees alike served the ends of the Communist Party.

All this they did with a pickpocket ingenuity, covering up their activities from the observation of the non-Communist members of these organizations, who were merely furnishing the bulk of the money and the personal service. Klaus Emil was told by his party that he must go abroad and finish his studies, because when the Nazis had been thrown out the Communists would need members with high technical qualifications to build up Soviet Germany; and he was first sent to France and then to England, where he was befriended by the Society for the Protection of Science and Learning, a body consisting almost entirely of non-Communists.

It must be emphasized that at no time did Klaus Emil have grounds for complaint against Great Britain. He never found it niggardly, or on the side of reaction. He was sent to Bristol University, where he got his doctorate of philosophy in mathematics and physics, and then to Edinburgh University, where he got his doctorate of science, and was given a Carnegie research fellowship. When war broke out between Great Britain and Germany the aliens tribunal, before which he appeared to show cause why he should not be interned, accepted his membership in the Communist Party as proof that he was anti-Nazi.

It is true that in 1940, when the Germans invaded the Low Countries and France, he was interned and taken to Canada. But this was the treatment which was applied both to refugees who were thought specially suspicious and those who were thought specially meritorious, and in his case it was certainly a proof that the authorities believed him worthy of being saved from a possible German invasion of Britain.

In 1942 he was allowed to return to Great Britain, where a position was waiting for him at Glasgow University. Soon afterward he was asked by Professor Peierls, a very eminent German-born refugee physicist, to come to Birmingham University to help him in some war work. This proved to be atomic research. In June of that year he signed the usual security undertaking, and applied for naturalization as a British subject a month later, taking the oath of allegiance to the King in due course, while at the same time he made arrangements to hand over all particulars of the research to couriers who he knew would deliver them to the Soviet authorities.

For the next 8 years he carried on this work in atomic research, first in England, then for 3 years with Professor Peierls in America, then again in England for 4 years as head of the theoretical division of the atomic energy project at Harwell. During the whole of this time he never flagged in his treachery. As steadily as the results were produced, so he handed them over to his Communist couriers.

His arrest on February 2, 1950, gave the thoughtful two reasons for terror. The first was the inadequacy it disclosed in the security measures taken by both Great Britain and America. The British should not have allowed Klaus Emil to take employment in a government atomic research project only 4 years after he had avowed to the aliens tribunal that he was a Communist and therefore repudiated all obligations of loyalty save those imposed by the Communist Party.

This act of carelessness should have been corrected when Gouzenko fled from the Soviet Embassy in Ottawa; for the name of Fuchs was scribbled in a notebook which was one of the exhibits in the Canadian spy-ring case. Moreover, in 1945, Elizabeth Bentley went to the FBI and told the story of her activities as a Communist agent, and this included an account of a courier system in which she was a link; and this system was intertwined with that by which Fuchs had passed to the Soviet authorities the results of the Los Alamos researches. It is a pity that no word about Fuchs crossed the Atlantic till 1949.

The second reason for terror was the statement Fuchs made. Here was one of the most gifted scientists of our time, with power to be part creator of lethal weapons transcending all the previous malice of mankind, and to be as dangerous in his work as a single-handed traitor, because of his rare and exalted gifts. And his statement read like the ramblings of an exceptionally silly boy of 16.

He was 38 years old. He was suspected of an appalling crime. He began by assuring the special branch officers of the happiness of his childhood, and went on to relate how brave he had been when a boy. It appeared that there was once a celebration at his school on the anniversary of the foundation of the Weimar republic; and as a protest many of the pupils arrived wearing the imperial badge, so he had put on the republican badge, and the other children had torn it off.

He recalled that; and he recalled, in the minutest detail, all his foolish and futile political activities at his universities. And in the course of this merciless recapitulation, which must have made the security officers groan aloud, he betrayed an unusual degree of political ignorance.

Every student of contemporary history knows that Communist strategy in Germany during the early thirties aimed at splitting the Popular Front and letting Hitler in so that he could be got out again by a revolution which the Communists would turn to their profit. Indeed, it is so well known that it would be virtually impossible for a non-Communist to write of those times without taking it as established historical fact, or for a Communist to write of them without attempting to disprove that assumption. But it is plain that Klaus Emil had never even heard of this interpretation of the events in which he took part. He wrote of them as naively as if he were still 20 and they had never been discussed.

Some of these tedious fatuities of his youth he recounted to the security officers for the sake of their moral, rather than their political, implications; and that, too, was a curious self-betrayal. Throughout the statement Klaus Emil expressed himself with extreme egotism and vanity. Even if we take into account the strong strain of self-satisfaction running through his father's writings, and remember also that he had spent all his childhood in minor industrial towns where his father was the unchallenged intellectual and moral leader, his sense of being an elect being must be pronounced extraordinary, particularly in a man of 38.

But it worried him, when what he had been doing was brought out into the open and he had to discuss it, that such a perfect character as his own should have been capable of practicing the continued deception, which, as he admits with an air of being fair-minded, had been a part of his treachery. He explained to the security officers at enormous length that this was all due to a mildly dirty trick he had played on some Nazi students during his campus intrigue in 1933. He had not given them fair warning that he was going to publish an attack on them for a course of action which, had they received such a warning, they might have abandoned. He had omitted to resolve this point in his mind, he said, and so he had set up a mental process which he described as "controlled schizophrenia." It was, in fact, plain lying and cheating, but these were too realistic terms to be used in the "Cloud-Cuckoo-Land" where he had made his home.

There was no limit to his sense that power should be his. At one point in his statement he rebuked the British authorities for not letting the internees in the Canadian camp read newspapers. He ignored the practical reason for this, which was the difficulty of keeping discipline and protecting the non-Nazi internees from the Nazi internees, had the news continued to be bad over any length of time. Gravely he complained that it had prevented him from learning the truth about the real character of the British; and it is implied that had he known more about them he might have spared them, might not have aided their enemies to drop A-bombs on them. Not for a moment did it cross his mind that perhaps it was not for him to smite them or to spare them.

As he demanded power, he showed why he, of all people should not have had it. This is not a superman claiming to govern the inferior masses, it is a subman who can only claim superiority to the masses in regard to special gifts quite irrelevant to government. His general ideas were childish; there is a passage on Marxist philosophy which would be considered poor at the least distinguished Youth Congress. He does not appreciate the material consequences of his treachery; he expressed concern that what he had done might "endanger" his friends, but he apparently meant simply that he might endanger their prospects of retaining their employment at Harwell, not that they might presently be blown up by A-bombs dropped by the U. S. S. R. But the fact about Klaus Emil which makes his appetite for government most appalling is one which might have been imagined to be a reassurance. He is not what he is supposed to be. He is feared as a fanatical Communist. But he is not even a loyal Communist; and therein lies his novel and terrible significance.

The statement shows that he is too infatuated an egoist ever to have given himself to any party. If he betrayed Great Britain and the United States to the U. S. S. R., it was only because they were the handiest objects for betrayal. He felt himself qualified to manage any society's business better than it could itself, and as he found himself in the center of the western society composed of Great Britain and the United States, he had to mind their business. Because he decided that the best way to exercise his supermanagerial powers was to attempt to destroy Great Britain and the United States, it is not to be supposed that he considered the U. S. S. R. worthy of survival. It would have to take its turn. He wrote in his statement:

"I came to a point where I knew that I disapproved of many actions of the Russian Government and of the Communist Party, and I still believed that they could build a new world and that one day I would take part in it, and that on that day I would also have to stand up and say to them that there are things which they are doing wrongly."

The word "also" is difficult to account for syntactically, but it would have to be a very obtuse reader who did not see what Klaus Emil meant.

What he was saying was that he had spent the summer of his days planning, in the cause of virtue, an unparalleled stimulus to the death rate of the Western world, and he intended to spend, still in the cause of virtue, the autumn of his days in rendering a like service to the Slavs. This holds out to us a far worse prospect than we saw before us when we regarded Klaus Emil as a fanatical Communist. In that case he would have aided the U. S. S. R. to impose a certain pattern on the world; a botched and loutish pattern, but still a pattern. But Klaus Emil's statement shows that what he meant to do was to invoke chaos; and at last we see just how serious a problem is propounded to us by the existence of the traitor scientist.

Till now we have looked at this new figure too exclusively in relation to ideological and international conflicts. We have considered the British and American traitor scientists simply as persons attracted to communism, and their opposite numbers in the U. S. S. R. and the satellite countries as persons attracted to democracy. This was a view which was naturally engendered by study of the cases of Dr. Alan Nunn May and the scientific workers involved in the Canadian spy ring. But the case of Dr. Fuchs reminds us that special gifts are sometimes found in persons of a low standard of general intelligence and character, and we see that a number of flibberty-gibbets might be engaged in atomic research who, like Dr. Fuchs, would indulge in treachery for the most trivial of reasons.

One can well imagine that an unbalanced egoist like Klaus Emil might decide to hand over the means of conquering the world to President Perón; and though Klaus Emil is certainly not insane, his statement may well make the prudent wonder where nature's recklessness ends, and whether we might not have a lunatic occupying a high position in some project. Many of us can remember a very famous pianist who carried on a long and arduous career while certifiably insane; and it does not seem impossible that a gifted scientist might decide to use the result of his researches to set fire to the world in order to please the Red Indian who is his spiritualist aunt's control, or the holy men in Mars whom his ouijaboard has indicated to him as waiting for the signal of the terrestrial flames to come down and bring us salvation.

It cannot now be argued that an individual scientist would be innocuous because he would have to work single-handed; he could cause vast destruction by sabotaging his own work in order to leave his employers defenseless before an enemy, or he could gather collaborators by the pretense of a saner mission. Considered internationally, we are all in peril in this situation, whether we are western democrats or Russian Communists. Considered nationally, we are all in peril in this situation, whatever our political views, however far we may be to the right or to the left; and it is to be noted that such is the injustice of the world that few people are in a more perilous position than the same scientists who work alongside their disordered colleagues.

Our civilization has, therefore, a new task before it. It has to reconcile the need of the community for protection from the maniacal use of science with the need of scientists for the fullest measure of freedom in their work. But it will be impossible to perform this task unless the nature of the problem is fully understood; and the case of Dr. Klaus Emil Fuchs should be studied in all its strangeness in order that we may realize how strange a passage of time we are now traversing.

The following article from The Sign magazine, August 1950, is reprinted by permission:

THE SECRET LIVES OF AN ATOM SPY

FOR AN IDEAL WHICH TURNED TO ASHES, DR. KLAUS FUCHS SACRIFICED HONOR, FRIENDS, AND A TRUSTING WORLD

(By Kurt Singer)

On the historic morning of November 7, 1949, a neatly dressed, tall man of most unhistoric appearance walked down London's Charing Cross Road to Cambridge Circus. No one notices as two men pass and pause, a cigarette is lit, a word exchanged, and they walk off together toward Trafalgar Square. No one notices as the taller man, lean, bespectacled, hands in pockets, shakes his head slowly and finally turns away. Perhaps only a single stranger is momentarily interested in the gesture of the shorter, dark man, who raises his hand to the arm of his companion. But the taller man does not hesitate. He hurries down the stairs of the subway. The shorter man looks after him for a brief moment, angrily, and then rounds the corner past the Nelson Monument.

This is the way, in our time, the secrets of our deadly alchemy are bartered for a price. But for Dr. Klaus Fuchs, sitting in the train looking impassively at the underground advertising, telling of beer, coats, and underwear, the fee, like the secrets, was not measured in a calculable monetary exchange.

Fuchs was not interested in the few hundred pounds thrust upon him by the professional foreign agent. This he accepted as the badge of his submission to Laurenti Beria, head of the Soviet Secret Service. The price Fuchs demanded was a world, one world, one Communist world—a world in which the boy of a war-torn childhood, the youth of Nazi terror and German discontent, the manhood of frustration and suspicion might all be finally synthesized in the maternal bosom of a great and secure world—Communist harmony. When Klaus Fuchs was 3 years old, the German Army unleashed upon Paris the forerunner of the atom bomb, a most terrible weapon, the big Bertha, firing giant shells a distance of 75 miles into the beautiful heart of the French capital.

His pacifist father hated war. Emil Fuchs was a Protestant minister, a religious Socialist, standing in the shadow of Tolstoi and Gandhi.

The boy Klaus in the provincial town of Russelsheim, near Frankfurt, was strictly forbidden to join the cheering of the soldiers off for the front. Little Klaus began life as the outsider, the observer, the nay-sayer. He had no close boyhood friends and except for his three elder brothers and sisters he lived in virtual isolation, shielded from the contagion of hysterical patriotism and living in an aseptic world of his father's making.

His father's house had emphasized values of brotherhood, duty, internationalism, peace, religion, but there was little flexibility or humor in the teaching. Life was gray, grim, earnest, boring, and there was no time for carefree joy or laughter.

The first war was followed by the annihilating inflation, and the roots of nazism flourished in the economic swamp which Germany had become. Even in the primary school, politics was an urgent reality, and the pacifist's son Klaus was the butt of soldiers' sons, who made fun of the timid, studious boy. The troubles of Germany turned his father inward to reflection and religious experience. He became a Quaker in 1925. Klaus found no comfort there. Instead, it was clear to him that the boys who fought back and did not fear the violent little nationalists were the Reds, the Communists.

Later, at Kiel University, when the Nazis were already a major political force, Klaus joined the Young Communist League. Against his father's quakerism he embraced the doctrine of the class struggle. But he was never a great reader of Marxist literature. His field was science and, like so many brilliant mathematicians and physicists, the experimental and analytical techniques he used so scrupulously in the lecture halls and laboratories, he abandoned completely when confronted with political argument. He accepted all the worn clichés of Communist propaganda. Russia was the worker's fatherland; all weapons were permissible in the class struggle; the Communists were fighting for a classless society; there was no such thing as absolute truth or objective science; art and science were class weapons; the artist and scientist who believed in communism were in uniform and must take part in the world struggle.

Klaus saw the Nazis seize power, he knew at close hand the terror they wielded. His father was sent to a concentration camp for 9 months and his sister, an artist driven to a nervous breakdown by Nazi persecution, committed suicide by leaping under a train.

His father urged Klaus to escape from Germany so that he might continue his studies abroad, but Klaus remained working in the Communist underground movement. It was not activity that appealed to him. The disorganized life, being hunted from pillar to post, the need to abandon organized studies, did not suit the young student-scientist.

After a short while he crossed the frontier into France and from there he came to Britain. He went to Bristol University, where he specialized in mathematics and physics and was awarded a doctorate in philosophy. His lodgings in Humpton Roads, Redland, Bristol, were the typical student's retreat, untidy, strewn with papers and books. It was a simple life and a happy one, on the whole Too happy, perhaps, for Klaus to justify his conscience, for his father was in Germany where he had chosen to stay, although American Quakers had offered him a chance to get out. Emil Fuchs had replied to them that his place was in Germany, in the fight against Hitler. Where, then, was the place of his son, Klaus?

Somehow, the student had to justify to his father that his departure from Germany was not a flight from fear, but a tactical withdrawal to a place from which he could renew his role in the struggle. For the first time, he was living in conditions of freedom and reasonable stability. Politics in Britain did not have the violence or the upsets that he had known in Germany. His fellow students were not consumed by bitterness nor deeply involved in doctrinal debate. Klaus, quiet and sensitive, emotional to an extent which his poker-faced appearance belied, was attractive to certain types of girls.

Lonely and abstracted, he aroused the maternal impulse, and during his years in Britain he was never without female friends who admired and fussed about him. At the same time, his studious, ingrown personality did not make him an exciting friend; his conversation did not often go beyond scientific small talk and university gossip. To his friends, Klaus was frankly a bore, but a nice bore.

Then his field of research widened. In 1938 he went to Edinburg University, where he took his degree as doctor of science. His original researches in atomic and nuclear physics were placing him in the forefront of the younger scientists, and he published papers in the Proceedings of the Royal Society. The refugee-immigrant was making a name for himself in scientific circles.

On September 1, 1939, Hitler invaded Poland and the war was on. Klaus Fuchs suddenly found that he was, finally, regarded as a German enemy alien. A few months later, despite his feelings against Hitler, his antipathy to nationalism, his years in British universities, he was told to pack a bag and get ready for internment. To him, British tolerance was a sham, as his Communist friends had told him. In the show-down, the British ruling classes were ruthless, heartless, barefisted—Fascist. The effect of internment on Klaus, the trip as an internee in the North Atlantic through waters infested by submarines, was to revive the Communist allegiance which had become quiescent. It also added the excitement of martyrdom to his essentially adolescent nature.

In his Nissen hut, in the Canadian camp, it was not difficult for Klaus to imagine that "fate" had pointed out to him the error of his backsliding ways. It is certain that he emerged from internment with his Communist faith renewed. Separated from his friends in Britain, surrounded by many of his countrymen who were grieved that though anti-Nazi they were treated as enemies. Klaus looked again toward the distant, greener fields of the Soviet paradise.

When he was able to resume his work, his old convictions were firmly fixed. In 1941, he was released from internment to continue his work, research which was to help in the development of the atom bomb. Although it was known that he was communistically inclined, so high was his qualification that he was allowed into the most secret consultations. Security officers, after careful screening, had reported that there was no danger he would become a foreign agent.

Meanwhile, British Military Intelligence was receiving reports of an extensive German plan to build a new weapon, an atom bomb, which would be decisive in the war.

In occupied Norway, secret underground agents reported the construction of strange, heavy-water plants, where hundreds of German scientists had been put on special duty. British-Norwegian commando teams went into action to cause as much physical destruction as possible for the new German production centers.

At the same time, a meeting was called in London to lay the plans for an answer to German atomic research. To this, the deepest secret of the war, Klaus Fuchs, the Communist, was given access. The Communist, now ready to conduct espionage for Russia, was given material to work with. At his trial, Mr. Curtis-Bennett, his defense attorney, said: "Anybody who had read anything about Marxist theory must know that a man who is a Communist, whether in Germany or Timbuctoo, will react in exactly the same way. When he gets information, he will automatically and unhappily put his allegiance to the Communist idea first."

Amazingly enough, although it was on record at the Home Office that he was a member of the German Communist Party, a year after being released from internment, Klaus Fuchs was naturalized as a Briton. The superb resistance of the Red Army to the German invaders, the atmosphere of allied amity, all made it easy for Fuchs to submerge the vestiges which remained of his British "conversion" while accepting its citizenship.

His work with Prof. Rudolph Peierls, one of the outstanding atomic research scientists, during 1941, showed that he was clearly a genius in his field, "more a candidate for a Nobel peace prize or membership of the Royal Society," as Mr. Curtis-Bennett said at his trial, than a likely traitor. Fuchs lived happily with Professor Peierls and his family in a large, detached house in Birmingham. The young scientist was a favorite with the children.

Dr. Klaus Fuchs was now close to the pinnacle of the atomic pyramid: the abstruse and most vital theoretical side of the bomb. The information which he acquired in this position, as well as his own brilliant discoveries, meant years of toil to a nation still young in atomic problems like Russia: what Fuchs had to offer meant the saving of possibly a decade of research.

As a member of the British atomic team, Fuchs was assigned to go to America to deal with their "opposite numbers." Despite the later recriminations of the FBI, the fact is that, not content with the British reports on Fuchs' reliability, the FBI did its own screening and passed him as suitable.

Fuchs was regarded as a dependable collaborator, a little "idealistic" perhaps, but nothing to worry about. The extensive Soviet espionage network in the United States was thus given a present of the man who was to be its most important link with the atomic mystery.

In England he had already been approached by Soviet agents in London and Birmingham. Now, assigned to Los Alamos, the atomic experimental center in New Mexico, his value was considerably enhanced. For nearly 18 months, Fuchs worked with the United States physicists and all the time a Soviet intermediary was never far away.

American atomic security was as highly organized as a stratoliner (FBI, Army, Navy Intelligence, Atomic Energy Commission, Civil Service Department, congressional investigations). Yet Fuchs, methodically and regularly, kept liaison with Soviet agents in two cities, Boston and New York, according to his confession, and probably more, according to the statement of Sir Hartley Shawcross, the attorney general.

In 1946, Dr. Klaus Fuchs returned to Britain, carrying the prestige of his considerable achievement in the atomic project. He was given the high post of head of the theoretical physics division of the Atomic Energy Establishment at Harwell. He was a scientist's scientist, devoted to the welfare of his colleagues, a steady contributor to the Proceedings of the Physical Society and of the Royal Society. He apportioned jobs, passed on the qualifications of applicants, selected people for promotion. As chairman of the Staff Association Committee at Harwell, he presided over matters affecting personnel with a fine impartiality, liked by his employers and associates.

At the peak of his career, Dr. Fuchs examined his course and decided that there was the possibility of a doubt creeping into his faith in communism: the sin of pride before the party in the Communist book of rules and regulations.

As a pledge of his subservience, Fuchs accepted a few hundred pounds payment from the Soviet agent. There had never been a road back for Fuchs; this was his way of demonstrating that he did not want one.

The fact is, however, that at this same time the first real doubts were creeping into his mind. He confessed later: "In the postwar period I began to have doubts about the Russian policy. During this time I was not sure I could go on giving information I had."

He participated in Harwell's social life, a little stuffily, unbending, awkwardly, but then genius has its mannerisms. It is nonsense to assume that his unmarked, repressed personality was a pose to assist his espionage. It was, however, a very useful weapon in the Soviet network. Fuchs, lonely, engrossed, inhibited, was actually alive only to a very small circle of intimates, who accepted the "flatness" as the hallmark of so many great scientists.

He said that he had divided his life into two compartments in his mind—the Communist and the British scientist. Some time in 1949, the wall separating these compartments broke under the pressure of his postwar doubts about communism, and his growing conviction that the life he was living among ordinary, decent, friendly Britons was, after all, more real and better than the Soviet paradise. The Communist ideal had receded into illusion, to be replaced by the reality of Russian imperialism haunting Europe. This change did not come easily with Fuchs. His friends knew that he was on the verge of a nervous breakdown.

The rest was inevitable. His defense gone, his Communist creed abandoned (and it is not likely that the meaning of his absences from his rendezvous was not understood by the Soviet comrades). Fuchs capitulated at the first interrogation of the intelligence authorities, somehow believing childishly that if he told what he knew he would "be allowed to remain at Harwell."

He was condemned out of his own mouth. "Is it right," Mr. Curtis-Bennett for the defense asked William James Skardon, the security investigator, "that before you took a statement from him there was no evidence upon which he could be prosecuted?"

"That is right," was the answer.

There was no road back for Fuchs, no way to redeem the betrayal of friends, no way to make his peace with the world he had bartered in exchange for the ideal which had turned to ashes.

Still confused, still not grasping the full meaning of the verdict, the balding, unhappy man made his last little accented speech in Old Bailey. "I have had a fair trial and I wish to thank you, my Lord, my counsel, and the Governor for their considerate treatment."

He received his 14 years, but somehow there was a different echo, a strange and curious echo; reminding of Budapest, Sofia, and Moscow, it hung in the air of the English courtroom. Then Fuchs was taken away, the man of many strange lives.

The following article from the New York Times magazine of March 12, 1950, is reprinted by permission of the New York Times Co. and the author, Stephen Spender.

The Inner Meanings of the Fuchs Case

(By Stephen Spender)

LONDON (by wireless).—To watch Dr. Klaus Emil Julius Fuchs at the Number One Court of the Old Bailey, seated in the prisoners' dock opposite Lord Goddard, the Lord Chief Justice (flanked by the city of London sword bearer and mace bearer in their traditional medieval costumes), was to witness the twisted forces of a world of immense destructive power, embodied in one man, confronting the ceremony, dignity, and decency of an older world calling him to justice.

One of the most powerful men in the world—the little man was described by his colleagues as "gentlemanly, inoffensive and a typical scholar"—who had perhaps altered the course of history, was sentenced to 14 years' imprisonment as a common criminal. The trial was a triumph of surviving institutions, human personality, and the law.

Nevertheless, it was haunted by formidable specters which dwarfed even the pageantry provided by the city of London; specters not only of communism, but also of the new forces of modern warfare, which could blow sky-high the whole world of values by which Fuch was tried (and perhaps the Communist values as well).

When Fuchs first appeared before a magistrate's court reporters described him as insignificant. At the Old Bailey he did not strike me in this way. He had, it is true, an abstracted, earnest, passive appearance—not at all flamboyant—the face of a student of affairs, with a very faint look of moral self-satisfaction concealed under the pale, attentive look of humility of the theological student. He listened to his own trial with a bowed attentiveness, sometimes shutting his eyes for minutes on end, at other times looking up with an expression almost of gratitude and always as if he understood completely the necessary and inevitable reactions of the court to himself.

The psychology of Fuchs may have been extremely abnormal. Nevertheless it was the product of a combination of circumstances which are disquietingly characteristic of Europe in the last 20 years: a complex of psychological, political, and tragic factors all acting on one man.

Fuchs is the son of Prof. Emil Fuchs, formerly a Lutheran pastor at Eisenach and later a professor at the University of Kiel. Emil Fuchs has been for a long time in close contact with English Quakers and it is through one of them that I have obtained some account of him. He is described as a man of great moral courage, a Tolstoyan Christian who, though disapproving of communism politically, regards Christianity as a Communist way of life. Thus young Klaus Fuchs grew up in an extremely ascetic, idealistic, and serious home environment. It is difficult not to conclude that Fuchs' relationship with his father plays an important role in his psychology. His father is a theologian, and Fuchs undoubtedly made a religion out of communism, by belonging to which he was acquitted of personal problems and automatically put on the side of the historic force which he supposed to be good, however much evil it invovled him in.

My informant, who knew Fuchs' father very well and had met young Fuchs and corresponded with him, described the son as "politically innocent." In addition, there was a strong strain of insanity in the family on the mother's side, and this may have tipped the scales to make an already fanatical young man accept the idea of treachery.

It brought grief to Emil Fuch that his son should have become engaged in war research. However, in 1949 when the father visited his son in England, their meeting was most cordial and Professor Fuchs went away convinced that his son's conscience was completely clear. My informant was certain that the father did not mean he knew his son to be a traitor.

Fuchs' childhood was passed in the Germany of the First World War and the postwar period of unrest. He was passionately opposed to the Nazis and when he was a student was a leading agitator among his fellow-students against them.

Shortly before Hitler came to power he saw how the Socialist Government of Prussia could be hustled out of office by six policemen without the Socialists or anyone else protesting. He decided then that the only effective opponents to the Nazis were the Communists. From this moment the whole of his idealism became attached to communism.

Meanwhile Fuchs' feelings were intensified by the persecution of his family by the Nazis. In 1932 he, himself, came to England as a political refugee; his father was sent to a concentration camp; a sister committed suicide.

The Germany from which Fuchs emerged was marked by the ruin of the middle classes by unemployment and despair; it was a Germany which seemed to offer no hope to its young people unless possibly by means of political revolution. It was a period when German students and even school children were largely politicalized and taught by their political leaders to interpret all situations in terms of their particular brand of political panacea. Children acquired from the politicians the terminology by which they judged all men and all things in abstract pseudo-political terms. This alone gave them a faith and something to live for which took them away from their own desperate situation. At the same time, by giving them, at an early age, an easily acquired ideology for judging the whole world, it deprived them of the incentive to mature intellectually.

Fuchs is also typical of the Germany of his time in his gift for self-dramatization. His confessional document shows amazing power to cast a role for himself, to analyze his own motives, to dramatize his actions, and to state his conflicts. The most effective of his self-dramatizations is his diagnosis of the state of mind in which he found he could betray the secrets of his friends while remaining on warm terms with them—a "controlled schizophrenia."

At his original Bow Street hearing, Mr. Travers Humphreys, the public prosecutor, was so impressed by this analysis that he described Fuchs as having a mind "possibly unique and creating a new precedent in the world of psychology." This description seemed to me so improbable that I asked two well-known psychiatrists and a brilliant young criminal lawyer to comment on it. The head psychiatric consultant at a great London hospital said that the divided personality of Fuchs was typical of cases he met every day. Essentially it was no different from the frequent one of the husband who is an excellent family man, but who also keeps a mistress of whom his wife knows nothing.

The criminal lawyer agreed with this diagnosis. In his work he often met attractive and reasurring persons who turned out to be blackmailers or forgers.

The other psychiatrist said that the split in Fuchs' mind could perhaps be described as dissociation, but not at all as "controlled schizophrenia." A schizophrenic does not with one part of his mind know and with the other part do. Fuchs consciously managed to keep what he knew and what he did in separate mental compartments.

This phychologist also said that atomic scientists, because they understood how to control enormous external powers, must be subject to the pressure of a great conflict if ever the question arose in their minds of whether the government for which they worked was the best trustee of these powers.

He drew a curious and interesting parallel between the control of the psychoanalyst over the dynamic interior forces of the individual subconscious mind and the control of the atomic physicist over external forces. "The psychoanalyst is subject to strains which often disrupt his own personality, and I imagine this might also be the case with the scientist if he begins to question the use to which his work is put."

One small point in Fuchs' defense seemed to me significant. His counsel mentioned that he had been interned in 1940, and wished it to be stated that he felt no bitterness over his internment. All he wished mentioned was that he had not, between 1940 and 1942, been able to witness the reaction of the British people to the threat of invasion.

Now the way in which this point was raised as one favorable to Fuchs is a clue to his mentality. He did not blame the British for interning him, for he understood that within their situation the Government was obliged to detain without judicial process foreigners who might, in the event of invasion, prove dangerous.

That is reasonable, but is it not a shade too much so? Might it not have been more human to assert that a slight sense of grievance had helped him to take the course he later took? And is it not characteristic of Fuchs that he should have made such a point of understanding a situation in which the British, through historical exigency, were forced to act unjustly?

For, by the same process of reasoning, he regarded all political and public action as a result of historic situations which have nothing to do with the feelings, the scruples, the humanity of individuals; and the very action which was criticized by the British themselves as unjust was fully pardoned by Fuchs. The question for him was not whether a society behaved badly, but whether it functioned efficiently and in response to circumstances dictated by the recognition of necessity; above all, whether it was a modern machine, adapted to run along the lines of history railroaded over human scruples by the Communist Party officials.

By the same method of ratiocination, Fuchs found himself in 1944 enjoying complete confidence in the Russian system, because he was convinced that the "Western Allies deliberately allowed Germany and Russia to fight each other to the death. Therefore, I had no hesitation in giving all information I had."

The sympathy for our Russian allies which was so much a reality in Britain and the United States at that time, the bombing of German cities by the Allies in preparation for a second-front invasion of Italy—these weighed as nothing against the magnetic attraction of an abstract, mechanistic interpretation of events. Fuchs found a kind of moral security in holding a theoretic view which repudiated all evidences of emotion conflicting with it. In a competition of the Harwell atomic station magazine, one writer hit on this side of Fuchs' character:

Fuchs'
Looks
An ascetic
Theoretic.

The innermost core of Fuchs' character was his wish to attain "freedom" by applying his theory of society to himself. "It appeared to me that I had become a free man because I had succeeded in establishing myself completely independent of the surrounding forces of society."

By this rather obscure statement he meant that he had brought his own personality in line with his theoretic view of society. He was dominated by a social theory and, therefore, on the level on which he acted politically, he was freed from all conceptions of "bourgeois morality," all personal loyalties. He could betray the country which had sheltered him from Nazi persecution. He could give away the secrets of his friends without even being embarrassed (at the time) by the feeling that this disturbed his warm relationship with them.

Another most important point is to remember the tremendous pressure of ideologies on the minds of scientists who have awakened conciences and are acutely aware of their responsibility toward history and the world. The confession by Fuchs is above all an indication of that pressure of two-world ideas acting upon the mind of an extremely intelligent, if neurotic, person. It would be deceiving ourselves to neglect the fact that there was moral pressure from both sides—the East and the West. Having first chosen the East, Fuchs then discovered "something decent in human nature" and made his confession to the West.

In his statement Fuchs expresses far more contrition at having betrayed his friends than at having handed over our most important secrets to a potential enemy. In court when the judge, before pronouncing sentence, asked him whether he had anything to say, he murmured in a low voice that there were other crimes he had committed, crimes against his friends.

At Harwell he had good friends, some of whom visited him at Brixton while he was awaiting trial. So presumably they had confidence in their personal relationship with him. He lived a quiet life at Harwell but obviously the few parties to which he went and the people he met in his work played an important, almost decisive, part in his development. The force of his self-diagnosis of "controlled schizophrenia" is that it was toward these friends that he felt disloyal, not toward the British Government. There is really nothing in his confession to show that he felt any real contrition of having betrayed our secrets to a potential enemy. Perhaps this is why he found it possible to have a clear conscience. As my Quaker friend explained to me, "He had a clear conscience when he was divided between Russia and his friends and now he has a clear conscience because he has made a clean breast of everything."

Such a state of mind seems extremely childlike and comes within what the psychiatrist calls "infantilism." There are only two possible explanations for the character of Fuchs: One is that he was diabolically insincere in everything he did, including his confession; the other that he was sincere. Of the two it is the more disturbing to think that he was sincere. But everyone who knew him seems to think that this was the case. According to his own lights, he was a man acting in accordance with a conscience all the way along.

Though not a "controlled schizophrenic," Fuchs was undoubtedly neurotic. His neurosis took the form of trying to escape from personal guilt into the abstract morality of a theory of history which freed those individuals who bound themselves to it from all loyalties based on relationships between individual human beings and on respect for the individual. As one of my psychologist consultants remarked, "the nominal reward of £100 which, on one occasion, he accepted from his Russian employers was a way of putting a seal on what he must have considered a pact with the devil." The devil for him, though, was the devil beyond-good-and-evil, the devil of history, the devil which knows only the difference between the right and wrong side of the "revolution." It was a righteous devil.

In order to understand the nature of the choice involved the reader must ask himself: If I had the power to press a button which would give victory to one side in the present conflict betwen ideologies and if I had to consider not just the future of my own country but that of the whole world would I be convinced that the side I am working for is the right one? The very existence of such a question would put one above and outside one's own nation. The atomic scientists are isolated in more ways than one. They are isolated in their secrecy, and also in their responsibility for uses to which their researches will be put.

And, although now the cry for more and more security is inevitable, it is well to remember that the only safety for the world lies in human beings developing a world point of view, which is the implication of the technical powers for contructing a new, or destroying an old, world that we now possess. If the Russians, at present, stand in the way of one world, then our hope of safety lies in developing a constructive faith which will challenge the nihilism of traitors like Fuchs at least as much as in police measures.

The following is reprinted from Parliamentary Debates, House of Commons Official Report, Session 1945–46 (422 H. C. Deb. 5s, page 2091, May 16, 1946):

DR. ALLAN MAY (SENTENCE)

44. Mr. W. J. BROWN asked the Secretary of State for the Home Department whether he will review the sentence of 10 years' imprisonment passed on Dr. Allan Nunn May.

Mr. EDE. I have considered this case, but I have been unable to find any grounds upon which I should be justified in recommending an interference with the sentence which the court thought it right to impose.

Mr. BROWN. In view of the fact that it is perfectly obvious from the evidence that this man is not a common criminal, and is not a traitor in the ordinary sense of the word, and that he has suffered a much heavier sentence than has been applied to many people who have sold this country for money; and further, in view of the fact that this whole issue of atomic bomb secrecy constitutes an extremely doubtful ethical area, ought not the right hon. Gentleman to have another look at this matter?

Mr. EDE. It was open to this man to appeal against the sentence. He would have run certain risks had he done so. It is still open to him to apply for an extension of the time within which to lodge such an appeal. I do not accept the implications of all the statements made by the hon. Member for Rugby (Mr. W. J. Brown). I can understand, although I could not condone, the attitude of a man who said he was willing to make knowledge which he had acquired generally available. This man did sell knowledge, which he had acquired in the service of this country, to a foreign Power, for their private and particular use.

The following is quoted from an address by Mr. Mackenzie King before the House of Commons, Dominion of Canada, on March 18, 1946. (Official Report of Debates, Dominion of Canada Parliament, House of Commons, vol. I, 1946, pp. 53–54.)

One other thing which is being said is that the alleged actions of the persons being detained were due to the secrecy with respect to the atomic bomb. I had a letter only yesterday from the secretary of one of the councils of Soviet-Canadian friendship, stating that all this had grown out of the fact that Russia was being denied information which the United States, Britain, and Canada had with respect to the atomic bomb. May I impress upon the house this fact—the disclosures of which I have been speaking tonight go back to 1943 and 1944. The organization for the purpose of espionage of which I have been speaking has been in existence for 3 or 4 years in this country, and the greater part of the information which it obtained or sought to obtain was secured before anyone knew anything about the atomic bomb outside of those in the immediate know. The attacks on Hiroshima and Nagasaki were, if I recollect aright, on the 6th of August and at a later date in that month, the 9th of August, I think. Documents in the possession of the Government are in large part prior to that time altogether. This espionage business has not arisen out of the atomic bomb in any way or any secrecy in connection with it. Now that the atomic bomb is a known factor, undoubtedly information is being sought in that connection, but the espionage net which has been referred to tonight has been in existence for a much longer time than the past 7 months. It was the 8th of September when the documents came into the possession of the Government.

The following is reprinted from Science, June 28, 1946, by permission:

The sentence imposed upon Allan Nunn May by the British Government has been strongly criticized by the executive committee of the Association of Scientific Workers of Great Britain in a communication directed to the American Association of Scientific Workers in the United States. The British association does not seek to justify Dr. May's breach of the Official Secrets Act, but feels that the sentence of 10 years penal servitude is out of proportion to the offense in view of the following facts. The maximum sentence in the proposed British atomic energy bill is only 5 years penal servitude; less severe sentences have been imposed upon persons who had actively aided the enemy; the person to whom Dr. May gave unauthorized information was a representative of an allied government; little consideration seems to have been taken of Dr. May's positive contribution to atomic bomb research; Dr. May was in a position to give fundamental scientific information only, having had no connection with the know-how of atomic-bomb manufacture.

[From the London (England) Times, September 3, 1946]

ESPIONAGE IN CANADA—WIDESPREAD AND HIGHLY ORGANIZED SYSTEM—REPORT OF THE ROYAL COMMISSION

(From a correspondent)

Much of the 733-page report by the Royal Commission appointed in Canada to investigate the espionage network revealed by a Russian cipher clerk, Igor Gouzenko, published in Canada at the end of June, is given to reproduction of the cross-examinations of the various suspects; but there are also several chapters summing up the evidence thus gained and analyzing the state of affairs that it reveals. The Commission found that the major part of the information was transmitted to the Russians through the agency of Canadian public servants, highly educated men, working under oaths of allegiance, and yet ready, after persuasion, to betray the secrets of their country to a foreign power. How was this accomplished and what does it portend?

The answer is summarized in the Commissioner's findings: "Membership in Communist organizations or a sympathy toward Communist ideologies was the primary force which caused these agents to agree" to carry on espionage at the behest of the Russians. The Commission also found that a fifth column exists in Canada, organized and directed by Russian agents in Canada and Russia, and that within this there are several spy rings. This organization is nothing new, but is the result of a long preparation by trained and experienced men using the Communist movement as the direct source of recruitment for the espionage network. Thus in every case but one, the Canadian agents working for Colonel Zabotin the Soviet military attaché in Ottawa, were members or sympathizers with the Communist Party. The Communist study groups in Ottawa, Montreal, and Toronto were, in fact, "cells," and as such, recruiting centers for agents.

THE MACHINE AT WORK

A good example of the machine at work is the case of the three Canadian scientists, Halperin, Durnforth Smith, and Mazerall, two of whom were members of a Communist cell which was largely composed of scientists employed at the National Research Council in Ottawa. Before the end of March 1945, no member of this group apparently contemplated espionage; then the Soviet Military Intelligence Organization expressed its desire for additional spies engaged in Canadian scientific research, and with a few weeks, and without the initiative of the scientists themselves, a political discussion group was transformed into an active espionage organization.

The report shows that the technique of this approach varied in individual cases, but that it was nearly always founded upon the basis of ideological sympathies. Money, at least in the early stages, played little or no part. The evidence before the Commission showed that within the framework of a Canadian political movement, the Labor-Progressive Party (Communist Party of Canada), "development" courses would take place, fostered by Canadians from the espionage network itself, and aimed at preparing suitable people for active participation in spy work. To begin with, certain selected sympathizers from among certain categories of the population would be invited to join secret "cells" or study groups, and to keep their adherence to the Communist Party secret from nonmembers.

One of the reasons for this secrecy would appear to be that it enabled the Communist Party to gain control, through the election of its members to directing committees or other positions of responsibility, in as many forms of public organization as possible—trade unions, professional associations, youth movements, and so on. A typical example is that of the Candian Association of Scientific Workers on the executive of which were several of the agents arraigned before the Commission, including Dr. Allan May and Professor Boyer, who was national president of the association.

Another, and still more sinister, reason for the technique of secret membership described above was to accustom the young Canadian Communist to what the report describes as "an atmosphere and an ethic of conspiracy," the gradual effect of which was to bring the subjects to a state of mind where they could throw off the moral obligations which they had accepted when entering upon their public duties. Once the victim was within the "cell," he was subjected to a course of study calculated to undermine his or her loyalties. But while this process was continuing, he or she might be quite unaware of the broader ramifications and real objectives of the organization. One of the agents, for example, Mazerall, testified that he was initially invited by a friend to join an "informal discussion group," and that for some time he did not realize it was a Communist cell. Indeed, in any of the small study groups, the extent of the secret section of the Communist Party would not be revealed to the junior members; the whole purpose of the cell organization as an operating ground for the fifth column would, as the report points out, have been frustrated had the rank-and-file members of these groups known the real policies and objectives of the conspiracy.

This principle extended even to senior members of the party, and to those actively engaged in espionage, who would be led to believe that their activities were exceptional, and left quite unaware of the scope and size of the general plan. Thus Lunan, one of the principal agents, testifying before the Commission, said: "I had no idea of the extent and scope of this work: I was amazed when it first became clear to me during my interrogation. I never thought of myself as being more than one person in a small group of five people." Only the leaders, such men as Sam Carr, the national organizer of the Labor-Progressive Party, or Fred Rose, the Quebec organizer, would have a full idea of the real extent and objectives of the organization which they helped to direct.

DEFECTION OF LOYALTY

The development course within the study group seemed to lead, as has been said, to a loosening of loyalties, then to a sense of internationalism, and finally to an acceptance of Communist doctrine and leadership as something transcending all national obligation. Professor Boyer must reacted well to this treatment, since he gave as his reason for imparting secret information to Fred Rose, in spite of his official oath of secrecy, that he thought his action would further "international scientific collaboration." Mazerall also had been worked upon in the same way: "I did not like," he said in his testimony, "the idea of supplying information. It was not put to me so much that I was supplying information to the Soviet Government, either. It was more that as scientists we were pooling information."

A further result of development was the inculcation of a sense of complete obedience to party doctrine, and especially to party leaders. This latter idea apparently extended to the point where loyalty to the leaders of the Canadian Communist Party could be shown to take precedence over national loyalty, and over official oaths of secrecy. Kathleen Willsher, the agent who was employed in the office of the High Commissioner for the United Kingdom in Ottawa, told the Commissioners that she believed the secret information she gave to Mr. Fred Rose was for the guidance of the national executive of the Communist Party of Canada. Her actual words are particularly interesting, since they illustrate the conflict of loyalties and the struggle with conscience which, in this agent, must have provided first-class material for the conspirators to work upon: "I felt I should contrive to contribute something toward the helping of this policy (that of the Communist Party) because I was very interested in it. I found it very difficult, and yet I felt I should try to help."

A "FIFTH COLUMN"

The picture that this widespread and highly organized system conjures up represents only one aspect, one spy ring, of the Russian fifth column in Canada. The existence of other parallel undercover systems, run by NKVD (the secret political police of the Soviet Union), Naval Intelligence, and the Central Committee of the Communist Party of the Soviet Union in Moscow, was revealed by Gouzenko in his evidence. In a statement which he made after he had put himself and his files into the hands of the Canadian police, and which is reproduced without comment in the report, he said: "To many Soviet people abroad it is clear that the Communist Party in democratic countries has changed long ago from a political party into an agency net of the Soviet Government, into a fifth column in these countries to meet a war, into an instrument in the hands of the Soviet Government for creating unrest, provocations, and so forth."

What appears to have happened is, in the Royal Commissioners' words, "a transplanting of a conspiratorial technique which was first developed in less fortunate countries to promote an underground struggle against tyranny, to a democratic society where it is singularly inappropriate."

The following article from U. S. News & World Report, November 24, 1950, is reprinted by permission of the copyright owners, United States News Publishing Corp.:

INSIDE STORY OF A NATIVE AMERICAN WHO TURNED SPY

Alfred Dean Slack seemed as nearly normal and average as an American could be. Now he is serving a term in prison for giving war secrets to Russia. And his friends and neighbors at Clay, N. Y., just outside Syracuse, are trying to figure out how it happened.

Until one day last June, Slack fitted snugly into the community at Clay. He merged easily with the crowd. He was 44, of medium he'ght, a little too heavy, like many others of his age. He wore rimless glasses, looked a little like a preoccupied college professor.

Slack had a good job. He had a new Cape Cod bungalow that he had built with his own hands. He was proud of it, and proud of his wife and two young children. His spare time went into work on the house. In idle moments, he liked to play the organ in his living room, or work at wood carving, or thumb through the chemical and scientific books in his little library. He was at home and loved it. He had been born within a dozen miles of the place where he lived.

Neighbors tabbed him as "a nice guy." One said: "He's a quiet fellow, but I like him." The justice of the peace called him "a home man." His grocer thought him "one of the nicest fellows I ever met."

This was the picture the community had of Slack when he climbed into his car on the morning of June 15, 1950, and drove off to his work as assistant production superintendent of a paint factory. A day later, the people at Clay knew Slack as a man who had given American war secrets to Russia. Two men from the Federal Bureau of Investigation had arrested Slack that morning when he reported for work.

Soon, the details came to the people in the home community. Six years before, while working at a war plant in Tennessee, Slack had told a Russian agent how to make a new explosive. He even had given the Russian agent a sample. And he had known the information was destined for Russia.

The neighbors at Clay puzzled over the story as they set about raising funds for Mrs. Slack and the children. The thing was hard for them to understand. Slack was not a parlor sophisticate or a college-bred Communist. He had not turned to communism because of joblessness. He was not even a member of the Communist Party.

All through his working life, Slack had worked at pretty good pay. He had no criminal record. He had been a quiet, well-behaved youth. There was nothing sinister in his background. He was just a quiet man who liked to potter about the house and play the organ.

On the surface of Slack's placid life, there seemed to be no clue as to how it could have happened. He had grown up in a self-respecting, middle-class family in Syracuse. He had a natural liking for chemistry. His father was a chemist. Slack had one brother and two sisters.

Young Slack had gone through school at the normal rate. He had finished North High School in Syracuse when not quite 18. Then had followed various jobs and two semesters at Syracuse University. Soon after he turned 21, Slack went to Rochester, got a job in the Eastman Kodak Co. laboratories and enrolled in night school. For two years he carried the double load of working by day and going to school at night.

Just before entering night school, Slack married. His work at the Eastman laboratories settled into permanency. He continued to dig into chemical and mechanical subjects in spare time at home after he finished school.

The great depression did not disturb Slack. All through this period, he had a regular job with the Eastman Co., growing in knowledge and responsibilities.

When the war came, Slack was one of the young men transferred to the Holston Ordnance Works of an Eastman subsidiary at Kingsport, Tenn. A new, superpowerful explosive was to be developed here. Slack became a department supervisor, with access to information about the development of the explosive. He worked here, and at another Eastman subsidiary at Oak Ridge all through the war years.

With the war over, Slack left Oak Ridge and war work. He tried engineering research, worked on various projects. Finally, he went back to Syracuse, took the job with the paint company, and settled back into his native environment.

His work history gave no clue to why Slack had turned spy. There had been good jobs—as chemist, engineer, plant manager—at fair pay. He had seemed to be happy. His first marriage had ended in a divorce in 1939, but this seemed to have left no scars. He had remarried. This happened to many men.

It is only in a study of Slack's friends that the pattern of intrigue begins to become apparent.

As an eager young student, working in the Eastman laboratories, Slack had met an older man named Richard Briggs. This new friend was a skeptic about the American economic system. This was in 1928. Briggs thought they were doing things better in Russia, the people's state.

Slack listened eagerly to Briggs. He felt much the same way. His own friendly feelings toward Russia, which were to grow through the depression years, already were beginning to flower.

In 1936, eight years after their meeting, Briggs left the Eastman plant and went to St. Louis. But he kept up his contact with Slack and soon was back in the East. And it was not long before he was calling on Slack again.

Slack was well on the upgrade now. He not only knew the Eastman processes, but by his outside studies of mechanics and general engineering he had picked up a good knowledge of many industrial techniques.

Briggs began to mine this vein of information. He asked Slack all sorts of questions: What is the way to do this? What is the formula for that? What are the processes for making this? He hinted that he needed the information for use in his own job. But some of the things Briggs said were vague. They set Slack to asking questions.

Briggs admitted that he was collecting the information for Russia. He was eloquent: Russia was the people's republic. It was behind the United States in industrial development. It would be a service to humanity to help Russia bridge this gap. Slack listened.

Soon, Briggs was suggesting that Slack might pick up some extra money for spare-time work. Slack could work out explanations and outlines of how things were done in the chemical field, with formulas and such things, and sell them to Russia. Briggs would put him in touch with the right man.

Slack was interested. Here was a chance to do something to help the people's republic. And he could pick up some spare money for doing it. At first he gave information to Briggs. Then Briggs brought a man named "George," who became a regular contact. "George" explained what he wanted and Slack worked out the information. He got approximately $200 for each report. Briggs died, but Slack went ahead with the work.

It all seemed simple. Russia was at peace with the United States. And this was industrial information, having nothing to do with weapons.

In 1940, about a year after the death of Briggs, Harry Gold took the place of the first Russian agent as a contact with Slack. The work continued.

Then America went to war and Slack tried to break off relations with Gold.

Slack had been picked for an important new job at Kingsport. He was married again, and happy. And he realized that there was a vast difference between giving industrial information in peace and providing military information in war.

There were constant reminders of this at Kingsport: restrictions on plant workers; security regulations; posters warning against giving information to an enemy. Russia was not an enemy, but Slack decided not to give Gold any more information.

Gold made several trips to Kingsport, demanding to know about the new explosive. Slack could tell him about it easily. But he refused, flatly.

Finally, Gold cracked down and began to threaten. He would tell about the other things Slack had done. No one would believe this work was as innocent as it sounded. Slack would be fired from the war plant, barred from work in any other, blacklisted everywhere.

Then Gold became persuasive again: Russia was an ally of the United States. It was up to Americans to help. He spoke of Stalingrad, and the stand before Moscow, and a devastated Ukraine.

Slack bent under the pressure. He brought a sample of the explosive out of the plant and gave it, with a sketch of the manufacturing technique, to Gold. The latter hurried it off toward the upper levels of the Russian pyramid.

That was in 1943. The crime lay on Slack's conscience for 6 years, through half a dozen different jobs, before it caught up with him in his home environment at Syracuse.

Because of the threats Gold had used, the Justice Department proposed a 10-year sentence for Slack. But Federal Judge Robert L. Taylor waived aside the recommendation. He said 15 years was not too much for conspiring to commit espionage for a foreign government.

And Alfred Dean Slack, a rumpled man with a worried face, wiped his rimless glasses, put them on again, and went off to prison.

From Security Manual, Manhattan District, United States Engineers Office, November 26, 1945:

SECTION II—SELECTION AND CLEARANCE OF PERSONNEL

1. General.

The purpose of personnel clearance is to assure the assignment to Manhattan District work only persons of demonstrated loyalty and the rejection and elimination of those who are potentially disloyal, disaffected, subversive, or who lack the character or discretion to protect the security of classified information disclosed to them.

2. Applicability of Personnel Clearance Procedure

a. Personnel to whom the District clearance procedure applies may be categorized as follows:

(1) Military personel.

(3) Civil Service personnel.

(3) Contractors' employees.

(4) Consultants.

b. *Definitions:*

(1) The term "classified employee," as used in this section, refers to all persons, excluding military personnel, engaged on classified work or having access to classified work areas of interest to the Manhattan District.

(2) The term "unclassified employee," as used in this section, refers to persons, other than military personnel, directly connected with work of interest to the Manhattan District but who do not have access to classified information or classified work areas. This definition is intended to include clerical personnel, construction workers, concessionaires, and other individuals at the Clinton Engineer Works, the Hanford Engineer Works, and other installations who, though not having access to classified documents, matériel, or equipment, are in daily contact with large groups of classified workers and consequently acquire, through observation and association, information which becomes classified through collection and association with unclassified information previously obtained. The term "unclassified employee" is not intended to include the great mass of workers who, by virtue of their employment by a private concern under contract with the District or one of its private associates, manually work on unclassified matériel or equipment which becomes classified upon final assembly, where the worker has no knowledge of the unusual character of the matériel or equipment, its potential use, or its ultimate consignee.

(3) The term "employee," as used in this section, refers to all classified employees and those unclassified employees to whom the personnel clearance procedure is applicable. (See sec. II, par. 5b.)

(4) The term "classified information," as used in this section, includes all information of a classified nature regardless of its form, substance, or mode of transmission, i. e., documents, matériel, equipment, etc.

3. Clearance Forms Required

a. *Proof of Citizenship.*—Prior to employment each employee (except aliens) will furnish proof of citizenship. Aliens will furnish proof of alien registration and will be processed in accordance with specific instructions hereinafter outlined. The following are acceptable as evidence of citizenship:

(1) Birth Certificate.

(2) Baptismal Certificate. indicating date and place of birth.

(3) Honorable Discharge from the United States Army, Navy, or Marine Corps providing same does not bear notation of alien status at time of discharge.

(4) Naturalization Certificate.

(5) Declaration of Citizenship, providing (1), (2), or (3) cannot be furnished.

Military personnel who are aliens are not subject to assignment to the Manhattan District. Citizenship of military personnel will be presumed unless the individual's Personal History Statement or Service Record discloses information to the contrary.

b. *Data Card.*—A data card will be prepared for each employee. Blank cards will be supplied by the District Security Office upon request.

c. *Questionnaire.*—Officers will execute Personal History Statement. All other personnel subject to Manhattan District clearance procedure will execute WD AGO Form 19–105, revised Personnel Security Questionnaire. The Personal History Statement may supplement this form within the discretion of the Area Engineer.

d. *Fingerprint Chart.*—All employees, including Civil Service personnel, will be fingerprinted on the National Defense Fingerprint Chart. The former policy of separately fingerprinting Civil Service personnel on the Civil Service Fingerprint Chart has been discontinued (War Department Personnel Circular No. 34, 26 March 1945). Military personnel will not be fingerprinted.

4. Screening of Forms Prior to Employment

Although desirable, investigation prior to assignment to classified work is in many instances impractical, in view of the volume of work involved and the immediate need for the services of certain types of employees. It is the responsibility of the Area Engineer, however, to provide the highest possible degree of personnel security after consideration has been given to all attending circumstances and operational purposes. Whenever possible, therefore, investigation should be made and the minimum standards of clearance set forth in *paragraph 5* should be applied prior to assignment of any employee to classified work. In all cases, however, all pertinent forms of a given individual will be carefully screened. No person who has been a consultant or employed on District Work will be re-employed in Category I or II (see par. 6c below) without approval from the District Security Section based upon a check of District files. No individual in the following categories will be assigned to classified work until the Area Engineer has assured himself, by reason of sufficient investigation, that the individual will not be a menace to project security:

a. Minors under 18 years of Age (Reference District Circular Letter (Pers. 34–19) dated 21 March 1944, Subject: Employment of Minors under 18 years of age).

b. Aliens.

c. Individuals not born in the United States or one of its territories.

d. Individuals whose background indicates that they may possess affinity for a foreign government. Indications to be considered are:

(1) Visits to a foreign country.

(2) Close relatives who reside in or are to owe allegiance to a foreign country.

(3) Service in the Army of a foreign country, etc.

e. Individuals having membership in organizations known to have been enemy sponsored or otherwise subversive or committed to the violent overthrow of the Government of the United States, or to adherence to the interest of any foreign power to the detriment of the interest of the United States.

5. Basic Principles of Personnel Clearance

a. The following are basic principles of personnel clearance:

(1) Each classified employee should be subject to careful and conscientious original selection.

(2) The identity of each classified employee should be positively established prior to his assignment to classified work.

(3) The loyalty, honesty, character and discretion or each classified employee should be established by investigation prior to his assignment to classified work.

These principles represent the desirable objective in personnel security; they should not be interpreted as the minimum standard for clearance. As a security objective, they should constantly be borne in mind and should be applied insofar as they are applicable to clearance of personnel at a unit or installation when all existing circumstances have been given careful consideration.

b. Where unclassified employees acquire some classified information through observation and association, and where criminal tendencies of employees of this group may adversely affect the work of the project, limited investigation is believed necessary. The extent to which the provisions of the personnel clearance procedure will be applied, if at all, to unclassified employees, is left to the discretion of the Area Engineer.

6. Minimum Standards for Clearance

The minimum standards for clearance are outlined below. The scope of investigation indicated under each category constitutes the minimum investigation required and should be supplemented by additional information when, in the opinion of the Area Engineer, further inquiry is believed advisable.

a. *Military Personnel.—*

(1) Clearance of military personnel will be based on a three-way central file check, routine investigation (mail investigation conducted by the Area Engineer Office or Service Command routine investigation), and FBI criminal file check. A Service Command special investigation will supplement or be conducted in lieu of the routine investigation for military personnel having access to TOP SECRET information.

(2) Inasmuch as the majority of the military personnel assigned to the District originally report to Oak-Ridge for duty, primary responsibility for their clearance is assumed by the District Security Officer. Area Engineers must take positive action, however, to insure that all military personnel within their areas have been cleared, and where clearance has not been granted by the District Security Office (where the individual is transferred directly to an area office) immediate action will be taken. For officer personnel, Personal History Statements will be forwarded to the District Security Office, which will take all necessary action. When enlisted SED and WAC personnel, transferring into the District report directly to area offices without having been cleared by the

District Office, all action to effect formal clearance will be taken by the appropriate Area Engineer.

b. *Clearance of Naval Personnel.*—Clearance investigations of Naval Personnel assigned to work of interest to the Manhattan District will be conducted by the District Security Office. Upon the assignment of a Naval Officer to the Manhattan District a Personnel Security Questionnaire, Form No. W. D., A. G. O. 19–105 will be transmitted immediately to the District Security Office where clearance will be initiated. Investigation will consist of a local agency check excluding ONI, a check of FBI, MID, and ONI central files and personal interviews with at least three unprejudiced individuals, be conducted by Manhattan District Personnel.

c. *Civilian Personnel.*—Civil Service personnel and contractors' personnel will be categorized as follows:

(1) Group *I*—Classified employees having access to (a) TOP SECRET information, or (b) SECRET information reflecting the over-all program of the District or of any major subdivision, or (c) detailed information concerning the processing and development of end products.

(2) Group *II*—All classified employees not included in Group *I*.

(3) Group *III*—Unclassified employees.

Minimum requirements for clearance are as follows:

Group *I*—Special Service Command investigation (Service Command or Manhattan District investigation), FBI subversive file check, and FBI fingerprint check.

Group *II*—Routine investigation (Service Command or mail investigation conducted by the contractor, or Area Engineer's office), FBI fingerprint check, and central FBI subversive file check.

Group *III*—Central FBI subversive file check and fingerprint check.

7. *Transmission of Records*

a. *Intra-District.*—

(1) Proof of Citizenship and Secrecy Agreement (see sec. III–A, par. 14) will be retained in the contractors' file subject to inspection. Declaration of Secrecy executed by military personnel upon transfer will become a part of the individual's 201 file.

(2) Intelligence files (201) of officer personnel will be maintained at the District Security Office and will not be forwarded upon interarea transfer. Intelligence files of contractors' personnel, Civil Service personnel, and enlisted personnel will be maintained by the Area Engineer's office having jurisdiction over the employing contractor or station and will be forwarded upon transfer of the individual to another Manhattan District station or contractor.

(3) A data card will be forwarded to the District Security office immediately upon assignment or employment of each individual and will reflect type of investigation initiated. When investigation of military and Civil Service personnel is completed the District Security Office will be advised so that the proper notation can be made on the master data card. It will not be necessary to notify the District office when the investigation of individuals other than the foregoing is completed unless derogatory information of a serious nature is developed, in which case complete information will be forwarded. Notification to the District Security Office that investigation has been favorably completed will not be necessary in the case of contractors' employees unless there is a change in the individual's clearance status, i. e., a restriction is placed or a previous restriction is removed. On all contractors' employees, this office will assume that initial clearance is in effect unless specifically advised to the contrary.

b. *To Outside Agencies.*—

(1) Upon transfer of District Civil Service personnel to other Government agencies, Intelligence files will be carefully screened and forwarded to the unit to which the employee is transferred. Upon the release or termination of contractors' employees or upon the separation, as distinguished from transfer, of Civil Service personnel, Intelligence files will be retained in the office of the Area Engineer and will not be forwarded to the Discontinued Projects Branch (District Circular Letter, Civilian Personnel 45–11, dated 6 December 1944)

(2) Part *I* of ASF Circular No. 403, dated 11 December 1944, requires that an entry be made on the Service Record of an enlisted man or woman or on the Qualification Card, WD AGO Form 66–1, of an officer or warrant officer, indicating the extent, results and date of loyalty investigation or check conducted. The notation required by this circular will be made on the Service Record or

Form 66–1 of Manhattan District military personnel when the individual is transferred from the District. All transfers from the Manhattan District are processed through the District Military Personnel Section and in each case the files of this office are checked and the proper entry is made. In view of this procedure, Circular No. 403 requires no action on the part of the Area Engineer. Intelligence files of military personnel *will not* be forwarded.

8. Transfer or Reemployment of Personnel

a. *Contractor's Personnel.*—Employees within group *I* (see this section par. 6c) will not be transferred from one phase of the project to another phase unless transfer of such individual is specifically approved by the Area Engineer having jurisdiction over the unit to which the transfer is being made and by the Area Engineer having jurisdiction of the unit from which the transfer is proposed. In the event of the individual's file contains derogatory information the approval of the District Engineer will be obtained before the transfer is effected. Request for such approval will be directed to the District Engineer, Attention: District Security Officer, and will include a recommendation by the foregoing Area Engineers as to whether approval should be granted, and if the individual is transferring from an installation other than CEW, this request will include a statement of the date and type of the individual's clearance. Decision as to transfer of other employees will be the responsibility of the Area Engineer having jurisdiction over the unit to which the transfer is to be made. Factors to be considered in approving such transfers are (1) the advantage to the District in accomplishing the transfer, (2) the hazards to security resulting from the individual obtaining information about another phase of the project, and (3) the information contained in the transferees intelligence file and employment record.

b. *Civil Service Personnel.*—Provisions of a. above do not apply to the transfer of Civil Service employees. *Prior to the transfer of a Civil Service employee, the approval of the Area Engineer and the District Executive Officer must be obtained.*

c. *Military Personnel.*—All transfers of military personnel emanate from the District Military Personnel Section which assumes full responsibility for compliance with general security policy.

9. Aliens

a. Paragraph 61, AR 380–5, 15 March 1944, provides that no alien shall be employed on a classified war contract unless consent has first been obtained from the Secretary of War. To expedite the granting of consent in cases of alien employment, this authority has been delegated to the Commanding Generals of the Service Commands. Consent for employment of an alien is actually granted by the Commanding Officer of the Service Command upon the recommendation of the Director, Security and Intelligence Division.

b. Aliens to be employed on classified Manhattan District work will furnish proof of alien registration and execute WD PMGO Form No. 301, in sextuplicate, which will be forwarded to the Director, Security and Intelligence, of the Service Command having jurisdiction over the employing facility, with a request that consent be granted for the alien's employment. Assignment to classified work prior to the receipt of this consent is prohibited.

c. Security of Manhattan District work requires clearance standards for aliens which are higher than those normally applied by the Service Command; consequently, letters transmitting Alien Questionnaires should specifically request that a special investigation be conducted and that a copy of the report of investigation be transmitted with the letter of consent to the requesting Area Engineers's office. Immediate arrangements to procure this additional service should be made with the Service Command inasmuch as the procedure is not in accordance with the Service Command's present operating policy.

d. It should be noted that consent for the employment of an alien by "X" company does not authorize the employment of the same alien by "Y" company within the same or another Service Command. When consent is desired for the alien's employment by a new employer, a new Alien Questionnaire must be submitted to the Service Command in which the new employer is located and new consent must be obtained. If however, consent has been granted for the alien's employment by "X" company in "X" Service Command, temporary consent can normally be obtained by TWX for his employment by "Y" Company in "Y" Service Command, if the name of the original employer, the date of the original consent and alien registration number are furnished.

10. Master Reference List

Each area office will submit to this office and to all other area offices lists of undesirables in accordance with letter dated 8 October 1945, Subject: Compilation of Master Reference List.

11. Termination or Separation of Potential Subversives

No action will be taken to terminate or separate any person for reason of their being potentially subversive without first submitting copies of reports of investi-gation and all information concerning the case to the District Security Officer for review and approval.

12. Types of Investigations and Investigative Channels

a. *Service Command Investigations.—*

(1) *Special Investigations.—*This type of investigation is based upon local agency checks and background investigation by investigators of the Security and Intelligence Division of the Service Commands. The extent of the investigation is determined by the Service Command acting as the office of origin as the investigation progresses. Appropriate leads in other Service Commands are included in the special investigation as a routine investigative procedure.

(2) *Routine Investigations.—*This type of investigation includes a local FBI and police check together with a limited verification of the data submitted by the subject on Personnel Security Questionnaire. Routine investigations are automatically converted into special investigations if derogatory information is developed.

(3) *Submission of Requests.—*Requests for Service Command investigations, accompanied by three copies of the PSQ, will be forwarded to the Security and Intelligence Division of the Service Command in which the forwarding area office is located. For investigation of military personnel, general Service Command policy requires the submission of loyalty check sheet forms, in quadruplicate, in lieu of the Personnel Security Questionnaire. However, exceptions have been made by the First, Second and Third Service Commands inasmuch as the PSQ is preferred. In every case, the local Service Command's preference should be determined and followed in the submission of these requests.

b. *Central FBI Subversive File Checks.—*

(1) This type of investigation consists of a check of central subversive files of the Federal Bureau of Investigation. In the absence of positive information, no report is received by the requesting area office. Where derogatory information is disclosed, a summary of information or a copy of the report is furnished.

(2) Requests for central FBI subversive file checks will be forwarded to the Intelligence Officer, Washington Branch Office. Personnel Security Questionnaires for each individual to be checked, bearing a red border stamped at the bottom of the form in order to insure priority in processing, will be forwarded with a letter of transmittal, alphabetizing the PSQ's submitted. Positive reports will be forwarded to the requesting office through the District Security Office.

c. *Three-Way Central File Checks.—*This type of investigation consists of a check of the central files of the FBI, ONI, and MID. Requests will be forwarded on loyalty check sheet forms, in quadruplicate, to Major E. M. Scherer, Office of the A. C. of S., G–2, Who's Who Branch, Pentagon Building, Washington, D. C., who will conduct all three checks. If immediate action is desired, the loyalty check sheet form should bear an EXPEDITE stamp to insure priority in processing. When expedite checks are requested by TWX, identifying information will include full name, date and place of birth and permanent address.

d. *Service Command Agency Checks.—*This type includes checks of the subversive files of the Headquarters, SID, District Headquarters, ONI, appropriate FBI field office, and local police. One copy of the Personnel Security Questionnaire should be forwarded for each agency to be checked. For military personnel, general Service Command policy requires the submission of loyalty check sheet forms in lieu of the PSQ. If there is an urgent need for an expedite agency check it will be requested of the nearest Branch Intelligence Office, with the facts indicating the urgency requiring expedite handling and sufficient identifying information. This type of check should be held to an absolute minimum.

e. *FBI Criminal Fingerprint Checks (Civilian).—*This type of investigation consists of a fingerprint check against the criminal files of the FBI. National Defense Fingerprint Charts will be forwarded by the Area Engineer to the Federal Bureau of Investigation, Washington, D. C., Attention: Mr. S. W. Reynolds. T-2 criminal records will be returned directly to the requesting office.

f. *FBI Criminal File Checks (Military).*—Criminal record checks for military personnel are available and requests will be forwarded to the Washington Branch Office for transmission to the FBI. Requests for criminal record check will state the individual's given name, middle initial or initials, last name, and Army Serial Number (original serial number will be given if the number has been changed). To facilitate expeditious handling by the FBI the 3″ x 5″ form (Exhibit I) will show in the upper right-hand corner the requesting office indicated as follows: Manhattan District _____ Area. Reports will be returned to the requesting office through the Washington Branch Intelligence Office.

g. *Investigations Conducted by Contractors.*—Mail investigations conducted by contractors and investigations conducted by credit agencies, when supplemented by the central FBI subversive file check or Service Command agency check and an FBI criminal check, will be considered equivalent to the routine investigation normally conducted by the Service Command.

h. *Investigations by Intelligence Division, Manhattan District.*—There may be instances when, because of the importance of the Subject, or the position, or because of information previously developed, it will be desirable to supplement a previous investigation or have the whole personnel clearance investigation conducted by the Intelligence Division, Manhattan District. In such cases the request, accompanied by any previous report of investigation, will be forwarded to the District Security Officer.

13. Company Clearance Procedure

It will be the responsibility of the Area Engineer to initiate the clearance of any company with which a contract is made to provide material or services, classified CONFIDENTIAL or higher. The Area Engineer will obtain sufficient identifying data from the Company to form the basis of a request for agency checks on the company and on its key personnel. (Dun and Bradstreet, Moody's Register, and Poor's Register will frequently be found helpful.)

a. Request for checks of the files of the Federal Bureau of Investigation, Office of Naval Intelligence, and Military Intelligence Division will be made on the company and key personnel in the same manner as agency checks are made for the clearance of personnel, and in the absence of derogatory information these checks will normally suffice for clearance. If, however, in the opinion of the Area Engineer, further investigation is considered necessary a request for such investigation will be directed to the office of the District Security Division, which in turn will complete the investigation.

b. Upon clearance of a company the files of investigation made will be forwarded to the District Security Division.

14. Distribution of Section II

It will be necessary for the Area Engineer to extract, supplement, and distribute pertinent portions of this section to instruct contractors in the proper methods of initiating clearance for its employees.

○